Biology of Insects

BIOLOGY OF INSECTS

By
MANJU YADAV
Lecturer
Department of Zoology
M.M.H. College
Ghaziabad (U.P.)
(India)

DISCOVERY PUBLISHING HOUSE
NEW DELHI-110002

First Published-2003
Reprint: 2011
ISBN 81-7141-739-6

Published by
DISCOVERY PUBLISHING HOUSE
4831/24, Ansari Road, Prahlad Street,
Darya Ganj, New Delhi-110002 (India)
Phone: 23279245 • Fax: 91-11-23253475
E-mail:dphtemp@indiatimes.com

Printed at: Dynamic Printers, Delhi

Preface

The fundamentals of Biology of Insects are presented within the framework of scientific discovery. Researches in Entomology have been made almost increadible strides in the past few decades. Consequently, existing concepts of Insect biology have been expanded. These has been a revolution indeed in this direction.

The text integrates the descriptive, experimental and biochemical approaches into a conceptual framework. All important points are illustrated diagramatically. The title is not intended to be comprehensive nor could it be at length, but it concentrates as putting across the basic principles of the subject as briefly and lucidly as possible. It does this with the aid of careful selected examples–some recent and other classic of the field and with numerous illustrations. The aim is to enthuse the reader with this active and exciting area of research and to lay a solid foundation on which further study of its various facets may be based.

The author has freely consulted various articles, discussion notes, reviews and extracts from the various scientific journals in the preparation of the present book, in able to make it comprehensive and upto date.

Though every care has been taken by the printer, publisher and me, it is quite likely that some errors might have found their way into the book but I hope these are of very insignificant nature. However, sugestions to improve book and pointing out of errors and mistakes, if any, will gratefully be appreciated.

The author expresses grateful to her friends and colleagues whose constant inspiration have initiated her in bringing out this book.

Special thanks are given to Mr. Vasan and staff of M/s Discovery publishers for their whole hearted co-operation in the publication of this book.

Author

Contents

1. **Introduction** **1–11**
Slowcoaches, On Three Legs, Feet First, High Jumpers, Long Jumpers, Flying By, In Water, Knowing Where to Go, Staying Alive.

2. **Insects Living in Water** **12–64**
Mayflies, Dragonflies and Damselflies, Stoneflies, Water Bugs or Water Striders, Alderflies, Dobsonflies and Spongilla Flies, Caddisflies, Aquatic Moths, Mosquitoes, Gnats and Aquatic Flies, Water Beetles.

3. **Plant Eating Insects** **65–115**
Radiation in Feeding, Chewing Insects, Sucking Insects, Gall Insects, Finding of Host Plant, Cues for Search a Suitable Habitat, Food Plant Acceptance, Dietary Acceptance, Artificial Diets, The Advantages of Polyphagy and Monophagy, Transmission of Diseases, Phytotoxemias, Plant Eating Insects in the Paleozoic Era, Pollinating Insects, Bee's Adaptations, Bee Flowers, Mouthparts, Flower's Attractants, Other Spectacular Adaptations, Methods of Pollination, eeding in Modern Plant Eating Insects, External Feeders, Internal Feeders, Co-Evolution of Insects and Plants, The Defensive Mechanisms, Feeding Behaviour, Orientation Behaviour, Diseases Carrying Insects, Diseases Caused by Bacteria, Diseases Caused by Fungus, Diseases Caused by Viruses, Diseases Caused by Mycoplasmal.

4. **Carnivorous Insects** **116–137**
Prey to Attack, Evolution of Entomophagy, Predaceous Insects, Finding Prey, Parasitoid Insects, Parasites.

5. Social Life 138–170

Honeybee as a Social Insect, The Workers, Ant Workers, Forms of Social Life, Swarming, Role of Males, Life Span, Slavery, Composite Nests, Guests, Thieves, Harvesting and Hunting, Walking Honey Jars, Ant Farming, Army March, Insect Communication, Language of Bees, Winter Swarming, Social Parasites, Termites, Broader Relationships, Classification.

6. Diversity of Insects 171–199

Evolution of Winged Insects, Origin and Evolution of Wings, Phylogenetic Relationships of the Pterygota, Origin and Function of the Pupa, Berlese's Theory, Novak's Theory, Poyarkoffs Theory, Heslop-Harison's Theory, Hinton's Theory, The Success of Insects, The Adaptability of Insects, The Importance of Environmental Changes.

7. Insect Hormones 200–323

Terminology, Metamorphosis, The Postembryonic Development, Development Stage, Moulting Process, Growth, Differentiation, Larva, Pupa, Hemimetabola and Holometabola, Abbreviations, BF-blastomo-factor, The Activation Hormone (AH), The Neurosecretory Cells of the Brain (NSC), The Corpora Cardiaca (cc), The Direct Effects of Activation Hormone (AH), The Indirect Effects of AH, The Control of AH Production, Chemical Characteristics of the AH, The Mode of Action of AH, The Moulting Hormone (Ecdysone) (MH), The Apolytic Glands (agl), The Prothoracic Glands (pgl), The Ventral Glands (vgl), The Pericardial Glands, The Peritracheal Glands, The Ring Gland (rgl), The Juvenile Hormone (Neotenin) (JH), The Corpora Allata (ca), Is There More Than One Corpus Allatum Hormone? The Control of JH Production, The Mode of Action of the Juenile Hormone, Chemical Characteristics of JH.

8. Insect Pheromones 324–352

Pheromones, Active Spaces, Sex Pheromones, Moth Attractants, Queen Butterfly, Aggregation Pheromones, Alarm Pheromones, Alarm Pheromones of Ants, Trail and Territorial Markings, Pheromones and Individual Differences, Primer Pheromones, Defensive Allomones, Blood-Borne Allomones, Allomones Produced in Exocrine Glands, Defensive Allomones of Social Insects, Allomones Promoting Associations, An Overview of Chemical Signals.

1

Introduction

You are guaranteeed to see insects in a garden or park on a warm summer's afternoon. The most noticeable insects are the most active: butterflies fluttering by, bees buzzing around flowers and wasps (yellowjackets) looking for any scraps of food. Lift up a few stones, and startled bettles run for cover. Look more closely at the plants and you may see caterpillars crawling along, munching away at the leaves. In a pond, mosquito larvae wriggle up to the surface to breathe air.

From even a casual afternoon of insect watching, it is obvious that insects have different ways of moving and a number of reasons for doing so. The butterflies in the garden could be searching for a mate or looking for food, either nectar for themselves or a food plant on which to lay eggs. During such activities, the butterflies may only need to fly from one garden to the next. Butterflies can travel great distances, powered by their own flight and carried by the wind.

Among the most famous examples are the monarch butterflies of North America, some of which spend the winter in large numbers in Mexico and then head north with the spring, eventually reaching Canada. On the outward trip individual butterflies undertake part of the journey, after which they stop, lay eggs and then die. The eggs develop into the next generation of butterflies which fly north again. The process is repeated so several generations are produced along the way. On the return journey, monarchs from Canada do fly the entire distance of over 2000 kilometres (1243 miles) to Mexico. The honey bees in your garden have flown from a hive perhaps as far as 6 kilometres (3.75 miles) away. A worker honeybee clocks up many kilometres (miles) during the six weeks or so it spends foraging for pollen and nectar to bring back to the hive. On average it does 15 trips a day, but when visiting some kinds of flowers an individual can do as many as 150 trips in a day. On a single trip a worker may visit as any as 1000 flowers. The nectar is carried in the bee's honey stomach while the pollen is collected in baskets on its legs. A worker may carry as much as 40 milligrams of nectar, up to half its own weight.

Common wasp workers also take nectar and other sweet foods but they eat these themselves. They bring scraps of meat and other insects, which they chew up in their jaws, back to the nest to feed the young. They are skilled flyers, easily able to defy our attempts to shoo them away. Most beetles can fly, but many take time to unfold their wings so when disturbed they often simply run away. Wings come in handy when looking for a new food source or a mate. In some years swarms of ladybirds (ladybugs) take to the air and then descend in great numbers on villages and towns. This happens when the numbers of their aphid prey have built up. Winged aphids appear and disperse on the prevailing winds along with their ladybird (ladybug) predators.

Wingless aphids often spend their entire lives on the leaf where they were born, and where their mother and grandmother were born before them. Some caterpillars hatch out from eggs laid on the food plant so only need to walk from one leaf to the next as they devour the foliage. Mosquito larvae are able to swim to the surface for air and forage for small particles of food to eat, but they are stuck in their pool or pond. The winged adults have the role of mating and flying to new areas.

Slowcoaches

The slowest movers in the insect world are larvae or grubs, such as those of some bettles which methodically chew tunnels through wood. They have no legs so can only wriggle along. Maggots of flies similarly have no legs and are slow movers. Just take a look at a can of worms which are in fact blow-fly maggots. They have no chance of escaping the tin or avoiding the angler's fingers. Yet the adult blow-fly could never be treated the same way because it is a fast flyer with a response time of one fiftieth of a second, twice the speed we can react to a stimulus. Fly maggots do not need to move much since they hatch from eggs laid in the food source, such as dung or rotting meat. Living in the food source makes maggots hard to see, so they do not need to move fast to escape predators.

Most caterpillars get about on their legs. In addition to the three pairs of true legs behind the head, they have a series of short stumpy legs, called false legs, on the abdomen. These have a ring of hooks to help them cling onto leaves. There is also a pair of claspers on the tip of the abdomen for extra grip. A wave-like motion passes down the caterpillar's body as each set of legs is released and reattached in sequence, moving the caterpillar forward. Looper caterpillars or inchworms have lost some of their false legs. They loop their bodies to brings the false legs at their hind end next to their true legs, then they release their true legs and stretch the body forward before looping again. Whether looping or crawling, stumpy legs are not designed for high speed. But caterpillars can walk a surprising distance. Tent-making caterpillars, which weave themselves a silken web or tent to shelter in, travel 2 or 3 metres from the tent each day to feed. Pine processionary caterpillars also live together in a silken web and travel out each night to feed in single file. When their food supply runs out they all leave in a long line in search of new pine needles to feed on. There can be over a hundred caterpillars in a procession, which stretches for 5 metres or more. Like most caterpillars, processionary caterpillars do their longest walk when leaving their food source and heading off for a place to pupate. They may descend to the ground from a nest in the top of a tree as high as 30 metres.

Caterpillars are not just restricted to walking, some can drop down on silken threads to avoid attack.Gypsy moth caterpillars have used this technique to gradually spread across the USA. Gusts of wind catch up the young caterpillars swinging from their threads and carry them several hundred metres (yards) into the air. When the winds abate, the caterpillars drift down to alight on a variety of plants which they are usually happy to eat since they are not fussy feeders. Leaf-mining caterpillars in comparison are stay-at homes, feeding within a single leaf. They do not have legs and simply tunnel along eating the inside of the leaf as they go.

On Three Legs

Many insects with proper jointed legs, such as ants, walk too quickly for us to see exactly in which order they place their feet. Among other techniques, high speed filming reveals that most insects balance on three legs at any one time, that is the front and back leg on one side of the body and the middle leg on the other side of the body. Not all insects follow this pattern. For example, butterflies and moths often hold their front pair of legs off surfaces. Brush-footed butterflies (nymphalids) never use their front legs for walking; instead they appear to be sensory and are perhaps used by the females to probe plants to see if they are suitable to lay their eggs on.

By increasing the speed at which the feet are placed on the ground, the insect progresses from a walk to a run. Unlike four-legged mammals, which may have four feet off the ground at once when running, insects always have some of their feet in contact with the ground. The main power for their legs comes from muscles inside the thorax which are responsible for lifting the legs and moving them forward. There are also muscles inside the spindly legs which are responsible for bending the legs joints.

By merely looking at a beetle's feet, scientist Nigel Stork can give a good guess where it lives. Most ground-dwelling beetles have few setae or bristles on their feet (tarsi or last-leg joints) and use their claws to get a grip as they move over the rough surface of the ground. Most beetles (including weevils) however, living on plants with smooth leaves, have broadly expanded feet with brush-like setae for gripping the surface, and walk with their claws held above the surface. Beetles living on plants with hairy leaves have comb-like claws which help them grip the individual hairs.

Feet First

The secret of how a fly can walk upside down across the ceiling also lies in its feet. In the same way beetles use the setae on their feet to cling onto smooth plants, flies use theirs to adhere to the smoothest of surfaces. The tips of the fly's setae come into close contact with the surface and are held there by a combination of secretions and forces acting to attract molecules. The pads of the setae can be peeled off and stuck on again rather like velcro. Flies, like many insects, appear to defy gravity, but walking on the ceiling is no problem because they weigh so little. In comparison, we would need such a strong adhesive to stick onto the ceiling by our hands and feet that we would not then have the strength to free ourselves.

High Jumpers

Relative to their own body height fleas outjump humans by far. Even aided by a pole humans cannot jump much more than three times their own height, compared to fleas which can jump at least a hundred times their own height. Fleas accelerate with a force of 140 g (accleration due to gravity) which is over 20 times that required to launch a space rocket. The jump is partly powered by muscles, but an elastic substance called resilin, in a pad on each side of the thorax at the base of the hind legs, really gives fleas their high-powered propulsion. The pad is the remains of a hinge, left over from when the flea's ancestors had wings. It stores energy when compressed by muscles in the thorax and this energy is released into the flea's jump.

The flea readies itself for jumping by putting the second joint of its rear legs in a vertical position and pushing the last leg segments against the ground as it tucks in its front and middle pairs of legs. As soon as the pressure on the resilin pad is released, the flea is flung into the air like a catapult. When it lands it can immediately bounce back again with even greater acceleration. Hungry fleas may have to jump hundreds of times to land on a host.

Another kind of spring is found in click beetles. They use this spring to help right themselves when they fall on their back, a time when they are vulnerable to predators. When on their back, click beetles curve their bodies so that a peg on the thorax spring into a pit. This makes the "click" sound and throws the beetle spinning into the air. They usually manage to land back on their feet. The jump itself is quite startling and may also help to put off predators. A click beetle can throw itself as much as 30 centimetres into the air. The energy required is produced by the action of large muscles in the thorax and is stored by the spring until the final moment of release.

Springtails, as their name suggest, also springs. They are small wingless insects which live in their millions in soil and among dead leaves.When disturbed they get away by releasing a forked "tail", which is normally held in place under the abdomen. As the tail is released, it hits the ground and the springtail is propelled into the air.

Long Jumpers

The other renowned jumpers in the insect world are the crickets, bush-crickets, and grasshoppers including the locusts. All of these have long back legs with powerful muscles. Grasshoppers also have a spring mechanism in their knees between the second and third joints. The muscles in the second joint (femur) of the back legs, work for several seconds building up energy which is stored in the spring and held there by a catch. When the back legs are flexed the catch is released, the spring is trigered and the grasshopper launches into the air. The muscles in a locust's back legs are a thousand times more powerful than an equal weight of human muscle. This is because like many insect's muscles they can contract more rapidly and at higher frequencies than human muscles. Locusts are able to jump up to ten times their own length.

The force grasshoppers exert on their back legs can be so strong that they can even break one of their own legs. However, an occasional broken leg among many grasshopeers is a small price to pay for escaping from predators.

Jumping is often a prelude to flight. When the grasshopper is air-borne, it spreads its wings to fly. Grasshoppers often jump and fly a short distance to escape predators. In mid-flight they often close their wings and drop down into the vegetation. Grasshopper nymphs cannot fly and have to hop from place to place. Bands of locust nymphs, called hoppers, travel 10 metres (11 yards) in a minute just by hoping along.

There are a host of other insect jumpers, including froghoppers, leaf hoppers, flea beetles and jumping plant lice. Though their jumping skills are well documented much less is known about the mechanisms involved in jumping. They all have skinny legs and the muscles responsible for jumping are in the thorax. It seems likely that the best of these jumper also have spring mechanisms in their legs.

Flying By

Whether for an insect or a person, flight is the ultimate way to travel. Flying gets you places faster and allows you to travel long distances. Dragonflies can do over 50 km/h in short bursts while locusts can fly longer distances at about 16 km/h. Insects were the first animals to fly. The first winged insects appeared over 300 million years ago, some 150 million years before the pterodactyls took to the air. No one is certain how wings evolved but there are several theories. It is known that some extinct insect groups had flaps extending from the thorax. These may have first served as surfaces to absorb heat, for camouflage, as extra gills or as sexual signals. It seems likely that once the insects had these surfaces, some kinds then experimented using them for gliding.

The many benefits of being airborne produced a strong selection pressure so that these insects evolved more elaborate wings and the means of controlling them so they could fly.

There are two kinds of systems to operate the wings. In insects, such as dragonflies and mayflies, the muscles in the thorax directly control the wings with one set of muscles contracting to raise the wings, and another to lower the wings. A nerve impulse is required each time the muscles work. In other insects, such as flies, the muscles change the shape of the exoskeleton of the thorax and this moves the wings indirectly. As the vertical muscles in the thorax contract, the top of the thorax becomes flat and the wings go up. As the horizontal muscles contract the top of the thorax arches and the wings go down. The muscles operate much faster because they do not need to be triggered by an individual nerve impulse. The process is aided by the wing hinges which are made of the highly elastic resilin, the same substance found in fleas. There is also a click mechanism like that in a light switch which ensures that the wings to either fully up or down and not to the middle.

Dragonflies and damselflies have two pairs of wings both of which are used for flying. This is aerodynamically a more difficult system to operate, yet dragonflies and damselflies are still master flyers. When the dragonfly is flying slowly the pairs of wings operate slightly out of synchrony with each other, so the front pair is ahead of the back pair. To create a burst of speed, or when gliding the pairs of wings are held together. The hawker dragonflies hunt on the wing catching midges between the hairs on their legs. A Costa Rican damselfly can even hover in font of a spider's web and pick off its prey, without getting caught itself.

Butterflies and moths also have two pairs of wings but their wings are linked together for flight either by a simple fold or a hook which attaches onto a patch of bristles. Not the speediest of flyers, they are proficient at launching themselves into the air. To prepare for flight, they close their wings together in vertical position above their bodies. Then they bring the wings down clapping them together below the body as they launch backward and upwards. The downward movement of the wings literally suckes them into the air by creating a vacuum above the body. Once airborne, they beat their wings slowly which gives them rather a fluttering flight.

Bees and wasps are good at manoeuvring in smaller air spaces than butterflies, having smaller wings and stouter bodies. They have two pairs of wings which when they fly are always joined by a row of hooks. Bumblebees are among the heaviest bees, and their weight combined with their small wings and round shape, make it a mystery how they can fly at all. Beetles sacrifice the ability to fly well for a pair of tough outer wing covers called elytra. These protect the delicate underwings as the beetle scurries along on the ground. The wing cases are held upwards, out of the way of the second pair of wings which beat up and down to protect the beetle through the air. But the wing cases are also important in providing some aerodynamic lift. The manoeuvrability of beetles on the ground is restricted by these wing cases. The smaller the beetle the bigger its wings have to be in proportion to its size to lift it off the ground and the bigger its wing cases have to be if they are to protect the wings when not flying. Beetles get around this problem by being able to fold their flying wings to fit under much smaller wing cases. A clever system of hooks assists in folding their wings, packing them away under the elytra. Earwigs have fan-shaped folds in their soft wings. These have to be coaxed out of their wing covers using the pincers on the tip of the abdomen.

Flying is a most efficient means of getting around but it does have high energy costs, limiting the distance travelled before refuelling is required. Locusts have fat reserves which can last up to 20 hours of powered flight as opposed to bouts of gliding. Honeybees eat honey before they leave the hive as fuel for their flight. Killer bees (an accidental hybrid, named for its ferocity) can fly 60 kilometres (37 miles) on one intake of honey. Another problem for the flying insect is the amount of heat generated by flight muscles. Some large tropical butterflies are restricted to flying in the shade because they cannot get rid of the heat efficiently. Night-flying moths encounter the reverse problem of needing to warm up their muscles before taking-off.

In Water

Walking on water is not a miracle for insects. There are a variety of true bugs, including pond-skaters and water crickets, which do it with ease. Unlike humans, they are so lightweight that the surface tension of the water can easily support them. The end of each leg is equipped with rows of water repellent hairs so they can skate on water without breaking the surface. The same system works for sea-skaters, which are also true bugs and can skate on the ocean surface. They are not completely in control of where they want to go and scores are sometimes washed up on the shore after storms. Both pond-skaters and sea-skaters use their first pair of legs to catch food, while they are propelled along by the middle pair of legs and the back pair of legs are used as rudders.

Insects that live underwater may swim, or crawl along the bottom. Water boatmen are true bugs which row themselves along with the back pair of legs fringed with hairs. Some kinds of water boatmen, also known as backswimmers, swim upside down. Their wing covers from a keel which helps them glide through the water. Beetles are also found living in ponds. Some, such as the silver beetle, just crawl around on the bottom, while dividing beetles are good swimmers rowing themselves along with their back legs. Other insects spend their time crawling along the bottom, such as caddisfly nymphs which carry with them their tube-like homes of sand and gravel. Dragonfly nymphs also crawl but put on a burst of speed by squirting water out of the abdomen.

Knowing Where to Go

A honeybee flies far and wide in search of flowers. Once it has found a good patch of flowers, it has two problems: how to find its way back to the hive and how to tell its nest mates where to find this source of food. Honeybees take their bearings from the sun, or from polarized light patterns (invisible to us but visible to insects) from a clear patch of sky, as well as memorizing key landmarks on their journey. They can also detect changes in the Earth's magnetic field whcih may help them navigate.

Once back in its hive the scout bee performs the classic waggle dance to give nest mates the direction and distance of the food source. If the flowers are within 15 metres (16.4 yards) of the hive the scout just walks around in a circle. In the waggle dance, the bee does a squished up figure of eight, doing a series of waggles with its abdomen where the two halves join. Where the bee dances on the vertical combs in the hive, the line it makes points out the direction of the flowers in relation to the sun. Straight up the comb means the food source is directly in line with the sun. The distance to the hive or the amount of energy a bee may need to reach the flowers (which will be more in high winds) is indicated by the number of circuits, and the frequency of waggling and buzzing as it dances.

However, the bees inside the hive cannot actually see what their informant is doing, so they must sense all the movements apparently through their antennae. Recent studies have shown that they interpret some of the information from the amount of times the scout bee beats its right and left wings. They also detect the smell of flowers and nectar carried by the scout through their antennae.

Ants also leave their nest to go in search of food. Since they run along the ground or other surfaces, they are able to mark their route with scent and retrace their step by smell. Ants, such as harvester ants which forage for seeds, create trunk routes from the nest by clearing away obstructions. This allows the ants to reach their destinations more quickly. The trunk route divides into branches and finally into lesser used routes. The smell for the main trunk route is so strong that it can last for days or even months. Ants do diverge from the marked routes especially at the sub routes or twigs. When they find a good supply of food, they lay down another kind of scent to attract their nest mates to forage there. However, ants are not unique in laying scent trails; many other insects make use of such hidden message for their nest mates.

Staying Alive

All is running smoothly in an ants nest on the African plains. The worker ants are tending the young, the queen is safe in her egg-laying chamber, and soldier ants are on patrol. Then there are ominous shuffling sounds, which get louder and louder as a large animal approaches the nest. A huge scaly nose pokes around the nest, sniffing and snuffing. Suddenly, a section of the nest is torn open by 6 centimetre (2.4 inches) long claws. The ants' nest is in turmoil as workers try to carry the young to safety and soldier ants rush to attack the intruder. Their bites have little effect on the scaly monster, known to us as the giant pangolin.

The giant pangolin is superbly equipped to eat ants. Its 40 centimetre long tongue is coated with copious quantities of sticky saliva and each time the tongue slithers out of its mouth, it can pick up hundreds of ants. In a night's attack, it can eat 200,000 ants weighing at least 700 grams altogether. Its scales provide protection against the biting ants, as do its thick eyelids. The pangolin closes its nostrils to prevent ants or soil from getting up its nose.

Insects are eaten everywhere and all the time. During the day many different kinds of birds pursue them in the air. Flycatchers perch until they spot an insect flying by and then take to the air in pursuit. Swifts and swallows swoop and dive to catch insects in their open mouths. These accumulate in a ball at the back of the throat stuck together with saliva. A swift can carry up to one thousand insects in a single mouthful. During the breeding season the swift population of an area the size of Gibraltar may consume as many as 18 million insects per day.

Insects are not safe sitting among vegetation or on the ground. Birds have excellent vision and can pick insects off leaves with their beaks, especially caterpillars. A pair of blue tits living in an English wood collect about 16,000 caterpillars in spring to feed their nestlings. Many lizards eat insects. Chameleon lizards live in the tropics (from Africa to Madagascar and India), where they clamber slowly along branches. They hunt by sight, lining up their eyes which move independently of each other to pin-point an insect, before shooting out their long sticky tongue. The tongue flicks out in one sixteenth of a second, quicker than a fly can react. Horned toads (which are lizards not toads) from the North American deserts and thorny devils from the Australian deserts concentrate on eating ants which they slurp up with sticky tongues.

Insects living underground are not safe either. Foxes and badgers are adept at digging up beetle grubs. The bateared fox from Africa listens for the noises made by dung beetle grubs as they gnaw away at the dung balls collected by their parents. The foxes then dig down to get at the beetle grubs. Ratels (also known as honey badgers, because they raid bees' nests to get at honey) have strong claws and also dig for grubs.

Freshwater fishes are certainly fond on insects. Many a trout has met its end by snapping at an angler's fly mistaking it for an insect flailing on the surface. Archer fish from the mangrove swamps of India and Australia take things a stage further by shooting a jet of water to dislodge an insect perched on a leaf hanging over the water's surface. They can also jump clear of the water to catch an insect as can the arawana fish from the Amazon. Aquatic insects are easier targets for fish. Large numbers of caddisfly nymphs are eaten by trout. Surprisingly, insects make up a large part of the diet of young alligators and crocodiles.

2

INSECTS LIVING IN WATER

Insects are six leggers invertebrates and are cosmopolition in their distribution. Insects are found in the clear, cool springs at the foot of glaciers and in the warm, yellow courses of tropical streams. They live in the wet moss of dripping cliffs, on the leaf surfaces of bromeliads high in the trees of tropical forests, in the exuding sap of trees, and in the digestive juices of insectivorous plants. Insects are at home in rushing books, in still water, in azure lakes, and in the morass of swamps. From the mountains to the oceans they inhabit all waters, forming in each kind the type of living community peculiar to it. Insects live and develop in hot springs throughout the world; larvae of certain flies and beetles exist in water whose temperature remains at 135ºF (55ºC) even when air temperatures drop well below freezing. And the larvae of the fly *Psilopa petrolei* live in the oil swamps, breathing by means of a tube that reaches above the surface and taking in food through a double-layered digestive tube whose inner wall holds back the petroleum but not the digestible matter.

Aquatic insects exhibit a multitude of clever refinements. In some groups, although the fully developed insects are much alike, the youthful stages differ completely in body form in accordance with their different environments and ways of life. But in other groups, exceedingly diverse ecological conditions are mastered by a single larval type, whereas the adults look very different from one another. Almost all the more important orders of insects are represented in the wet element. Only a few species spend their lives uninterruptedly in water. Some live out of water only as pupae.

But most pass through all their developmental stages in water and take to the air as adults. To the latter type belong those four orders-mayfiles, dragonflies, stoneflies, and caddisflies–that may be regarded as the "most genuine" aquatic insects, since all of their members without exception are bound to the water. Among the other orders only a very few species or groups generally are water insects

According to one school of thoughts it is believed that land insects such as cockroaches, praying mantids, walkingstick insects, grasshoppers, and crickets are the descendants of orginally aquatic species. In actual fact, swimming cockroaches, grasshoppers, and crickets have been discovered; and walkingstick insects, whose bodies have certain features adapted to life in the water, have been found clinging to submerged stones. What is more, the stoneflies, which belong to the most typically aquatic insects, seem to be distantly related to the grasshoppers. The first prerequistie for living in water is breathing. Water insects have two types of respiratory systems: closed and open. In a closed system, oxygen from the water is taken up through the body surface or through special extensions of it, such as the gills. But in an open respiratory system, air passes from the atmosphere or from the air passages oi water plants through openings such as spiracles directly into the air tubes. During its various stages, which frequently are subject to different conditions, a water

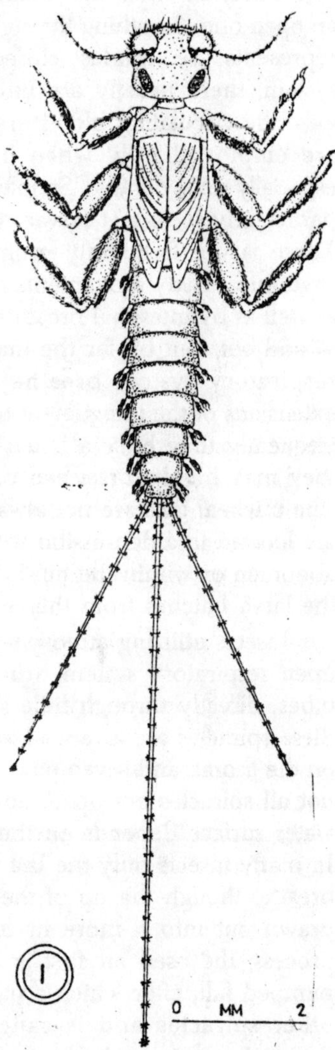

Fig. 2.1. Caenis–a common Ephemerid.

insect may at times use a closed respiratory system and at others an open one. Breathing through the skin without special adaptations represents the simplest closed system. Beneath the skin, in this system, there usually are little integumentary sacs, called blood gills, the circulating blood in which takes up oxygen: often these are employed only when breathing conditions have become especially unfavourable. Simple integumentary breathing is possessed particularly by insect larvae that live in highly oxygenated water. These larvae frequently promote the circulation of oxygen with rhythmical body movements that increase the circulation of water, as well as by intestinal breathing, in which the rectum pumps water in and out. But by far the majority of water insects with a closed respiratory system breathe through tracheal gills. These are extensions of the integument that are richly supplied with air tubes: frequently they have a beautiful shape, like leaves or clusters, or they may highly. branched or in the form of bristles or threads. The tracheal gills are not always freely visible, but in many larvae are located in a depression with a movable cover at the end of the abdomen or within the hind gut. Often gills, are still lacking when the larva hatches from the egg, and first appear in later stages.

Insects utilizing atmospheric air show a wide variations in the open respiratory system, whereby air enters the tracheae, or air tubes, directly through little spiracles. As is normal in all insects, these spiracles are arranged along the sides of the body, two pairs on the thorax and seven pairs on the abdomen. Yet in water insects not all spiracles are open, and the method of taking air in at the water surface depends on the position of the functional spiracles. In many insects only the last pair is operational, and these insects breathe though the tip of the abdomen, which in many larvae is drawn out into a more or less lengthy respiratory tube. In the process, the used air first is expelled, and then the tracheae are pumped full, after which a part of the air flows out again through other spiracles and is caught on hair patches or in various depressions of the body surface, very often beneath the wings. Frequently, there are also air chambers inside the body, such as thoracic sacs, tracheal vesicles, and parts of the gut. The air thus stored outside the tracheal system serves as a reserve, used especially when the insect is obliged to stay under the water for a long time or even to pass the winter beneath the ice.

In addition to respiration the air also has an another important function as a hydrostatic device that helps in regulating body density, and therefore buoyancy. Water insects that live predominantly on the bottom carry little or almost no air in their respiratory system and are heavier than the water. But most become lighter than water by virtue of their air store if they are not to be carried to the surface, they must hold themselves down by clinging to submerged objects or must work their way down by expending energy. By voluntarily displacing its mass of air, an insect can alter the position of its body in the water, as is important when a certain part of the body must be elevated to the surface for taking in air. Many insects, by regulating the pressure of their air store, are able to rise or fall slowly, or to float at a desired level. The ideal situation where the body weight is exactly equal to that of the displaced water has astonishingly seldom been realized by insects. It is also surprising that only a very few forms live in the open sea.

The water surface has a special feature. Its uppermost layer is a fine, elastic skin, that enables many water insects to suspend themselves from it, to brace themselves against it from below, or to walk upon it. As suspensory and bracing device, bristles, hairs and hair circlets, spines, and integumentary pegs are used, and in mature insects the feet are particularly useful. Various insects are able to walk on the water surface–that is, on the surface film–because their legs and frequently their entire underside are rendered unwettable by a dense feltwork of hairs. The springtails that live on the water surface differ in two respects from terrestrial springtails. First, their leaping organ is expanded like a beaver's tail. Second, from their venter they are able to push out into the water a peg that, unlike the rest of the body, is wettable and holds these excessively light creatures to the surface, lest they be too severely buffeted by the wind. Complete or partial unwettability is of vital importance to many water insects, since they would find it impossible to breathe and would drawn if they got wet. But they remain dry thanks to a thick coat of hair or scales or to fatty

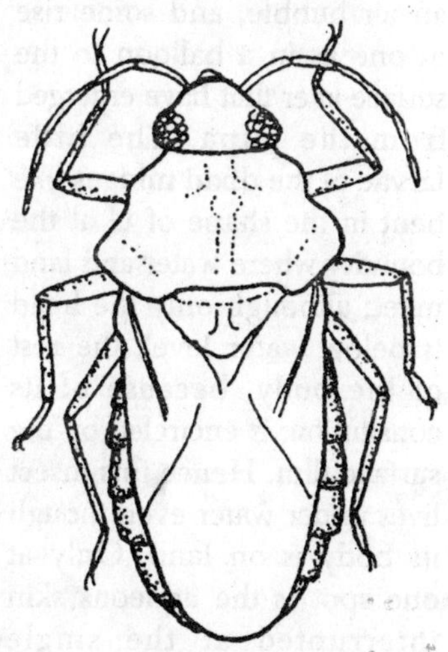

Fig. 2.2. Hebrus pusillus Fallen.

materials exuded from glands. In some caterpillars that liver under water and in parasitic wasps that attack various water insects, special structures of the body surface prevent them from getting wet. The unwetttable parts of insects are coated when under water with a thin film of air, and therefore gleam like silver.

Certain water insects spend the winter or pass through the pupal stage inside an air bubble, and some rise in one as in a balloon to the surface after that have emerged from the pupa. The little larvae of the dixid midges like bent in the shape of U at the boundry where water and land meet; although only the head is below water level, the rest of the body, because of its consitution, is encircled by the surface film. Hence this insect lives under water even though its body is on land. Only at one spot is the aqueous skin interrupted at the single unwettable part of the larva where the respiratory organ is situated. The hairs, as already noted, are important not only as organs of touch and other senses, but also as clinging and bracing devices and as a covering that permits air storage and keeps the insect dry. And they find still further applications in water insects.

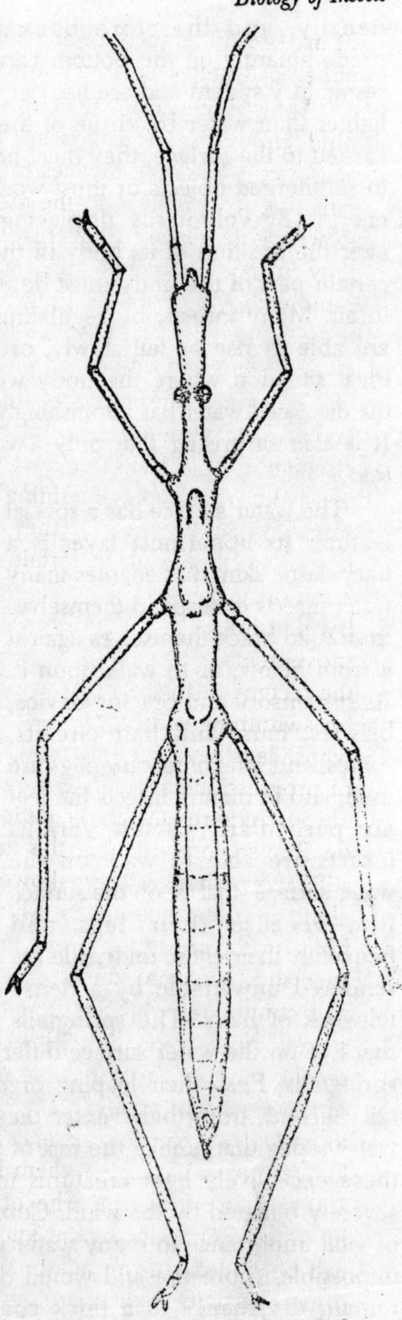

Fig. 2.3. Hydrometra vittata Stal.

Many breathing aperture are ringed with hairs that hinder the entry either of water or of dirt particles.

Various cleaning devices or brushes on the mouth, legs, or hind end of water insects are built of hairs, as are the sievelike snares or prehensile devices for catching food. Hairs also serve as fans that swirl food in to the mouth or even move the whole body slowly along. Hairs frequently support larvae and pupae so well they do not come into direct contact with the substratum; and in rapidly flowing streams fringes of hairs fasten to the substratum those flattened larvae that are continually in danger of being swept away by the current.

The movable organs of the body, the gills and legs of many larvae, are brodened by means of hairs, thereby promoting circulation of water. Fringes of hairs also can serve as devices that assist the insect in maintaining a constant depth by reducing the speed of sinking. Finally they increase the propulsive surface of legs, swimmerets, or, in certain larvae, the rear end. Actually, only a minority of water insects are swimmers. They propel themselves by paddling movements of the legs, by writhing and beating movements of the body, occasionally by expulsion of water from the rectum, and rarely by rowing motions of the wings. The capacity for swimming is most highly developed in certain beetles and bugs whose legs have been modified into oars. The most frequent means of progression in water, however, is crawling, which by no means always requires the participation of the legs. Thus, the body may be shoved forward with the help of projecting warts or segment stiffened perhaps with swallowed air. Or the body may be dragged along after the mouthparts, which repeatedly take a more advanced hold on the substratum. Or, the hind end, endowed with various specialization such as hooks, or the extruded hind gut, can be used as a continuously advancing pint of support, in which action sticky secretions occasionally may play a part. Certain larvae are nearly sessile, some of them in motionless tubes, others attached with silk of fastened with hooks to objects under water, or possibly clinging to rocks by means of suckers. The water insects include the oldest known insects. Traces of them have been found from a time about 150 million years ago. The original ancestors of all insects lived in water. But as far as the known water insects are concerned, it can be duduced from anatomical characteristics of the respiratory organs that in earlier geologic eras these insects lived through long periods as land insect and must have reentered the water only much later and become equatic.

Mayflies

After striking the light of the sun on the horizon is broken up by the gray-green webwork of the willows along the shore. The last rays wander over the ripples of the slowly flowing stream, here quivering in the rings made by a leaping fish, there vanishing quickly in transient whirlpools. Occasional insects appear at the surface, slip from their skin, rise from the water on shimmering wings; and others follow, more and more of them. Like summoned spirits, hundreds, thousands, millions, arise from the water. Over the wavelets there is a twitching, a hissing and a rustling, a moving mosaic of bodies, wings, and empty skins. A cloud, like whirling snowflakes, rises and grows denser and denser, now ascending like a mighty column of smoke toward the heavens, then wafted abroad

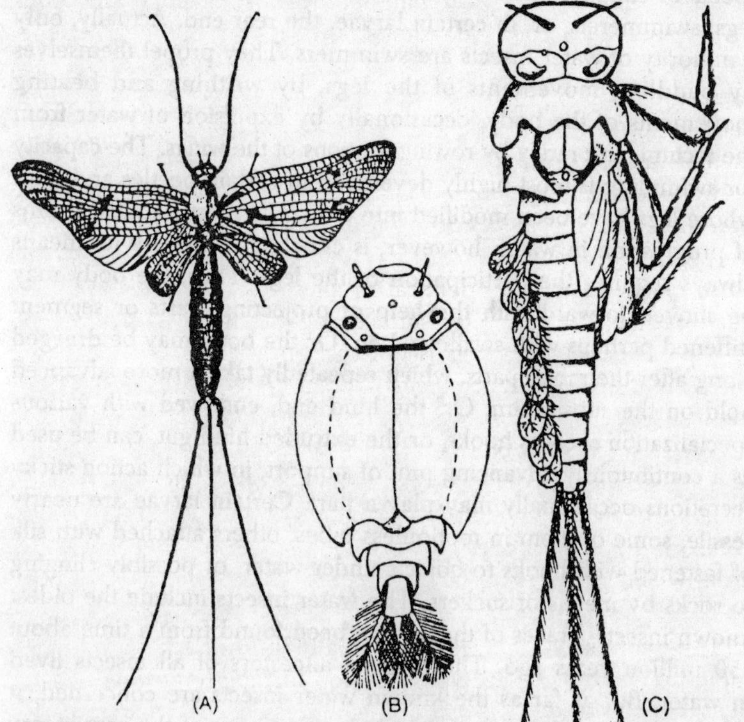

Fig. 2.4. The mayfly. A–Ephemera adult, B–Prosopistoma nymph and C–Heptagenia nymph.

by the gentle evening zephyrs. In this unforgettalbe drama of nature, the principal players are the mayflies (order Ephemeroptera), those images of the transitory that long ago were celebrated in song by the ancient Greeks. For one, two, three, or even four years these insects exist as nymphs, striving to meet their vital requirements in the water and perhaps shedding their skin as many as twenty times, all in seeming preparation for a dance of only a few hours– the nuptial flight that marks their adulthood.

In adults, the mouth and digestive tract become exclusively and aerostatic apparatus for regulating the center of gravity of the body in the air. Driven by rapid wing movement, with the body held almost vertically, at one instant they shoot nearly ten yards upward. Then, turning horizontally and beginning to hover, they sink slowly, borne up as though with parachutes by the outspread wings and caudal filaments, only to rise once again. Up and down they go, in form and movement picture of perfect grace. Their eyes which seem to gleam are among the most beautiful and most capable in the insect world. Almost always only the males dance thus. But one female after another flies into the swarm.There, she is seized by the long forelegs of a male, which at other times are stretched out in flight like antennae. Female and male drift downward together, mating in flight, and the female lets her eggs drop into the water either one after the other or in a single big package. Soon thereafter she dies, and the male returns to his companions. The eggs, numbering from 100 to 3,000, sink into the water. The course of events is not indentical in all speices. In *Palingenia* the female remains at the water surface and the male flies down to her. Or the package or eggs is not cast off at once, but sticks for a while to the abdomen or to the caudal filaments of the female. A few species from more rapidly flowing streams go beneath the surface, cloaked in a fine mantle of air, and glue their eggs to rocks; the *Baetis mayfly* even submerges in foaming water falls. And a very few are viviparous. *Cleon dipterum*, for example, hangs hidden in the thickets along the bank for ten days to two weeks, during which the eggs in the abdomen develop into tiny nymphs that only then are borne to the water.

The nymphs of mayflies differ greatly in form and their living conditions. Very flattened types develop in rapidly flowing water; their body and legs are broad and compressed, the separate parts pieced together like the plates of a knight's armor. By means of their almost knife-sharp contours and the frequent trailing fringes of bristles, theses insects are enabled practically to weld themsleves to the rocks to which they cling. The violent current finds no surfaces to seize upon, and instead of washing the body away presses it more forcefully against the substratum. A few species make themselsves fast with gills that act as suction devices, and others use the labium (lower lip) as a suction plate. In the quieter streams nymphs of a more cylindrical build sit on the bottom or on plants. Hairy species creep through the mud, sometimes at depths as great as 30 feet. Others excavate passageways along the shore with forelegs modified into digging implements. A species in Indonesia bores into wood and has been known to damage wooden water conduits. Many nymphs swim along in open water in spurts, assisted by their caudal fans, formed of three elegant bristles broadened with fringes of hairs, and by two series of gills of various leaflike shapes; the gills, moreover, are in constant waving motion as respiratory organs. In an escape from danger, the first swift movements may be speeded greatly by the expulsion of water from the rectum.

Most of mayfly nymphs are vegetarians however, a very few lead predatory life. They scrape algae from stones or leaves with their mouth, or whisk it off with their forelegs, which bear hairs especially adapted to the job. The *Oligoneura* mayfly sifts plankton through such hairs, and *Isonychia* catches plankton in a basket beneath the mouth, a basket open against the current and formed from rows of hairs on the forelegs. Toward the end of its development the mayfly nymph accumulates air beneath its outer skin and rises to the top of the water. Floating on the surface or clinging to some object, it swells by swallowing air; suddenly the skin bursts and in an instant there seems to be a mature mayfly fluttering along the water. But it only seems so, for the too-short caudal filaments, the too-heavy, lusterless wings, and the eyes betray their immaturity. On a plant at the water's edge, or on the water surface, or may be even in free flight, the true final molt soon occurs. The pre-adult form, inserted between life in the water and life in the air, is unique in the world of insects. It is only after the final molt that the mayfly attains the degree of completion essential for life ink the new, quite different medium, the air. This achievement, however, entails very great sensitivity to even the least amount of water.

The most extreme form of an air-bound life is exhibited by a south American species that falls over and dies if it touches the ground; instead of legs this mayfly has only little useless stumps, except for the male's forelegs, which are essential for seizing the female. Mayflies belong to the most ancient of insects and are distributed over the whole world; about 1,400 species are known. North as far as Alaska and south as far as New Zealand and patagonia, they fly above fresh water, especially in the forenoon and evening, but also at night, when they frequently are attracted to electric lights. They dance in small groups, large groups, and in swarms, and in many lakes and rivers for unknown reasons there occurs a nearly simultaneous mass emergence of millions. In such numbers they can force ships to heave to, can darken light buoys, and can extinguish campfires, and on the next day cover the shore and roads in deep layers. They have been spread as fertilizer on the fields. In the economy of nature mayflies are of crucial importance, for countless fish live almost exclusively on them and their nymphs. And they supply the basis of life for many birds, as well as for many other water insects. Year after year life manifests itself in uninterrupted abundance in these delicate, frail creatures.

Dragonflies and Damselflies

In summer the heat lies over the pond. Water birds doze in the half-shadow of the scarcely moving reeds. The gentle buzzing of innumerable insects seems woven softly into the midday silence, like an audible veil. An odor as of the primeval world, or wet plants and heated mud, saturates the air. This is the kingdom and the time of the dragonflies and their smaller, more fragile relatives, the damselflies. As curious testimonials to their resurrection after a long under water existence, their empty nymphal shells hang everywhere on the plants, grotesque in form and position, as though the life that now goes sliding away, ghostly and silent, decked in kaleidoscopic color and borne on shimmering wings, had not entirely left them.

Like flashing needled the small species crisscross the luxuriant tangle of swamp vegetation. Larger ones whir glitteringly through the reeds and rushes. And huge dragonflies dart mutely, life an artistic mosaic of festive colors, over water and countryside. On the shore is a fisherman. In spite of his careful tread, the carpet of moss squishes and the water quivers beyond the heavy belt of grass. For a few moments, a big dragonfly hangs motionless above

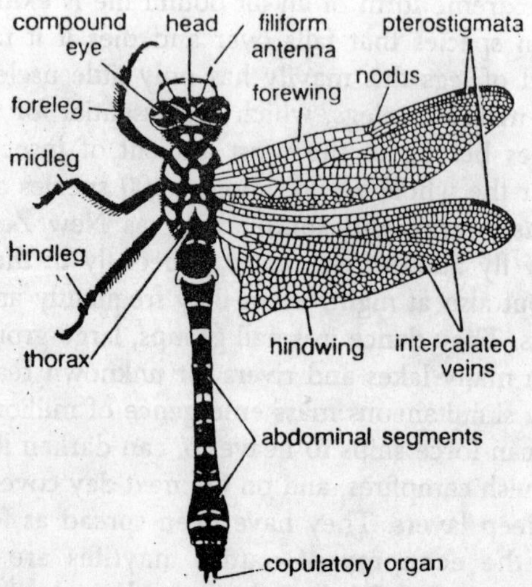

Fig. 2.5. Dragon fly.

the man's head, and its rattling wing beat draws his attention. The fisherman's actions seem odd, for on one occasion he hangs a dragonfly on his hook, and another time only a dragonfly's wing, and instead of fishing in the water he lets the line dangle in the air. We have guessed the riddle: he is a naturalist who is experimenting with dragonflies. The study of behaviour, of animal psychology, has been extended even to insects, and besides the social insects dragonflies and damselflies seem especially suitable for such attention. In some dragonflies, not only does an individual or pair take possession of a territory, but also the sexes cooperate to an astounding degree in the process of egglaying. Big dragonflies, such as *Aescbna*, take charge of a hunting preserve with definite boundaries, and seldom tolerate trespassing by another dragonfly, least of all one of the same species. Individual domains abut, but frequently it is possible for an interloper to force his way between two occupied zones. Then three smaller preserves are formed. But apparently such subdivisions are permitted only up to a certain degree of saturation of this imaginary mosaic of hunting areas.

In damselflies, this staking out of territory is also common *Calopteryx*, which flutter up and down, almost like butterflies, along streams and brooks of the Old World. Their satiny dark blue, violet, or purple wings gleam in intense reflections through the dark green shadows and the scattered rays of sunlight in the midst of the

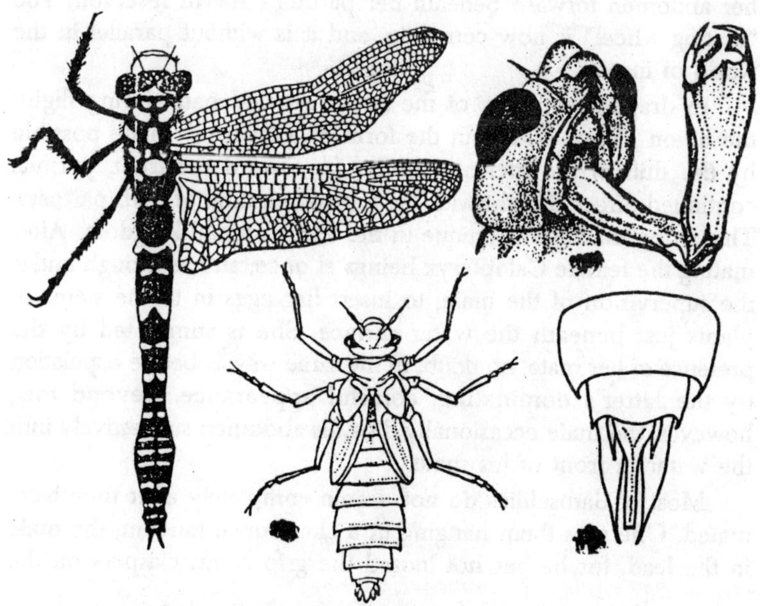

Fig. 2.6. The Dragonfly Cordulegaster. A–Nymph, B–Mouth parts of nymph and C–Terminal end of abdomen of a nymph.

bushes along the shore. A male will defend an area of about 15 square yards all day long against penetration by his companions. Frequently the males come together in small companies for sleeping, then next morning take over new hunting preserves, entry to which is granted only to females.

By use of the casting rod, it has been shown that the females recognize males of their own kind by the distinctive color of the wings, which differs from species to species. If both sexes are in the right mood, the male begins a dancing flight, turning hither and yon with quicker wingbeats and coming nearer and nearer to his prospective mate, who waits quietly. Perhaps on one occasion he drops quickly into the water surface with the tip of the abdomen uplifted: the abdominal underside is a luminous light color or red. Finally he drifts down over the female's wings, which are held together above the back, and runs quiveringly forward to her head, bending the tip of his abdomen forward beneath him and filling the sperm reservoir that is situated there. He then seizes the female's neck with his terminal claspers, and stretches out again. While the two are resting thus connected, the female, who is behind, bends her abdomen forward beneath her partner's sperm reservoir. The "mating wheel" is now complete, and it is without parallel in the world of insects.

In dragonflies many of the larger species mate during flight; the union of two bodies in the form of a "wheel," made possible by the shift of the sperm reservoir far toward the front, permits continued directed flight with the full cooperation of both partners. This flying machine is unique in the entire animal kingdom. After mating the female Calopteryx beings at once, and as though under the supervision of the male, to insert her eggs in to the stems of plants just beneath the water surface. She is stimulated by the presence of her mate, no doubt in the same way as before copulation by the latter's dominating, colorful appearance. Beyond this, however, the male occasionally dips his abdomen suggestively into the water in front of his mate.

Most of damselflies do not disjoin completely after they have mated. One sees them hanging in a chain or in tandem, the male in the lead, for he has not loosed the grip of his claspers on the

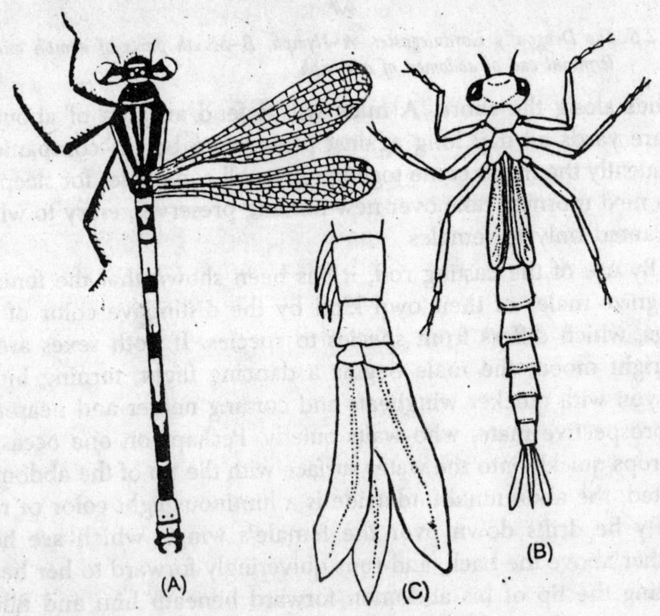

Fig. 2.7. The damselfly Coeagrion. A–Adult, B–Nymph and C–Terminal end of abdomen of the nymph.

female's neck. In such a way pairs of the genus *Sympetrum* seesaw up and down, while the female brushes the water with her abdomen on each excursion and lets eggs drop into it. Aroused by this spectacle, other males often fly up and join solo in the seesaw flight, or may even attach themselves to the pair as a third member. In the American genus *Tramea*, the "tandem" hovers over water; the female is released, drops to the water surface and deposits eggs, rises vertically, and once again is seized behind the head with astounding assurance and carried off in flight by her partner, who meanwhile has been hovering nearby. In the majority of species, which deposit their eggs under water, the male accompanies his attached partner, who descends backward. While she is inserting the eggs one after the other into a long plant stem, the two vanish gradually beneath the water surface, cloaked in a delicate, sliver mantle of air.

Among broad-wings damselflies (Agrionidae) the male stands still and vertical on the female, attached to her neck only by the claspers at the hind end of his body, frequently letting his wings quiver or buzzing them excitedly, and stimulating has companion to oviposit. Often whole groups of such pairs assemble, and a series of them lined up on a stem at the water's surface resembles a beautiful ballet of breathtaking delicacy. Another enchanting scene is presented by an *Agrion* air on a floating water-lily leaf. The female sticks her posterior end into the water through a hole previously eaten in the leaf by beetles, and bending it forward inserts the eggs one another the other into the leaf from below. Meanwhile she is slowly rotated like an hour hand by her mate, who is standing on her neck and either whirring perpendicularly or creeping along on the surface of the leaf. When the first circle of eggs is finished, the female sinks her abdomen deeper, and the procedure begins anew, with an increased radius. For the third circle, the entire abdomen disappears into the water. The final result is eggs inlaid in circles into the underside of the leaf, like a handsome piece of embroidery.

The dragonflies and damselflies are incomparably older than man and are one of the oldest groups in the animal kingdom. Their typical form, completed in primeval time, seems inalterable. We are acquainted, of course, with some aberrant ancient types, yet we see no deep-seated changes since the Jurassic. Dragonflies and damselflies have in part highly specialized characteristics, including numerous peculiarities found only in this group of insects. The dragonflies and damselflies comprise the order Odonata, which we divide into three suborders. The suborder Anisozygoptera is an especially old group, a remnant of which survives only in a single living Japanese species. The suborder Zygoptera includes long, slender insects whose very similar fore and hind wings are held erect during rest and whose eyes are widely separated. These are the damselflies. The third group, Anisoptera, contains the dragonflies. Their wings always are held outspread after maturation, and the hind wings usually are wider than the forewings.

The flying machine of dragonfly-damselfly may well be the best on our planet. The center of gravity lies between and below the bases of the wings, which, in contrast with those of nearly all other insects, are connected directly with the very strong flight muscles. Consequently dragonflies and damselflies are masters of every type of flight, even when the position of the body makes flight almost impossible. They can fly backward, move vertically like helicopters or even like rockets, or stop and turn in the midst of the most rapid progression, as if they had been rammed into. So violent are these maneuvers that it is surprising the wing membrane is not torn to shreds. Yet it withstands these strains, for it is not smoothly stretched out, but is ribbed with a bountiful system of veins that make it elastic. The speed of a big dragonfly may reach about 60 miles an hour; and flight is noiseless, since the wings beat too slowly to produce a hum.

In small species there are some 50 to 90 beats per second, but only about 15 in certain damselflies. The latter have an alternating wing movement, in that the hind wings are elevated while the forewings are depressed. But other dragonflies and damselflies move both pairs of wings together, only alternating while hovering, at which times the wing margins brush against one another and cause the rustling that differs so slightly and yet so insistently from the sound of reeds rubbing together in the wind.

Including procreation most of species spend their entire lives in flight, with the exception of nightly pauses or those compelled by bad weather. For the most part the insects spend these intervals in trees, suspended by the legs. These limbs, unsuitable for walking, are raptorial (adapted for seizing prey) appendages situated for forward and armed with spines; their potential is increased by the uniquely movable first thoracic segment. With them the prey is ensuared during flight as in a basket and is at once conveyed to the mouth, which is equipped with sharp, toothed, knifelike mandibles. All dragonflies and damselflies are voracious predators, and many are cannibalistic.

Some species catch mosquitoes almost exclusively, but larger ones prefer butterflies and motlis, as well as smaller members of their own group. These predators have been seen chasing moths in the dusk in the company of swallows and bats. Some dragonflies, like their enemies, the kingfishers, even plunge into water in the effort to catch prey. The little damselflies snatch aphids from leaves as they fly past, and some of the large darners (*Aescbna*) occasionally may be seen fluttering along treetrunks or cliffs, probing every cleft for resting moths or other insects. Once a resting underwing moth (*Catocalanupta*) was seen to be rouse in this way by a darner, but due apparently to its size and strength was able to escape, which few prey succeed in doing. It is astonishing that the darner discovered this completely camouflaged lepidopteran, in asmuch as experiments have shown that large dragonflies ordinarily respond to objects in motion.

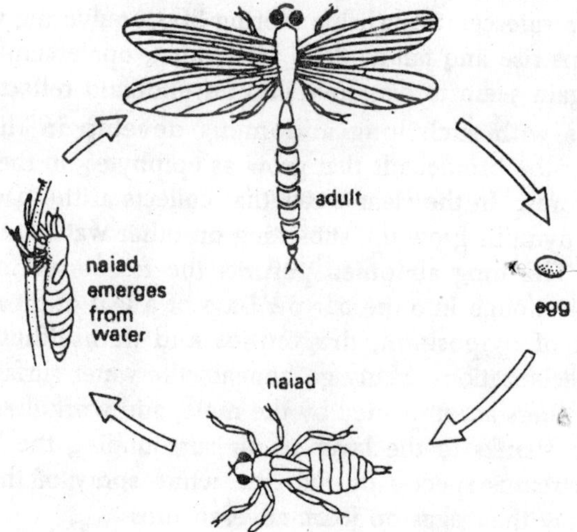

Fig. 2.8. Dragonfly (Odonata). Stages of life history exhibiting hemimetabolic metamorphosis or to say incomplete metamorphosis.

The eyes of dragonflies and damselflies are magnificent iridescent and probably are the best in the whole world of insects. According to the species, one eye is composed of about 10,000 to

30,000 facets. As a further aid to vision, the head is exceptionally movable, as may be observed in dragonflies that lie in wait for their prey at the tips of branches or on the stems of reeds. They can twist the head sideways 180°, backward 70°, and forward and downward 40°. This gives an enormous visual field for the eyes, which already have a large visual field, since they cover almost the entire face. Moreover, a highly refined system of joints in the neck compensates for all involuntary movements of the body–for instance, those caused by the wind shaking the perch–so that any object that has been fixated visually is held precisely in focus. The actual field of observation of these insects, lies above them and in front of them, so they always attack their prey from below.

The eyes of dragonflies and damselflies are capable of seeing at night and may hunt until late evening and in almost complete darkness. In Java there lives an exclusively nocturnal species. South American damselflies of the family Pseudostigmatidae sail almost invisibly through dark parts of the forest in spite of their size. These are ghostly creatures, venerated by the natives more or less as the souls of the departed. In the deep shadows of a luxuriant vegetation, where the smell of mold holds away, one sees transparent, iridescent wings shimmering like a quivering veil. The four wingtips rise and fall as deep blue-violet, opalescently-ringed spots, or again seem to dissolove into an undulant reflection.

Species with such long abdomens develop in the water reservoirs of the bromeliads that grow as epiphytes on the trees of the virgin forest. In the clear water that collects at the base of the leaves, the nymphs grow up, subsisting on other water insects that live there. The long abdomen permits the female to insert her eggs deeply enough into the narrow base of a leaf of these plants. In this act of oviposition, dragonflies and damselflies exhibit peculiar specializations. Many go beneath the water surface, some alone and others accompanied by the male, and work there for up to an hour, thanks to the layer of air surrounding the body. In mountain streams species fly into the white spray of the raging water and lay their eggs on foam-covered moss.

Many dragonflies and deamselflies and damselflies bore their eggs into living or dead plants, or even into wood, with the help of a cutting device at the tip of the abdomen. For each species there is a typical arrangement or grouping of the eggs, and some species make attractive incised designs. Others bury their eggs in sand, mud, or moss in shallow water or along the shore. In doing this, the huge dragonflies *Cordulegaster* hover erect on the tip of the abdomen and force themselves down ward rhythmically once a second, by means of special wing movements, each time boring an egg into the ground. Another species hovers in a horizontal position, and with corresponding wing strokes hammers its eggs into the earth with a sharp ridge, directed downward, on the tip of the abdomen. A great many dragonflies and damselflies tap their eggs into the water with an up and down movement while they are flying, or place them in this way onto floating foam, wood, or the damp bank. A few species dart through the surface film, hissing like speedboats and in this way brush the eggs off.

An *Aeschna juncea* female frequently creeps into a dark hole in the bank and lays her eggs, while the males patrol outside. The females that insert their eggs into plants produce several hundred, of rather large, elongate form. On the other hand, those that oviposit freely produce thousands of smaller and rounder eggs, most of which are laid in series or in large batches. These eggs are enclosed in a gelatinous layer that swells strongly in water and that frequently takes the form of strings that twine about water plants. At first ensheathed in a membrane llike little mummies in their wrappings, the hatching nymphs squirm forth form the eggs. If the eggs were deposited on shore or in over hanging twings, the nymps work their way into the water with fishlike undulations. But most nymphs hatch in the water, free themselves at once from their first skin, and then, as much more finely articulated nymphs, take up their predatory, aqueous life. This lasts in rare instances only a few weeks, but mostly from one to three years, or even for four, with perhaps 10 to 15 molts.

The nymphs are well camouflaged by their colour, which changes according to that of the bottom. They live there on submerged plants, or in mud or sand, frequently deeply buried. They eat everything they can, including little fishes and slightly smaller nymphs of their own kind. Lurking motionless or creeping slowly, these uncanny creatures suddenly shoot out their raptorial tongs, previously folded almost invisibly beneath the head, and grasp the prey. The respiratory organs of the nymph are in its expanded rectum, which contains up to 24,000 gill leaflets arranged in six double rows. A strong intestinal musculature pumps water in and out. But atmospheric air also can be taken in, as happens when the oxygen content of the water falls too low. Thus such nymphs change without difficulty from aquatic to aerial respiration; this also is the case with those neotropical species that live on damp pillows of moss, on rocks at the age of gusing waterfalls, or on dripping crags. In addition to getting rid of the remains of digestion and taking in oxygen, the rectum of many nymphs has a third important function as an organ or locomotion. It can expel with great force the water taken in. This provides rapid movement.

The nymphs of damsefly frequently also have at the posterior end three handsome gill leaflets; these, however, have only a very limited respiratory function and serve for the most part as oars. In the species that lead a buried life the up of the abdomen frequently is lengthened into a breathing tube that extends out of the mud into free water or even beyond into the air.

The nymphs of all dragon fly and damselfly do not live exclusively under water. An Australian species digs passages in the banks to just above the water's surface, where it hunts at night. In Hawaii, some species come ashore when food is scarce, and one even lives solely on land beneath plants. Many nymphs can withstand for more than two months the drying up of a puddle; they dry out in the mud and reawaken when it is moistened a new. When the time has come, the adult dragon fly or damselfly is formed inside the nymph. The nymph, as a type fully co-ordinated and shaped for a creeping, underwater existence, produces the basis for a completely and in comparably different being. From the one comes the other, without transition. The nymph spends its last days at the water surface. In front on the thorax, above the level of the water, spiracles have opened, allowing air to enter. The former intestinal respiration no longer functions. A thin layer of air already of has freed the nymphal skin from that of the adult, which lies beneath. This double creature in its nymph shell climbs yet higher and finds something to cling to. It is strange that in this grasping reflex the nymph seems to live on in the empty skin, even after the adult has freed itself at the front end and, exhausted by this effort, remains for some time bent back and hanging downward. It then frees itself entirely, it wings and body expanding as blood is pumped through them, and again rests, hanging while it hardens a little, until it finally soars into the air.

Still soft and completely helpless, the new dragonflies or damselflies frequently fall prey to their enemies, especially to birds.

They are easily visible because of their still light colored, almost white skin; the final coloration first appears with sexual maturity, which, according to the genus, may require at least several hours, but more likely days or even weeks. The length of life also varies greatly, from two to three weeks in some three to six months in others, and as long as ten months in the over wintering *Sympecma species.* Dragonflies and damselflies by no means always stay near water, but frequently hunt far away in woods, over fields, and even in the middle of cities. These insects are distributed over the whole world, as far North as the Arctic Ocean, but the majority of the 4,500 species live in the tropics. A few develop in salt water lakes and in brackish pool, others in hot springs.

Stoneflies

There are about 2,000 species of stoneflies distributed throughout the world, the stone flies (order plecoptera), which lead somewhat concealed lives, are familiar to few persons besides fishermen. Dependent more exclusively than many other insects on water rich in oxygen, their nymphs with few exceptions inhabit only brooks, rivers, or clear lakes; they are absent from waters

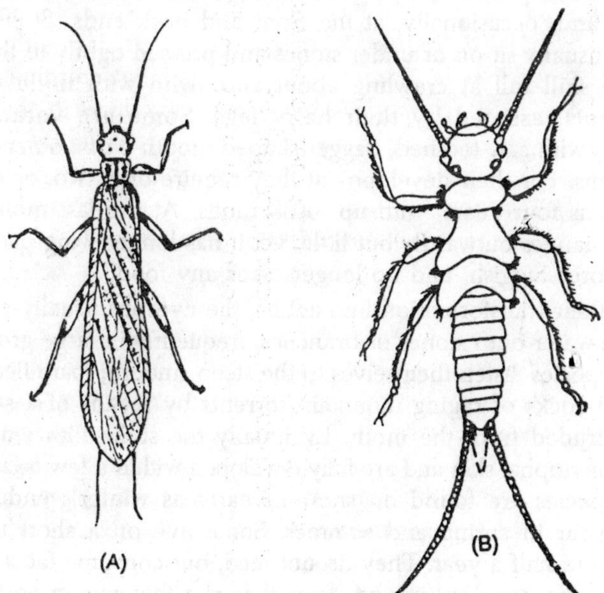

Fig. 2.9. The stonefly, Perla. A–adult and B–nymph.

polluted by civilise man. Stone flies, sometimes called perlids, choose cold, flowing water, and ascend mountain streams to the foot of glacier; many are looked on as relics of the Ice Age.

Morphologically the stoneflies are elongated insects with parallel sides, slightly reminiscent of the distantly related grasshoppers. With the exception of a few tropical species, they lack striking colors; from a large head, long antennae protrude. Their wings are folded back against the abdomen and extend beyond it. The second pair of wings is folded fanlike beneath the first pair. In spite of having large wings with strong veins, the stone flies are poor flier; the two pairs of wings beat slowly and asynchronously, producing a characteristic fluttering flight. Shortly before the female lays her eggs, she brings them together in sticky clumps beneath the abdomen, or in the form of strings on its upper surface; then she flies low over the water until the eggs are washed off by the waves, finally to adhere to submerged stones.

The nymphs are mostly flattened and resemble the nymphs of mayflies. But since oxygen uptake occurs predominantly through the skin and perhaps also through the rectum, gills frequently are lacking or are present only as rudimentary clusters at the base of the legs and, occasionally, at the front and back ends. Stone fly nymphs usually sit on or under stones and pressed tightly to them; they are skill full at crawling about and swim with undulating movements, assisted by their hairy legs. Some are voracious predators with saw-toothed, dagger-shaped mouth-parts; others are vegetarians. For their development they require one, two, or even as much as four years, with up to 33 molts. At the last molt the nymph changes outwardly but little, yet it has longer wing pads, is much more sluggish, and no longer takes any food.

For their transformation into adults, the nymphs usually crawl form the water onto stones or branches, frequently in large groups. Certain species fasten themselves to the steep smooth, foam-flecked cliffs and rocks of raging mountain torrents by means of a sticky crop, extruded from the moth. Eventually the stone flies emerge from the nymphal skin and are fully developed within a few minutes. A few species are found on snow as early as winter's end, but most appear in spring and summer. Some live on a short time, others up to half a year. They do not feed, but consume fat stored up during the nymphal period. Stoneflies of a few genera perform mating dances, during which they beat the abdomen against stones so hard one can hear it.

Water Bugs or Water Striders

In day time when sunlight bewitches the brook, casting the reflection of sky, thickets, and trees in cool colors onto the surface, and making the water below luminous with its warm light, here and there it traces on the bottom silhouettes composed of symmetrical spots. At one instant they are still or flowing slowly along, then suddenly they dart jerkily away in ghostly, disembodied designs over the golden yellow of the bottom, the copper of sunken leaves and the shining green of the water plants. Soon we recognize the cause as insects flitting over the surface. These are the water striders, bugs of the family Gerridae. Their legs, kept dry by a dense felt work of oily hairs, indent the uppermost layer of the water a little; these impressions cast the striking silhouettes, which are much larger than the insects and seem hardly to belong to them. The body of the water striders is always dry, too, thanks to an oily, silvery coating of hairs, if it got wet these insects would drown. There is indeed one North American species that dives to the bottom and catches its food there, but otherwise the water surface exclusively is the environment of the water striders. Upon it they rest with their long, slender legs extended crosswise like the outriggers of a boat; upon it they dart forward for perhaps as much as a yard with a single stroke of the middle legs, or perhaps proceed in jumps as high as 4 inches. Their claws are not at the tips of the legs, as in other insects, but are located in a depression a little above and to the side. In this way rupture of the surface film is avoided.

The hind legs serve as rudders, and the forelegs are borne bent over the head, above the water. With forelegs the water striders size their food, which they suck dry. This consists of little insects that have fallen into the water; their struggles on the water surface are perceived at a great distance by sensory hair on the water striders' legs. Among the water striders, as among many

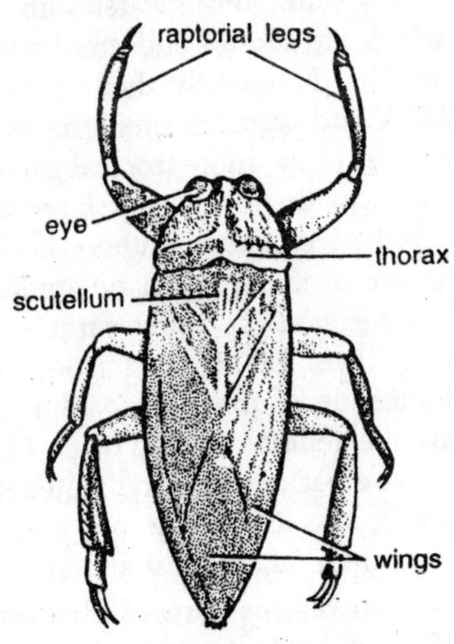

Fig. 2.10. Belostoma (Giant water bug).

other bugs, a species may have winged individuals and others that either are wingless or display various degree of wing reduction. Why this is so has not yet been satisfactorily explained. Water striders spend the winter under moss or leaves on the bank or further inland, and are in the water again early in spring, when little open channels first appear in the ice. They lay their eggs on plants or on wood just below the surface, and the hatching nymphs first sink to the bottom and only later swim to the top. At first the nymph is normal in form, but then for a while seems to have lost its abdomen, which becomes only a very small and compressed protuberance at the end of the large thorax. But the abdomen is extended again after the third molt. Water striders are sociable insects, and at times hunderds may assemble in the shade of a bush.

There are several species of bugs that live at sea, certain ones often hundreds of miles from land, laying their eggs on floating object is such as pieces of seaweed, feathers, or snail shells. In shallow water close to shore, among all sorts of standerd plant parts, lives another odd family, the water measurers (Hydrometridae). Walking along slowly and as though uplifted on stilts, these needle-thin creatures, with eyes that seem to be attached to their neck, may be seen seeking their prey on the water surface. Their eggs are sculptured works of art.

Veliids are active at low temperatures and the bugs of the family walk on the water with alternating movements of their legs, which are shorter and more strongly flexed than those of the water striders. Frequently they anchor themselves with two legs to the bank and wait for small insects carried along by the stream. The species of the more tropical genus *Rhagovelia* even walk across raging rapids with the help of large swimming fans near the tip of the middle leg. These feathery circlets, which are kept folded up in a crease of the leg, can be spread at need and act like oars pulled just beneath the water surface. All the veliids dive often, cloaked in a layer of air, and frequently run about hanging form the underside of the surface film. In contrast to the surface dwellers are the genuine aquatic bugs. The best-known representatives surely are the back swimmers (Notonectidae). With the keel-shaped back down, the flat venter up, and the hind legs widely spread, like oars, these insects rest and swim close beneath the water's surface.

On various parts of their body and beneath the wings, a layer of air is held; the greatest store of air lies on the ventral surface

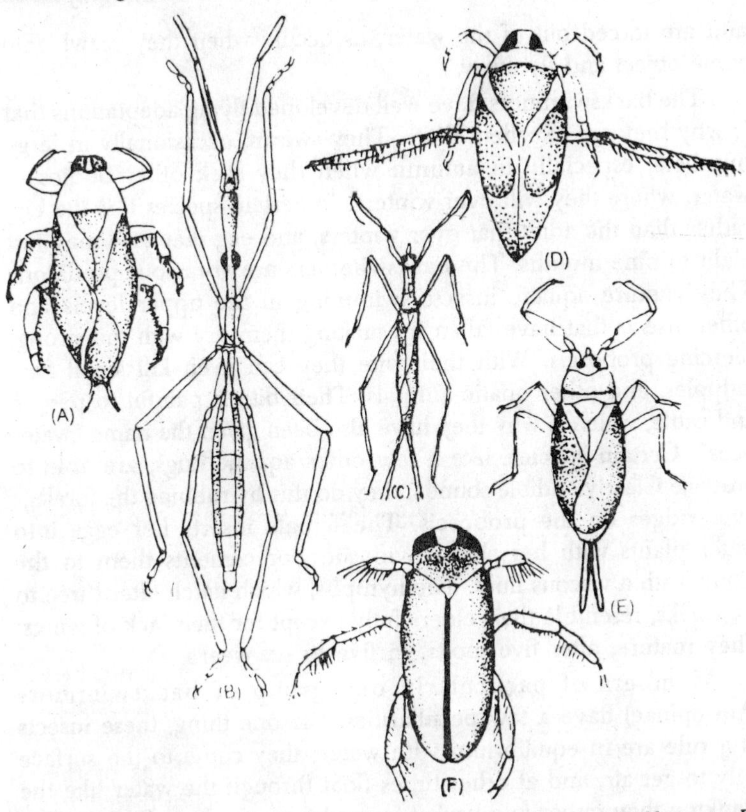

Fig. 2.11. Aquatic bugs. A–Belostoma, B–Hydrometra, C–Ranatra, D–Notonecta, E–Nepa and F–Corixa.

in two channels formed of four dense rows of hairs that are unwettable on their inner faces. This is the cause of the side- down position of these bugs. Being less dense than water, the back swimmers are carried to the surface unless they drive themselves downwards with rowing movements of the hind legs, which are especially adapted by means of hair fringes as swimming organs, or unless they hold themsleves down by means of the four front legs while the oars, directed forward, hand free in the water. The back swimmers seem to be suspended from the water surface by means of their four front legs and hind end, but actually they are big pushed upward against it, so that at the five points of contact the water surface bulges outward slightly. But if the back swimmers come up with the back on top, they break through the surface film and are forced out of the water, as occurs when they crawl onto some object and fly away.

The backswimmers have well developed flying adaptations that is why they are excellent fliers. They swarm, occasionally in large numbers, especially in autumn when they seek other bodies of water, where they will over winters. In certain species it is the egg rather than the adult that over winters, and egg diapause may last eight to nine months. The backswimmers are voracious predators. They capture aquatic insects swimming in the upper levels and other insects that have fallen in, sucking them dry with the strong, piercing proboscis. With their bite they can even kill small fish, tadpoles, and other quatic animals. Their bite is painful to people and cattle, which is why they have also been given the name "water bees". Certain species, like a few other aquatic bugs, are able to produce Clearly audible sounds; they do this by rubbing the forelegs over ridges on the proboscis. The female inserts her eggs into water plants with her sharp ovipositor, or cements them to the plants with a viscous fluid. The nymphs, which hatch after three to six weeks, resemble their elders fully except for their lack of wings. They mature, after five molts, in five to six weeks.

Members of particularly one group of backswimmers (Anisopinae) have a few peculiarities. For one thing, these insects as a rule are in equilibrium with water; they come to the surface only to get air, and at other times float through the water like the plankton they rather in a basket formed from the legs. For another, these bugs have red blood, which makes the ventral surface frequently look red. A family at first glance much like the back swimmers is that of the water boatmen (Corixidae). But these do not swim on their back and do not live on the surface. They spend more time on the bottom, gathering their predominantly vegetable food with their forelegs and sucking at it with their weak proboscis.

Especially at night, when they are attracted to electric lights the water boatmen fly frequently. Their takeoff is swifter than that of most water insects. Much lighter than water, these bugs shoot away from the bottom as soon as the grasp of their abnormally long middle legs is released. They hurtle through the surface film into the air, where their very rapid course may pass over directly into flight. Their method of taking in air when in water is very different from that of the backswimmers, for the water boatmen

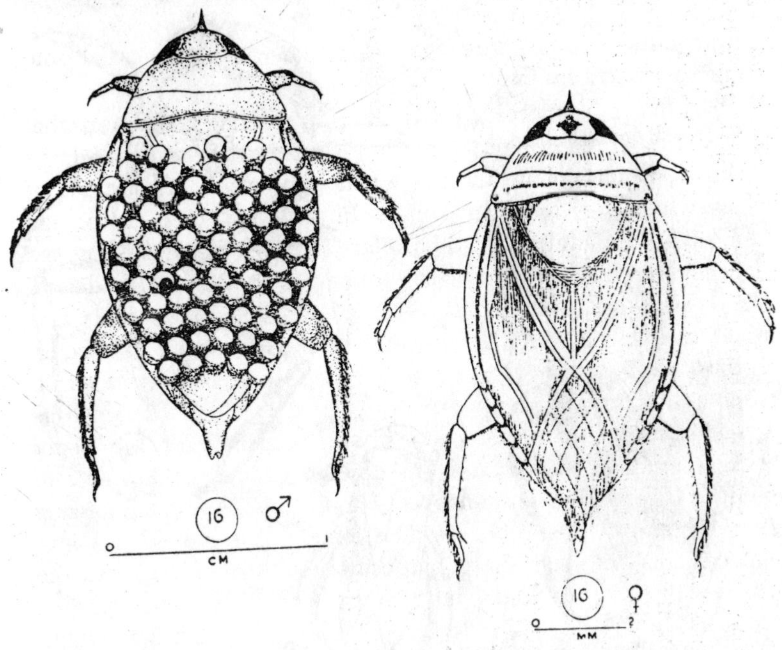

Fig. 2.12. Diplonychus rusticum (Fabr) male and female.

have no special ventral channels. They hold the greater part of the air mass beneath the wings, and a thin layer of air by means of hairs on the ventral surface; the two air supplies are connected between throax and abdomen. And the water boatmen do not come to the surface hind end first, like the backswimmers, but stretch head and thorax out into the air and also frequently leap out of the water like fish.

They also produce a peculiar sound these bugs chirp by means of the proboscis and forelegs, particularly at mating time, and hence have received the designation "water cicadas". The eggs are glued usually to plants, but by certain North American species even to living crayfish. With about 300 species, of which many have a delicate transverse pattern of wavy lines, the water boatmen are the largest family of the water bugs. They have worldwide distribution, occurring in all possible bodies of water up to 16,000 feet above sea level, and even in brackish estuaries. Occasionally they appear in great swarms. Their eggs are used as food by the poorer people of Mexico and Egypt. Broadly oval, like beetles, are

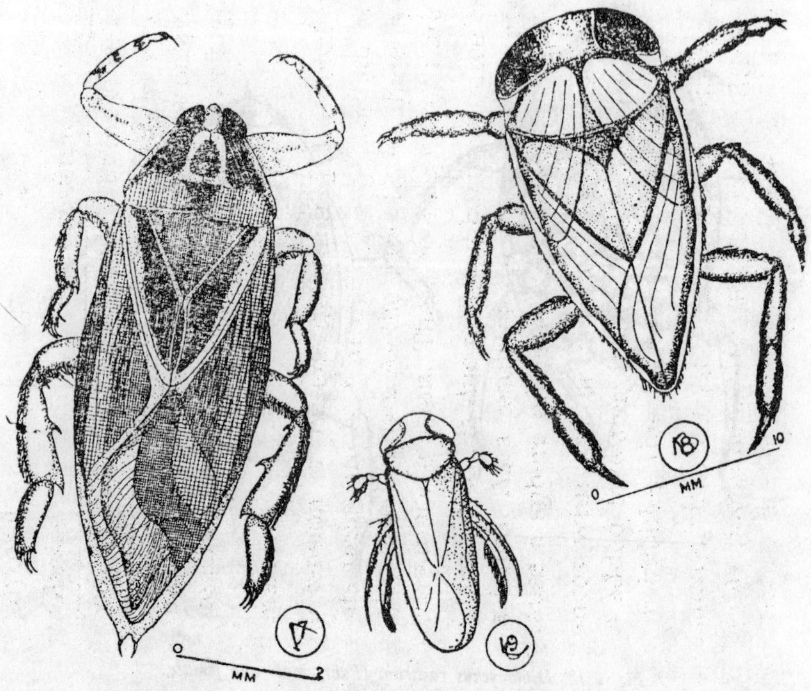

Fig. 2.13. A–Lithocerus indicum (Lepel. and Serv.); B–Corixa sp.; C–Sigara pectoralis Fieber.

the creeping water bugs (Naucoridae). They are conspicuous because their forelegs have been transformed into prehensile hooks with long, dagger-sharp points. With these the living prey is seized, then pierced with the poison bearing proboscis and the juices sucked out. The bites of these bugs are very painful and sharper than those of bees and wasps.

The creeping water bugs have hair-fringed hind legs and thus they can swim very well, and also fly very well, even though only very rarely. Air for breathing is taken in at the water surface with the posterior end and stored as a layer on the underside as well as beneath the wings. These bugs swim with the back uppermost, as well as in the reverse position, and frequently walk belly upward beneath the water surface. The eggs are inserted into plants or merely stuck to their surface. Similar in structure and way of life to the creeping water bugs are the giant water bugs (Belostomatidae), known particularly in America as "electric light bugs", because they frequently fly to electric lights, and as fish-devouring "fish-killers". This predominantly tropical and subtropical family contains about 200 species. Some are more than 4 inches long and hence among the largest of all insects. These insects have become especially well known because of their care of the young, though of a rather involuntary sort, has been observed in a few species. For, after the male has fought against it for hours and has made innumerable vain efforts to escape, eventually he is overpowered by the larger female and forced to hold still while she at her leisure plasters his whole back with eggs. To no avail the male rolls and scratches afterward, for his weighty gift it glued on too firmly.

The first stage nymphs hatch after only ten days, freeing the father of his burden. But usually he is fallen upon anew, and the female polishes the empty eggshells from his back, making room for a new clutch. In flight the giant water bugs breathe through big thoracic spiracles. In the water these become non functional, and the insects inhale by suspending themsleves from the surface and taking in air through a short respiratory tube protruded from the posterior end into the air like a periscope. Air is stored both on the ventral surface and in the thorax. With their poisonous proboscis the giant water bugs kill insects, frogs, tadpoles, salamanders, and fish, and their bite is very painful to man. In certain South American and Asiatic countries these great bugs are enter by the natives, and some are used also in the manufacture of medicines.

Water scorpions are the oddest forms among the water bugs belonging to the family Nepidae. These distantly resemble real scorpions because they have two grasping legs far forward on the body and a long respiratory tube at the rear. They may be seen crawling about on floating plants or on the bottom in very shallow water, or resting half buried in mud, while the tip of their breathing tube sticks up through the surface. The tube is composed of two trough-shaped staves. Like almost all water bugs, the water scorpions are predatory, stalking their prey slowly or lying in wait for it, seizing it like lighting with their raptorial tongs, and killing it with their toxic beak. Their bite is painful to man, and they are not called "toe biters" for nothing.

There are two subfamilies of water scorpions, Nepinae and Ranatrinae which are similar in structure and way of life, but are different in appearance. Members of the Nepinae have an oval shape, and those of Ranatrinae have long, thin, stick like form. The grasping legs of the latter group are much like those of praying mantids in function and attitude. The Nepinae seem hardly ever to fly despite the large wings, which, retaining a store of air beneath them, cover the curiously gaudy, red back, but they are able to swim slowly, even though their legs do not seem particularly well-suited for doing so. The Ranatrinae, on the other hand, fly frequently and have been seen even in migratory swarms. Water scorpions pass the winter on the bank under stones and cast up remnants of plants, or in the water beneath the ice, where they can live for a long time almost with out oxygen. They insert their eggs at the water surface into rotting plants, moss, or algal carpets. The eggs of Nepinae have on the upper side a wreath of short threads that serve as respiratory tubes. The eggs of the Ranatrinae, which are laid in rows, have only two of these tubes, but they are much longer.

Alderflies, Dobsonflies and Spongilla Flies

A small minority of nerve winged insects (order Neuroptera) live in the water, but only during their larval stage. These include the alderflies, dobsonflies, and spongillaflies. In the spring the shores of many bodies of water are swarming with small brown thick headed gnomes. On their four large prominently veined wings they fly ponderously and shakily over short stretches between the vegetation or across the water from one twig, post or stone to another. Over the ground they run in pairs, with wings laid back against the body and the

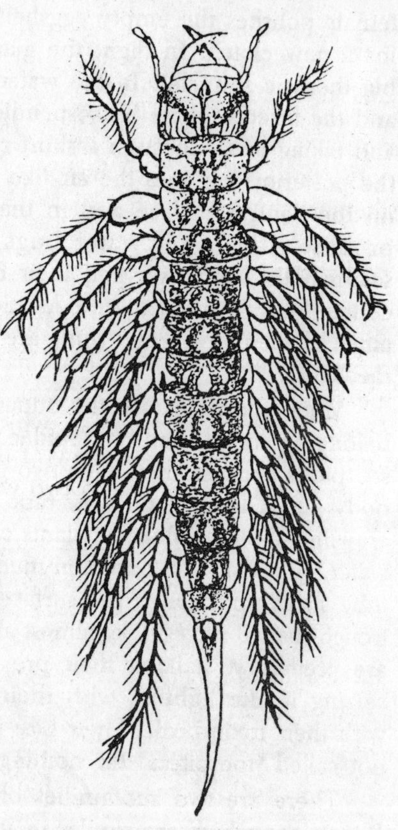

Fig. 2.14. The larva of Sialis.

antennae extended forward, the larger female leading the male.

During their adult life of but a few days the alderflies of the family sialidae, they mate and lay their eggs. They take little or no food, possibly nibbling a bit at flowers. The female glues her approximately 1000 eggs in several groups on plants, wood, or stones, close together and at an angle with the vertical. The hatching larvae burst the eggshells with a special tool situated on the head, drop into the water, and swim off, rowing with the legs and undulating the body. The larvae are predators that dismember their preywith their sharp jaws. In their later stages they usually creep about in mud on the bottom. They breathe through long gill chambers on sides, seem to be independent of vegetation, and penetrate to depths of up to about 65 feet. Molting nine times, they lives as larvae some two years, and then go ashore, frequently very far inland, where they pupate. The pupae, which lie on their side in a cavity in the ground, are able to crawl if they are disturbed. The adults are fully developed in one to two weeks.

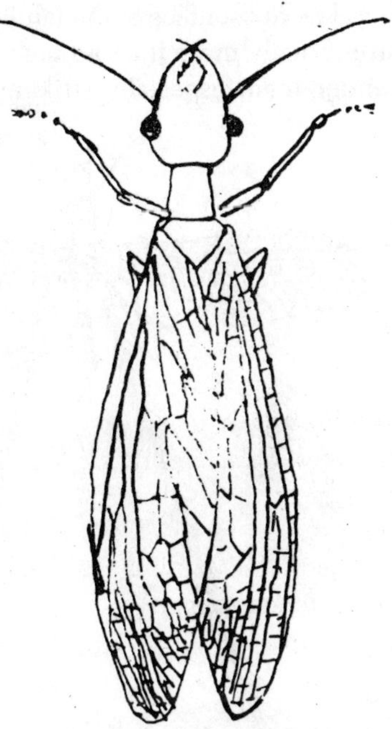

Fig. 2.15. The Dobsonfly, Corydalus.

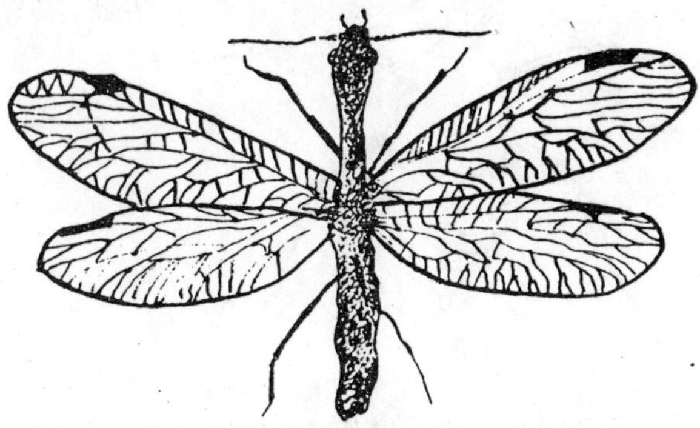

Fig. 2.16. The Snakefly Raphidia.

The dobsonflies of the family Corydalidae, and of extraordinary size (usually more than an inch long) and quixotically shaped head render them especially striking insects. Huge, crossed mandibles

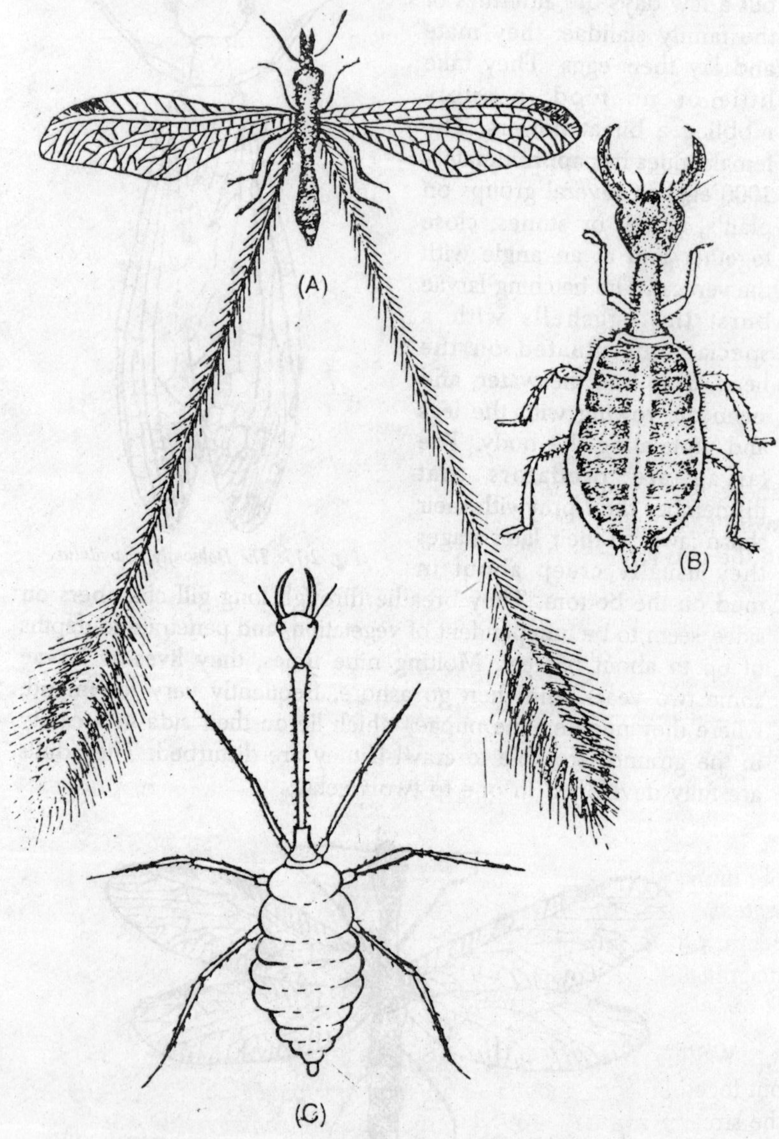

Fig. 2.17. Croce–(A) adult (B) larva, and (C) larva of Pterocroce.

just out like sabers from the head of the male, but dangerous as they look these are hardly of any use in biting; rather they probably serve as claspers in the act of mating. The dobsonflies, live predominantly near swift flowing water, and with their very large, frequently spotted wings, fly at dusk, often to electric lights. Eating little or no food, the male live but a few days, the female perhaps one to two weeks. The female fastens as many as 3,000 eggs to leaves, frequently three to five yard above the water. The larvae creep about as voracious predators for two to three years, and are able to swim backward as well as forward. In North America they are known as "crawlers" or "hellgramites" and are often used as fish bait. A close relationship to the water also is shown by the spongillaflies (sisyridae). They are distributed over the whole world except for the Arctic and Antarctic regions. Near the water, on branches or other objects, sit these little dull colored insects, whose wings are hairy on the margins and on the veins.

During breeding season the females fasten their eggs to plants above the water. After hatching, the larvae drop into the water and swim away with kicking movements, buoyed up by an air bubble in the gut. Eventually they develop a series of abdominal gills that look like the opposed legs of a millipede. The larvae hunt out fresh water sponges, no doubt homing on the gentle currents that emanate from them, and suck on them with two long, slender, independently movable oral tubes. By moving over the sponges in a special manner, the larvae avoid being gashed by the spines sticking out in every direction. The two long antennae, serving as crutches, are moved forward alternately step by step and, with the help of the propelling legs, lift the soft body over the sharp points of the sponge. Another peculiarity of these larvae is that they vomit the indigestible parts of their food. Pupation takes place out of the water in a cocoon spun on plants from threads that flown from the rectum. The cocoon of the American genus Climacia is a magnificent net woven with hexagonal meshes.

Caddisflies

Wading in a brook, we may have seen tiny houses, artistically put together from pebbles or all sorts of rubbish, on the bottom of the stream. We may also have seen the protruding, caterpillar like fore part and a six tiny legs of the little artisans that pulled their houses along after them. These are the larvae of caddisflies (order Trichoptera), and their skill at construction arouses our admiration.

The whitish larvae, like hermit crabs, have to hide their soft bodies inside a house. They make it by spinning a tube, open at both ends, from a silken thread that flows from the mouth and that is manipulated with the help of specially adapted forelegs.

Most species disguise the outside of these cases, into which they are anchored with two hooks on the posterior end, with all kinds of materials found in the water. For instance, they use minerals such as grains of sand and pebbles, all possible parts of plants, bits of wood, or animal matter such as snail shells and the shells of little bivalves. By means of sticky silk these things are mounted on the tube length wise, crosswise, or spirally.

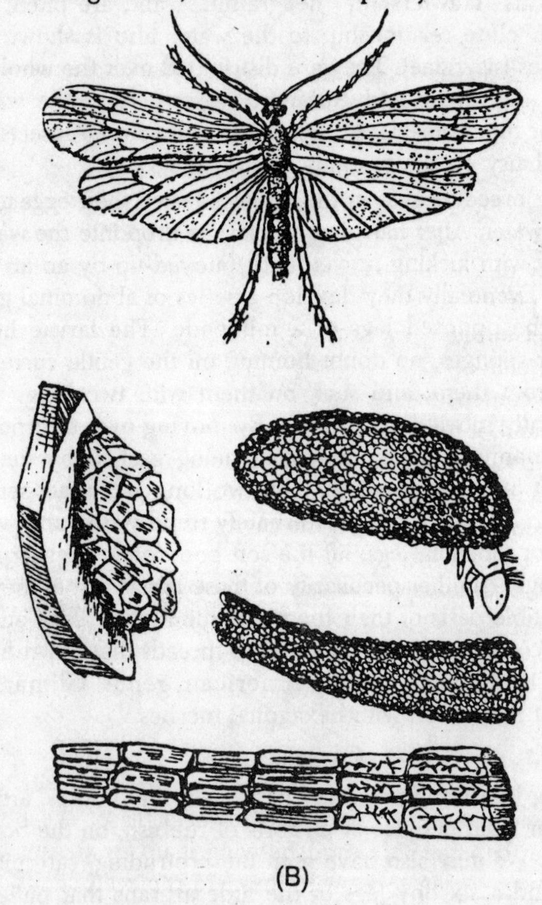

Fig. 2.18. The caddisfly, (A) Halesus (adult) and (B) cases of tricopteran larva.

The cases themselves may be cylindrical, flattened, four cornered, or triangular in cross section. They may widen conically from back to front, in correlation with the growth of the larva, which builds on at the front end and requires a more capacious dwelling as it increases in girth. Or they may have a regular cylindrical form when the posterior part of the house keeps wearing away or is removes. Furthermore, the house may be straight, curved, or twined like a snail shell; and at the rear it is narrowed down into one or more apertures for the circulation of water. In many instances the individual species may be recognized from the style of construction land the material employed. Often enough, however, these matters give difficulty even to the experienced scientist, since many species use different material or vary the style of construction according to the environment and both things may happen simultaneously. The particular virtue of these dwellings resides in their adaptation to environment and way of life.

The beautiful, regularly cylindrical case of species of the family Phryganeidae is a spirally twisted ribbon of bitten off plant stems of equal length, apparently measured by the larva against the fore part of its body. This form of case combines lightness with the potentiality of slipping unhindered through narrow passages and a tangle of plants. In thus affords the larva, which is predatory on other water animals, the requisite motility. The European *Crunoecia irrorata* also lives in a specially regular structure. After the larva has been content for more than half its life with a simple case made of sand, it then construct one out of many small, transversely placed parts of plants, arranged in four rows that meet at right angles. Each individual bit of plant is gnawed out circularly on the inside in such a way that the cavity of the case makes a smooth tube.

Very intriguing also are the structures put together out of snail shells, especially where larger and larger snails are used toward the front end, corresponding to the growing size and strength of the caddis larva. Certain phytophagous species, which in their youth live on the bottom, rise in the course of the year, toghether with the growing vegetation, and finally live for a while on the surface. These larvae then cut large pieces from green leaves and affix them crosswise to the upper and lower sides of the house, thus giving it a flattened form and buoying it up by the air contained in the leaves. When these will, they are replaced again and again with new ones. In the autumn the larvae once again go to the bottom, and their case now is given a wholly different appearance through the lengthwise attachment of tiny dead bits of plants.

Oxyethira has a case made of transparent silk; like a spider this caddis larve pulls a thread along behind it at all times. It also stretches about it abode a criss crossed network of threads, and does acrobatics on it like a rope dancer, or perhaps lets itself down to the bottom supported by a thread. A few species simply cut off a length of reed, line it with silk, and so have their finished case. An American species gets along similarly, boring its way into a little bit of wood or twig, hollowing this out, and then carpeting it with silk. Let us now visit turbulent waters- rushing brooks and rivers, or the shores of large lakes pummeled by wing whipped waves. Here the animals have the task of not being swept away, and frequently utilize the current in obtaining food. They are forced not only to orient-their body against the flow, but also to move about as little as possible, and the ways and means of their battle for existence are fascinating. Many species build flattened cases that offer fewer surfaces for the water to attack and that are especially well adapted to hugging the substratum; frequently these have leaf like expanded margins, made of sand or little pebbles fastened with silk or with sticky material, that surround them and lie close against the bottom. The apertures for water circulation that are essential for respiration are constructed by a few species in the form of two or three big, pyramidal, stone chimneys. Other larvae weight their stone houses additionally with larger fragments, frequently with a large flat stone on each side, the firm attachment of which seems almost impossible for such a small creature.

Outriggers, which function as brakes, in the form of piecedin, long bits of wood or stems, are found in the cases of other species, and frequently the entire structure is interwoven with rough staves, crisscrossed and sticking out in all directions. Houses with cylindrical or smooth constructions are curved like horns, and hence are less easily rolled over by the current; others are anchored by a strong rope woven of silken threads. The cases of certain species stand out like pointed teeth from the stones to which they are hold by suction. The head of such a larva fits exactly into the case's opening, which lies close against the substratum, and is with drawn slightly so that a little space of lowered pressure is produced and effectuates the strong, sucking attachment. All these caddis larvae feed on algae. Certain species that lack cases gnaw long grooves into the surface layer of rocks which consists of soft lime and algae, and at the same time spin a tunnel like covering for their feeding grooves, including in it particles of lime. These tunnels are migratory, since they are constantly, being broken away in back and added to in front.

One group that lives in swiftly flowing water and builds no cases subsists principally on small aquatic life and water insects. A few larvae of this group move about freely in the water with the help if their swimming legs, or creep assuredly with the aid of a thread. When not moving, they anchor them selves with their two posterior hooks. Many lead an existence almost like that of spiders, catching their prey in nets. Special brushes on the forelegs and in some species on the posterior end serve to sweep together the plankton that is caught, as well as to clean the net. The nets mostly are irregular webs on the bottom or on plants, thickened medially into a funnel that narrows below or at one side into a silken tube that leads finally into a hiding place. When prey has strayed in to the net, the larva seizes it. But there are much cleverer plankton nets that have been used by these insects for millions of years, nets of a type that man has made for only slightly more than half a century. These are spun tubes, broadened in front into the shape of a funnel and stretched between plants or stones with the opening facing the current. The net of the species *Neurecolipsis bimaculata* is curve like a trumpet. The owner sits at the bottom of the funnel or in an attached side tube. Masterpieces are provided by the species of the genus *Hydropsyche*; between stones or plants they build a chamber whose entrance widens on the upstream side, and the rear portion of which is closed off by lattice work made of two- ply spun threads that are bound together with sticky material at the intersections. The artisan sits in front of this net, hidden either in its case, which is fastened to the bottom, or in a dwelling tube, and has nothing to do but to collect whatever is left behind by the water that streams through, and to clean the filter now and then. Several of these structures frequently are found lined up in a series, constituting a system of traps built by an insect.

The larvae of caddisflies live for the most part in rather shallow water, yet they have been found in depths of over 20 feet, even as deep as 130 feet. They respire either through the entire skin, or , more often, through gills, which run along the body as two or more rows of threads of tufts; at times these are only on the lower surface. In their houses these larvae circulate the water by an almost constant oscillation of their bodies. Effectiveness is promoted by lateral lines, such as rows of hairs on both sides, that increase the moving surface. The body floats freely within the house, supported fore and aft by humps on the back. During their growth, the larvae molt from five to seven time and ultimately pupate, frequently gathered together in large groups. Thus, by including pupal stage, their metamorphosis is a complete one, in contrast with such insects as dragonflies and mayflies. Case-bearing larvae fasten the cases to the substratum, perhaps also weighing them down with rather large pebbles, and close them off except for small openings for circulation at both ends. The spun closures are variously constructed covers that differ according to the species. Some have only a single rather large, round or slit shaped aperture, and others have several little ones, up to about 30.

Species of the genus *Heliopsyche*, which live in little snail shell shaped cases in the thin layer of water on dripping cliffs, fasten down only the closure before they pupate. Before emerging as a mature insect, the pupa bites off the cover, and the house is carried to calmer water. Many larvae in strongly flowing water spin an additional leathery cocoon inside their case and pupate there; in highly oxygenated water respiration is possible is spite of a completely close cocoon. Some larvae that are free- swimming or that live in webs build themselves spherical house of sand for pupation, other spin an enclosure between leaves, and any merely bury themselves in the bottom or hide in hole or in clefts in rocks. Thus may the beetle galleries in the branch of a tree, which for years there after may have served many a small bee or a wasp as a dwelling place, in later years eventually become useful under water to quite different insects. But certain caddisfly larvae themselves bore deep hole into wood and may damage old bridges or vessels that are laid up for long periods.

Ancient scientist such as Aristotle and Pliny designated these larvae as "*destroyers of wood.*" The pupae, which already resemble the mature insect slightly have free antennae, legs, and wings, and frequently free jaws, too. Like the larvae, the pupae maintain a circulation of water in their house by means of constant movements. The appropriate opening in the sieve are kept clear at all time with the aid of specially adapted cleaning instruments on the front and rear ends of the body. These take the form of brushes, pegs, or shears, and are present only in this pupal stage, which lasts scarcely more than two weeks. At emergence time, the oral shears serve also to cut open the front closing membrane of the house. As additional organs peculiar to it, the pupa has on its back special hooks that enable it to make the forward and backward movements essential for cleaning. When the time of emergence has come, the pupa frees itself from the cocoon or hiding place, and crawls upward on a plant or on any other object. Many pupa can swim upward with strong strokes of the middle legs, which are equipped with swimming hairs, and others are borne by air to the surface. There they support themselves by mean of the hairs of the lateral lines or else by grasping something with their jaws, until the mature insect emerges and flies off or walks ashore on the surface film.

The mature caddisflies ultimately display few differences among themselves. Rather monotonous in shape and drab in colouration, they resemble in form an attitude little moths, to which they are distantly related. Their wings, however, are covered with recumbent hairs rather than with scales. With few exceptions hidden through out the day, the caddisflies appear in clumsy, reeling flight toward evening, when they are attracted to electric lights. The caddisfly mouth is best suited for lapping, and many of the adults take no food whatever. The nuptial dances of caddisflies are enchanting; frequently they circle with whirring wings above the water surface, paddling vigorously with the middle legs, the female in advance and the male following. Or other species dance in a vertical posture in the air, with the very long antennae diverging. The flight period may last three or four weeks, or for some species only a few days.

The eggs some 20 to 800, according to the species are thrown in a gelatinous, swelling mass into the water, scraped off in flight by the surface film, glued to twigs or to other vegetation along the bank, or deposited deep beneath the water. Eggs fastened outside the water are protected by a gelatinous mass from drying out, and the first stage larvae that hatch can afford to wait there to be washed into the water by rain or dew. The life cycle of caddisflies mostly is an annual one, but frequently there are two generations a year. In a few species, the egg is the overwintering stage, but in most the larva overwinters on the bottom.

Caddisflies develop in almost all fresh water of the world. The larvae of various species also live in damp moss on trees and cliffs, in salt and brackish water, in the icy water of grottoes, and in the water collectors of bromeliad plants growing on the trees of tropical forests. They even are able to develop in the fluid of a tropical insectivorous pitcher plant (Nepenthes), apparently protected by their thick skin from the fluid's dissolving action. Occasionally mass swarms are observed, and then even in cities the house fronts along whole streets facing a river may be covered with these brown insects. Caddisflies are very ancient insects, known as early as the Lower jurassic period and very richly represented in the Tertiary. Over 150 species have been preserved in amber, but none of these exists today. Present-day caddisflies are divided into two sub orders with more than 20 families and over, 3,000 species. Their great importance in the pattern of nature is as one of the most important sources of food for fish.

Aquatic Moths

Water moths probably are not generally known. In fact, only a vanishing minority, with representation from a few families, have become adapted to the water. The degree of adaptation varies widely, with all kinds of intermediate forms between land and water insects. Thus some caterpillars of the family Arctiidae feed on plants extending over the water and, in doing so, occasionally crawl partly or wholly beneath the surface, and others of the same family dwell under water, swimming and crawling along with air carried in their hairy coast and constantly renewed at the surface.

Caterpillars of different families, such as noctuids, cossids, and pyralids, burrow in plants below water level, often clear down to the roots, without coming in contact with the water itself except perhaps when migrating from one plant to another. For doing this, moreover, certain cater- pillars bite off a piece of the plant and use it as a float. Another species builds itself a little boat out of a short, hollow stem, closing the two openings with silk. More highly specialized water insects are found among the micro moths of the family pyralidae. These sometimes appear on summer evenings over bogs, ponds, and little lakes in swarms like undulant white veils.

The individuals, like snowflakes, may chase after one another over the water surface in graceful nuptial dances. Their larvae build themselves houses of varied form, typical for each species. These houses in the water are made from leaves spun together, or simply from a piece of hollow plant stem, mostly they are filled with air, but by some species with water. Some float and some are dragged along as if by caddisfly larvae. Many larvae, especially those that live in flowing water, spin nets on stones or plants and live in them inhaling air bubbles that get stuck in the nets and likewise eating algae that are caught there.

For pupation, appropriate pupal cradles frequently are constructed. *Caraclysta* simply uses the caterpillar's case, namely a piece of hollow stem, and attaches it with spun silk at the water surface at right angles to a plant; or else makes a floating, oval, air-filled chamber out of the tiny leaves of rushes, carpeting the inside with silk, on the chamber's bottom the caterpillar lies enclosed in a silken tube.

Similar houses of certain American and Chinese species do not contain air, but oxygen-rich water flows through them without being able to wet the pupa resting within. The cocoon of *Hydro campa*, covered with two large pieces of leaf and fastened below the surface to the skin of a plant, is a silken web inside which the pupa stands upright. The web is filled with air that has escaped from holes bitten into the plant stem by that caterpillar; in other genera, on the contrary, the caterpillars themselves extrude air into the case from the supply stored within the body.

Before the moths emerge from cacoon, air is concentrated beneath the pupal skin. At the moment when the mouth breaks out of the shell, the air is stowed a way under the wings, as though in a diving bell, and lifts the insect to the surface. The pathway remains visible for a while as a vertically suspended with column of wax particles in the water, which have been brushed off the wings during the swift ascent. This waxen coating is the water proofing that keeps the insect fully dry until it reaches the surface, crawls ashore, and unfolds its delicate wings. Many caterpillars that live their whole lives under water breathe through gills that cover the body in tufts. They promote the circulation of water through their webs with oscillatory movements of their bodies. Whereas most water moths dip only the abdomen into the water when depositing eggs on plants, a few species with an unwettable garment of scales go completely beneath the surface for example, the female of *Acentropus niveus*, which lives only in water, swims with the help of legs fringed with hairs and with wings reduced to little oar-shaped stumps. The males have normally developed wings and go whirring over the water in search of some female that comes briefly to the surface. Strangely enough, in many regions there also are females of this species with complete wings. Thus the species has developed two different female forms, one adapted to air and the other to water.

Mosquitoes, Gnats and Aquatic Flies

Mosquitoes are among the most common and widely distributed insects. Rare is the person who has not had the experience of drifting contentedly off to sleep only to be awakened by the insistent buzz of a mosquito. Mosquitoes have earned a worldwide reputation as torturers of man and animals, and as disease carriers they have helped shape the world's history. They have made regions of the earth uninhabitable and have extinguished thriving cultures. Malaria-carrying mosquitoes helped free ancient Rome from the besieging hordes and likewise brought her people into great danger. Mosquito-transmitted yellow fever forced the French to give up construction of the panama canal after the loss of 20,000 lives. Of the thousands of different species of mosquitoes (family Culicidae), only a few are vectors of the dreaded tropical disease, such as elephantiasis, yellow fever, and malaria, the last of which at times penetrates far into the North.

The disease-causing mosquitoes are among the best-investigated animals in the world, for in combatting them a gigantic amount of scientific work has been done and is being done in many countries. Nevertheless, much remains to be learned, since the subject is a complicated one. Among the malaria mosquitoes, for example, some species bite only animals, and others also include man, but these and other characteristics may change from locality to locality. The activity of mosquitoes is affected by climate, light, and temperature. Thus they have two generations a year in the North, but up to ten in the South; and a northern European female produces 100 to 350 eggs, and Italian female perhaps, 1,300. The female of the most important European malaria mosquito, *Anopheles maculipenis*, live an over winter, so far as has been observed, mostly in the stalls of cattle, where at night they suck the blood of the animals, preferably from their eyelids.

Almost all female mosquitoes require blood meals, since it is only thereafter that their eggs can ripen; the males are not bloodsuckers, or only in exceptional cases. In many localities, the Anopheles females go into dwellings, too, infecting people with malaria; to transmit malaria, a female must have previously imbibed

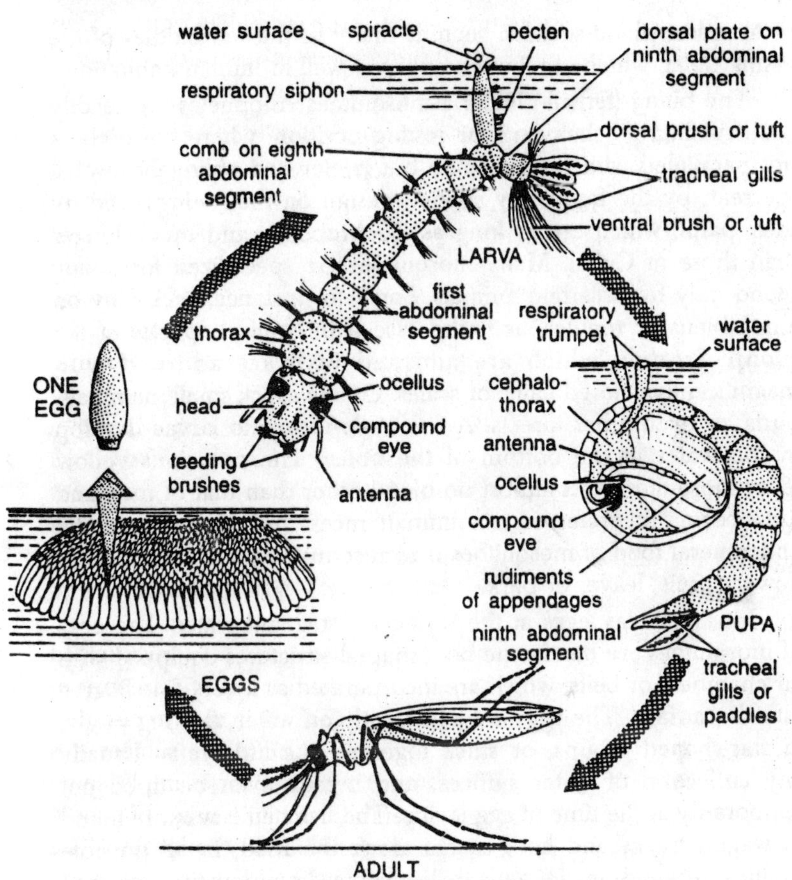

Fig. 2.19. Culex. Life cycle.

blood containing the malaria parasite in a quite specific stage from a persons with the disease. For another tropical disease, yellow fever, the mosquito *Aedes aegypti* is chiefly responsible, yet in the South American virgin forest the disease also has been transmitted from apes to man by species of the genus *Haemagogus.* The proboscis of mosquitoes is a tube formed from the labium (lower lip). It does not itself penetrate, but serves as a sheath enclosing a combination of six stilettos. These are sharp as knives, some of them saw-toothed and others with tips as smooth as daggers. When these pierce the skin the mosquito's saliva flows at once into the puncture, and the victim's blood is sucked in as a continuous stream by the pumping action of the digestive tube. There are multitudes of harmless blood-suckers, such as many of the mosquitoes of the genus *Culex*, which are known only too well in human habitations.

The biting (female) malaria mosquitoes Anopheles are readily distinguished by their angular resting position, where the body is not paralleled with the substrate but is elevated obliquely toward the rear, by the frequently spotted design on their wings, and by their palpi, which are as long as the proboscis and much longer than those of Culex. Many mosquitoes are specialized for taking blood only from certain animals. Some for instance, suck only on amphibians or reptiles, as is the case with tropical species of the group *Sabelini*, which are adorned with are adorned with magnificently colored hairs, or scales. Others attack small mammals, birds, or even other insects. A few *Aedes* mosquito larvae develop in the water at the bottom of the holes. The notorious yellow fever mosquito wants almost no blood other than that of man and even seems to prefer certain human races. But the fundamental and general food of mosquitoes is almost any vegetable fluids from flowers, fruit, leave, or bark.

Mosquites lay eggs at the surface of stagnant water. The eggs of mosquitoes are handsome boat-shaped structures equipped with air chambers or belts, which are incorporated as a very fine filigree into the surface. The eggs are laid mostly on water, floating singly, in star-shaped groups, or stuck together like little rafts. Usually any collection of water suffices, and it may even be dried out temporarily at the time of egg-laying. The hatched larvae are found in wagon tracks and hoof marks along the road, in all possible cavities in wood, in the axils of leaves, in discarded tin cans, and in flower vases and fonts. A few species even live in the digestive juice of carnivorous plants, feeding on the remains of digested insects or predatory upon other water insects that likewise are adapted to this extraordinary milieu. And yet others are at home in pools of salt or brackish water and in holes in the rock.

In the far North the eggs laid in the summertime in dried-out puddles remain at rest for eight to ten months, frozen solid in the winter. Also preserved in ice for the winter, in the tubular traps of the insectivorous *Sarracenia* plants in North America, are the larvae of the mosquito *Wyeomyia smithi*. The larvae of mosquitoes are just as striking as they are characteristic. The majority, including those of *Culex*, hang stiffly in groups beneath the water surface, with the head extended; again and again undulant wagging movements come over the whole assemblage. The side of the short, thick fore body and of the cylindrical, clearly segmented abdomen bear tufts of long bristles that sand stiffly away body and the rear end is divided into a cone-shaped propulsive organ bearing swimming fans, and a somewhat longer respiratory tube with a tiny, star-shaped wreath at its tip. By means of this wreath the body is suspended from the surface film of the water.

The larvae of the fever mosquitoes *Anopheles* float horizontally, fastened to the film along the back by means of special bristles and warts and by means of the respiratory shell, into which open the two principal air tubes of the body. This shell replaces the respiratory tube. The food of mosquito larvae consists of little particles in the water, especially of algae, but also of animal

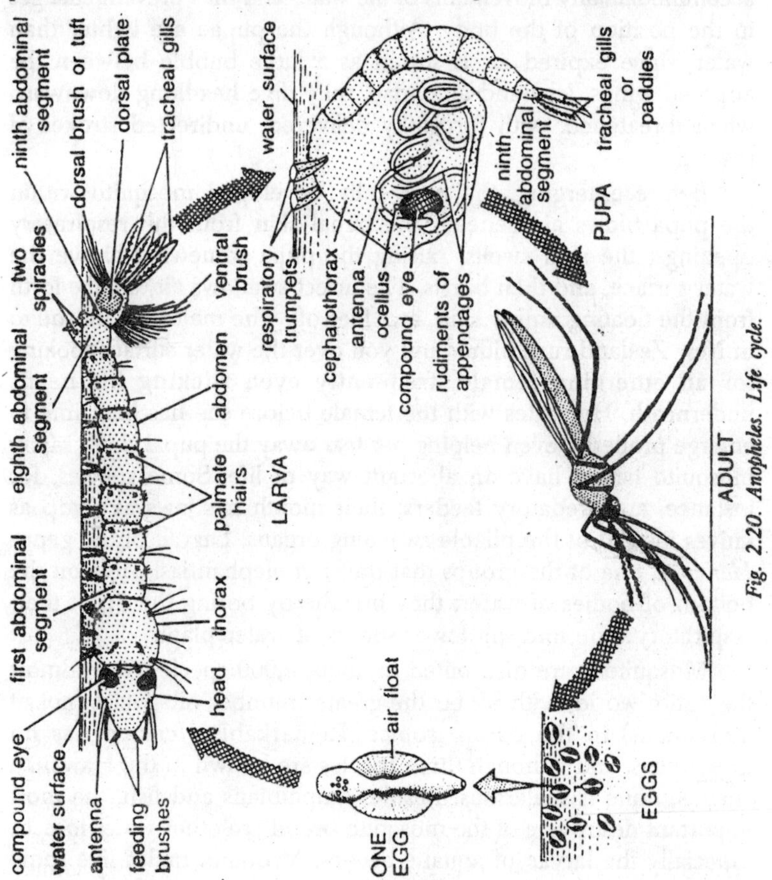

Fig. 2.20. Anopheles. Life cycle.

substance, that are swept in by means of a current engendered by two large, hairy fans, beating in opposition, on the outer mouth parts. Before the accumulation is swallowed, it is sorted over with other oral parts, and balls of what is deemed unsuitable thereafter sink to the bottom. The water thus filtered by a single larva in a day may amount to 2 quarts or more. On the fourth molt the mosquito larva transforms to the pupa; frequently the massive numbers of pupae cover entire ponds with a black layer. These creatures, slightly less motile than the larvae, have a thick, lumpy forebody, an abdomen curved halfway beneath it, and leaf like caudal fins. The pupae also hang from the surface film, fastened to it in the fore part of the back region by two short, continuous respiratory tubes, that, because of their basal articulation, accommodate any movements of the water and thus prevent changes in the position of the body. Although the pupae are lighter than water, since expired air is stored as a large bubble between the apposed wings, legs and antennae, they dive headlong downward when threatened, with forcefully delivered, undirected strokes of the tail.

Before emerging, the new, fully developed mosquito within the pupa blows air beneath the pupal skin from the respiratory openings; the skin swells, raising the pupa somewhat above the water surface, and then bursts. The insect swallows air, climbs forth from the floating empty skin, and flies off. The male of a mosquito in New Zealand runs hither and yon over the water surface looking for an emerging female, frequently even sticking his heads underneath. He mates with the female before she has had time to emerge properly, even helping her tear away the pupal shell. Many mosquito larvae have an aberrant way of life. Some species, for instance, are predatory feeders; their mouth has jaws as sharp as knives instead of the pliable sweeping organs. Larvae of the genus *Mansonia*, one of the groups that transmit elephantiasis, live on the bottom of bodies of water; they breathe by boring the tip of their respiratory tube into the lower shoots of water plants.

Mosquitoes are distributed in about 2,000 species over almost the entire world, with by far the greater number of species (not of individuals) in the damp tropics. Remarkably, Iceland has no mosquitoes, and although three species are known in the Hawaiian Islands, none is malarious. Besides amphibians and fish, the most important destroyers of the mosquito brood are other water insects, especially the larvae of aquatic beetles. Mosquito males live some two to three weeks, the females three to four months, and under favorable conditions development from the egg to the adult maybe completed in less than two weeks. The tropical genus *Psorophora* even does it within five days; the egg hatches on the first day, and a molt occurs on each day following.

The tiny flies known as gnats also suck blood. Of these, the biting midges, or punkies (Ceratapogonidae), in spite of a body length of only 0.04 to 0.08 inch, sting more painfully than mosquitoes. Some are vectors of disease. Many species attack only amphibians, reptiles, or insects; one of them sucks mammalian blood from the abdomen of a mosquito that has ingested it. Others visit flowers, and certain tropical species are useful as pollinators of cocoa and rubber trees. The larvae of many gnats, especially the bloodsucking ones, develop in water. Shaped like·worms or wires, they move with undulations, seeking prey on the bottom or in the mud; their body length of about 0.8 inch is larger than the gnat that arises from them. The females deposit eggs in gelatinous masses, at times in long strings.

In certain years, many places in the world are besieged by the tiny, dark, humpback black flies (Simuliidae). Called buffalo gnats in America, black flies in the Canadian woods, and Columbian gnats in Europe. They fall upon animals and people in swarms of millions. They penetrate beneath clothing and into the nose, ears, and eyes; their bites, containing a poison that affects, are nerve centers leave red, bleeding spots. Black flies kill buffalo, cattle, reindeer, deer, and horses, by the thousands; human beings have died also for 20 to 30 bites may cause a fever of as much as 104°F. These gnats develop only in flowing, and especially in cold, water where their wedge-shaped larvae cling to plants and rocks in part by a suction plate and in part by spun threads at the rear of the body. A few larvae walk along the threads of a web they have spun, by putting out a drop of sticky material from the mouth, by flexing the body and attaching the rear end to the drop, and then by stretching out forward again and setting the next attachment point on the thread. Black-fly larvae have two large fans of bristles spread out in front of the head that act as a sieve. Algae or other particles carried in by the current stick to these bristles, are swept off from time to time with special oral bristles, and are swallowed.

The pupae, by means of threads and numerous hooks, are caught fast in a basket of sticky webbing; this basket, with the opening facing the current, is hung on plants or rocks and comes to a point at the rear. Out of this house there project upward from the pupal head two bundles of long, thin, silvery threads, through the entire surface of which oxygen is taken up from the water. Towards the end of development, the pupa stores air beneath the skin, and this air carries the emerging gnat to the surface. Certain moth flies and sand flies (Psychodidae) also bite painfully and in the tropics and subtropics transmit various diseases. These tiny insects hold their broad, hairy wings in a moth like position. Many of their larvae live in the zone between land and water, where they feed on rotting plants. In the water they take it through open spiracles at the rear end. Certain larvae live in flowing water, holding onto objects with their stiff hairs, and a few tropical species that inhabit waterfalls are fastened to the rocks with a series of suction cups.

The bites of the much larger horse flies and deer flies (Tabanidae) are pure torture. Although in Africa and India these insects transmit certain diseases of cattle, elsewhere they cause annoyance only by means of their itching bites, which may cause appreciable losses of blood. Only relatively few of the larvae live in water; these are predators, either on the bottom, in lower layers of mud at the surface, or frequently on the banks. They are single-shaped maggots that inhale through a protrusible by short breathing tube formed from the last segment. Among the most interesting water insects are the larvae of the family Corethridae, whose members are closely related to the mosquitoes. Respiration take place through the skin of these freshwater predators, for the tracheal system has been reduced to two paired, curved air vesicles that act exclusively as a hydrostatic apparatus. Living in little puddles as well as in big lakes, and of all insect larvae probably the best adapted to a pelagic way of life, they float at any desired depth in the water, wholly independent of atmospheric apparatus. Living in little puddles as well as in big lakes, and of all insect larvae probably the best adapted to a pelagic way of life, they flat at an my desired depth in the water, wholly independent of atmospheric air or of any point of support, as completely planktonic organisms. They swim with the help of a large fan of hairs, directed downward at the posterior end, and eat all possible living aquatic things; these are seized with the two antennae, which are lengthened with stiff, curved tufts of bristles, and then cut up with the razor-sharp mandibles.

The *Corethra* species, which live in large numbers as deep as 120 to 160 feet in big lacks, bury themselves by day in the bottom mud, perhaps more than 70,000 larvae per square yard, and then ascend at night. Astonishingly, a pressure change from 4 atmospheres to 1 atmosphere within the space of perhaps a single hour is thus no problem to these organisms, which look as though they were made of the most delicate glass. Thousands of their pupae float vertically in the water and apparently rise or sink automatically to the level with which they are in equilibrium. When the gnats emerge from the pupae and their swarms fill the air, frequently with a high-pitched buzzing but often without sound, the empty pupal cases are washed ashore by the waves and look like a broad band of white foam.

A few peculiarities are displayed also by the water-dwelling larvae of soldier flies (Stratiomyidae). Those of the genus *Stratiomys*, which are long, legless maggots drawn out posteriorly, hang from the water surface as though glued to a single point by means of the slender rear end, but with the body constantly undulating or writhing. At the posterior tip of the abdomen, where the two spiracles open, there is a beautiful outspread wreath of hairs; the wreath lies on the water surface. At a sign of danger it is clapped together in the shape of a bell, containing a large bubble of air; with this the larva glides to the bottom.

In spite of their tininess, the mouth parts are astonishingly diverse and complex; they not only shave the algal layer from leaves, but also sweep in free-floating food. Moreover, they are able to move the whole body forward slowly by means of their constantly moving rows of "cilia." The integument of these larvae is very thick because of the large amount of incorporated lime; and it gets much thicker still before pupation. Ultimately this protective cost of mail conceals a much smaller cocoon containing the resting pupa. In gorges as far north as the glaciers, live the flattened larvae of the net-winged midges (Blepharoceridae) during eleven months of the year. Amid the strongest currents they are fastened almost immovably to rocks by means of six big suction cups along the middle of the underside; these are among the most perfect retaining devices in the animal kingdom. By pulling in a fleshy cone beneath the cup, a space with reduced pressure is formed, producing the suction; the cups also give off a sticky material. Grazing on the algal layer, these larvae move both forward and sideways by alternately releasing the suction in the individual cups.

Among the strongest currents, the pupa is glued to rocks slightly below the surface by means of special plates. During the instant of emergence, the gnat clings with its long hind legs to the pupal skin while the forelegs rest on the water surface; then it flies off. In torrents this manuever frequently fails, and hence adults are rather rare. Their wings have few or almost no veins, but are crossed by a network of folds; and the eyes are divided into an upper zone with very large facets and lower zone with little ones, a division that enables the insect to see when it is light and when it is dark. The majority of females suck the blood of insects, especially of other gnats, but the males frequent flowers. The water insects that most universally have mastered the most various environments are the larvae of the shore flies (Ephydridae). These are thick-skinned creatures of simple, maggot like form, lengthened posteriorly into a forked respiratory tube. They crawl by means of eight pairs of spiny warts, or swim with violent twitching movements of the body. They live in salt lakes and salt bogs, in salt pits, in hot springs, and even in oil swamps, where almost no other living thing is able to exist.

One and the same species seems to live just as well in warm as in cool water, or in fresh, brackish or sea water; and it lives in pool on the seashore no matter whether these have dried up today, are filled with rain water next week, or later are inundated again with sea water. These pools, which are evaporating and being refilled constantly, finally attain a much higher salt content than the sea itself, and may reach temperatures approaching 104°F (40°C). As is well known, other water animals die when changed suddenly from fresh to salt water and vice versa. When large numbers of these flies all deposit their eggs on the water at one time, they frequently develop swarms that move along the shores of salt takes as compact, black belts visible miles away. Peculiar salt-water flies are the little Pacific *Pontomyia* (Ephydridae). Using their middle and hind legs and their oar-shaped wings, the males run along the surface, forelegs held aloft and bent toward the rear. The females, larval in shape, have only the two posterior pairs of legs, both greatly reduced, and they live under water in mud tubes they have built. Some of the larvae of another family have the ability to live both in fresh and salt water. This family is the midges (Chironomidae), which resemble mosquitoes. A few, as a very great rarity among insects, live exclusively in the ocean. Where there is surf they live in the canal within sponges, and are found at depths of over 90 feet. The mature female gnat of one species herself has only a maggot like form, almost without legs, antennae, and wings; she does not leave the canal within which she developed, but mates by stretching her rear end out of the water. Some of these gnats walk on the water with the hairy soles of their feet, or fill the air with dense swarms.

Of all water insects, the midges are thc family that probably is richest in number of species. Their larvae show the greatest diversity of ways of living, and yet almost all have an identical worm like form. They live on land in plants and rotten wood, in hot springs, in icy lakes at elevations above 15,000 feet in the Himalayas, in subterranean caves, in waste water, in wells, and in the ocean. The best-known midge larvae no doubt are those red worms of the *Chironomus group*, which cover the bottom of lakes like a carpet. Their mass surpasses the combined masses of all other living organisms in these bodies of water. Like few other insects, they have red blood. Protruding vertically from their tubes buried in the mud, they wave back and forth, catching particles sinking from above.

The pupae likewise remain in the tubes, but rise to the surface when transforming to the adult, and do so in such masses that the emergence of these insects from the water makes an audible rustling. In buzzing clouds that fluctuate in shape from steeples to mushrooms, the midges fill the land scape for miles around, slipping into the eyes, ears, and nose of persons and animal. Meanwhile the empty pupal shells on the shores are heaped together into broad ribbons by the waves. Midges and their larvae, like mayflies, are an important source of food for water insects, fish, birds, and other animals.

Many larvae build dwelling tubes from webbing or gelatinous masses. These tubes differ from species to species. Tubes that lie in or are stuck into mud mostly are curved, so that both open ends protrude into the water. Frequently tubes are fastened to plants, some by threads that let them wave freely back and forth in flowing water. Many species include sand or algae in their spun structures, and carry these houses around with them like caddisfly larvae, others build on rocks masses of adjacent tubes of webbing, which become more and more heavily petrified by depositions of lime.

Before pupation, the various houses are enlarged at the front, forming a chamber, and are closed over with spun lid. A few species bore passage ways lined with webbing into colonies of sponges or moss animalcules (Bryozoa), and others eat grooves into the upper sides of green leaves and close these over, like tunnels, with spun material. Many mine all the way into plants, and such mines sometimes are filled with air; more often, water is circulated through them by continual oscillatory movements of the larvae. A North American species spins a net in its passage way about every ten minutes, and then eats the net together with the plankton caught in it.

Water Beetles

Beetles show many gradations from land insects that like moist conditions to the most highly perfected water insects. In the boundary zones between land and water, in wet sand that at times is inundated, in mud or moss or pools brooks, lakes and seas, underneath stones washed by the water, or on plants sticking out of the water, there is abundant beetle life. By far the majority of the species that have settled in such places nonetheless have no particularly close relationships to the wet element. Yet some show in at least one developmentsal stage the beginning of an adapation to water. Certain larvae, for example, though much like land dwelling larvae, nevertheless respire through gills located on the rear end and live in a chamber that can be closed by a lid and from which the body can be protruded. Among beetles the first adaptive stages toward a life in water show in the form walk on the water surface or a hair covering on the body, in which a layer of air is taken along when the insect lays its eggs under water. The pupa of certain heloids is bedded down in an air bubble under a stone in the water, and this air bears the emerging beetle to the surface. A few weevils (curculionidae) that crawl on floating leaves and on under water plants near the surface have larvae that mine in stems below water level and that breathe air from the plants.

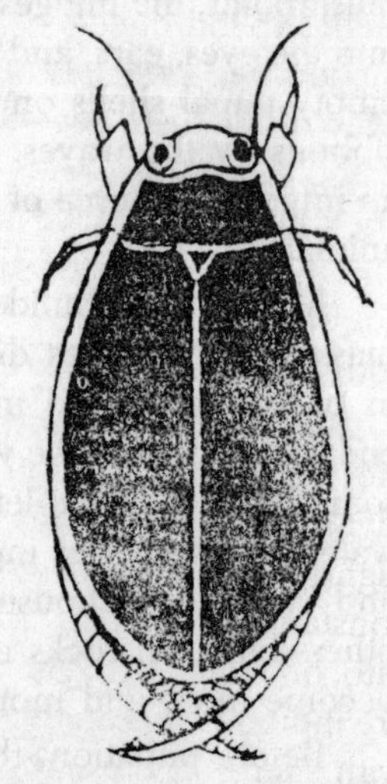

Fig. 2.21. The water beetle Cybister.

The larvae of the long horned leaf beetles (donaciines) of the family *Chrysomelidae* inhale air through two big posterior spines hooked into plant roots, while the used air escapes through special spiracles. Perhaps never moving more than inch or so away, these larvae mostly hang motionless throughout their two year life, sucking sap from a hole they have bitten out and keeping it watertight by forcing in their head an first thoracic segment. When ready to pupate, the larvae builds a house by enclosing its body and its support in a wax like secretion and then expanding the enclosure to the proper shape by inflating itself with air. In addition, the larva curls up into a humpbacked position with the support of a cushion of air it has collected beneath its body. Afterward, the larva collapses again, becoming much smaller, and ultimately points over the whole inside of the house with a layer of varnish made of water repellent saliva and of sap from the gut therefore it pupates.

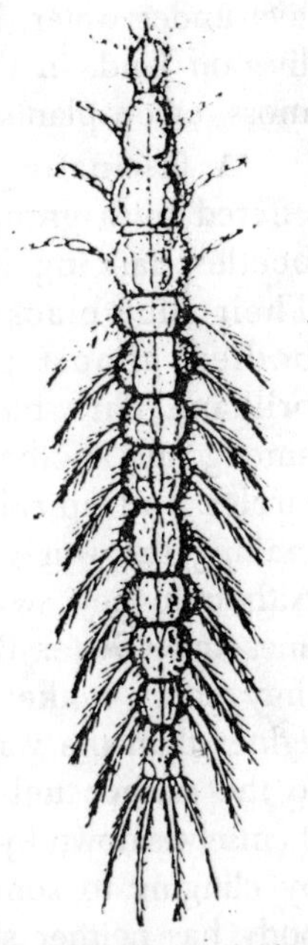

Fig. 2.22. The whirling beetle Gyrinus-larva.

In one genus of these long horned leaf beetles, namely *Haemonia*, even the adult beetles spend their whole life deep under water without coming to the surface. Their slight oxygen requirement is satisfied by a thin layer of air held beneath the hairs of the body. The little beetles of the family *Helmidae* are equally perfected for a life in water. They likewise remain in the water during all stages, scarcely ever coming to the surface at all. They take up little bubbles of oxygen from algae and rocks with the mouth and conduct them by way of body groove to the spiracles as well as to the lower body surface.

Generally, the term "water beetles" refers to the families of the whirligig beetles (Gyrinidae), the predaceous diving beetles (dvtiscidae), and the water scavenger beetles (Hydrophilidae). Their highly polished, broad, and frequently flattened body, kept greased constantly by mean of countless glands, seems as though molded into one single, streamlined figure. The individual portions of it are indicated merely by finely engraved lines. The legs are flattened, with sharp edges, and the hind legs in particular are modified into oars, most of them further expand by means of borders of hairs.

The mature beetles and the larvae live under water, but the pupae live on land, in the soil, under moss, or on plants.

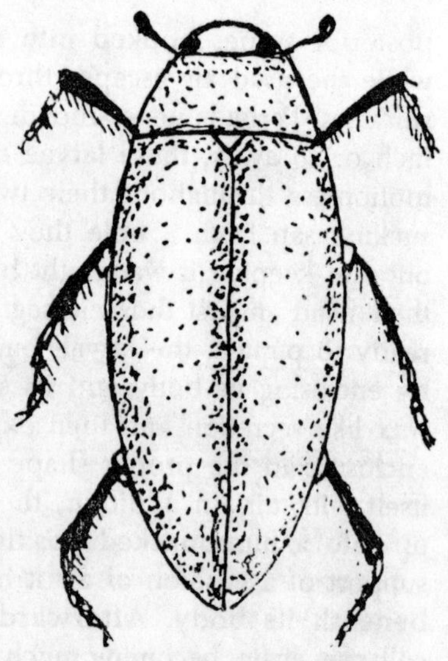

Fig. 2.23. Water scavenger beetle Hydrophilus.

A fascinating spectacle is offered by a group of whirligig beetles dancing on the water. Their blue-black streamlined bodies, almost penetratingly brilliant, dark back and forth among one another in turbulent circles and spirals, seemingly making the water surface boil. Although they swim in straight lines under water. Because the air they carry makes them much lighter than the water, they bob to the surface unless they keep themselves down by swimming or by clinging to something. Their well lubricated and always dry body has neither sharp edges nor projections that might rupture the water surface when they go tearing along. The form and co operative action of their four swimming legs are impressive miracles. Their eyes, too, are fantastic; one pair is above the water into the interwoven shadows of the bottom. The whirligig beetles represent a very ancient group, belong to the most highly specialized of water insects, and are geniuses at almost everything. They swim and dive, sit on rocks on the shore in the sunshine, and in the evening and at might fly around for great distances in the vicinity. And in spite of the unimpressive appearance of the legs, the beetles progress quite well on dry land, even hopping when necessary, and the species that prefer swiftly running water swim quite fast against even a strong current. On land and on the water surface the whirliging beetles breathe through spiracles like land insects, but under water they carry on their rear a bubble of air that is renewed from time to time. These beetles over winter both under ice in a big air bubble on plants, and on land under stones.

3

PLANT EATING INSECTS

A number of insects macerate plant tissue or imbibe plant fluids digest these tissues to utilize them for energy, growth, and reproduction. This ability of insects is called *phytophagy* or *herbivoury*. Oddly, insects have limited abilities to digest cellulose, the major substance of which plants are constructed (although some have symbiotic micro-organisms that assist in its breakdown). Plant tissue is a source of sugars, proteins, fats, salts, water and vitamins. Most plant tissue provides adequate nourishment for insects, although different species, do have different nutritional requirements. That insects are to varying degrees discriminating feeders reflects a long and fascinating history of plant-insect coevolution, which we shall explore briefly in this chapter and the following two. Insects feed on leaves, buds, stems, roots, fruits, and seeds, as well as on plant tissue in various stages of decay. They feed externally or internally, a borers or leaf miners; they produce galls and others distortions; sucking insects imbibe plant juices and may cause weakening and yellowing. Many insects, even non-phytophagous species, also use plants as shelter, and some make still other uses of plants. Insects like leaf-cutter bees, use pieces of leaves to line their nest cells and leaf-cutter ants harvest bits of leaves that they use as a substrate for growing fungi in their nests. Thus insects indirectly make use of plants.

Every plant is attacked by variable number of insects. Clover is fed on by some 200 insect species; corn, by over 300. Over 200 species attack citrus trees; 400 or more, apples. Among forest trees, elms harbor at least 600 species, while oaks hold the record with nearly 1500 insect enemies (many of them gall formers). Out of its normal habitat, a plant may sometimes be relatively free of insect pests. Eucalyptus grown in Mexico or California, for example, shows little evidence of insect attack; yet in its native home in Australia, eucalyptus harbours a great number of insects. Obviously its foliage is perfectly edible for adapted insects; but insects elsewhere have not evolved mechanisms for overcoming, the repellency of eucalyptus oils. Ginkgo trees are seldom if ever subject to insect attack whereever they are grown(they no longer exist in the wild). These trees are relics of a very ancient group of plants that have evidently survived as a result of repellent substances in their tissues that deter both insects and disease. The degree to which a plant species is immune to insect attack is, in general, a reflection of the defenses it has evolved and the evolved abilities of insects to overcome these defenses.

Insects are specific in their habitats. Some insects inhabiting a particular plant, some may restrict their feeding to that plant species alone and are said to be *monophagous.* Others may be general feeders and include a diversity of plants in their diet; such insects are said

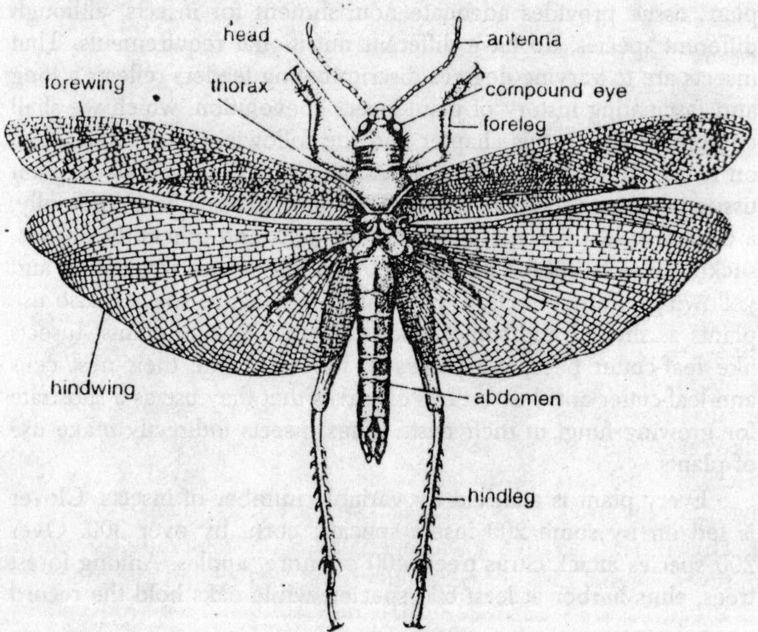

Fig. 3.1. Locust (Dorsal view).

to be *polyphagous.* White oak, for example, plays host to several species of gall wasps that not only form gall on white oak alone but even form a distinictive type of gall on only one part of the tree. White oaks may also be attacked by gypsy moth caterpillars, which are decidedly polyphagous insects, attacking a wide variety of broad-leaved trees and even, at times, conifers. Under artificial conditions these caterpillars have been reared on over 400 plant species; but they reject certain plants (such of shade and forest trees. Some of the Orthoptera are strongly polyphagous; Hungry migratory locusts will gnaw on wooden fence posts, and crickets will consume clothing. Indeed crickets, many ants, and some other insects are essentially omnivorous, including both plant and animal tissue and sometimes detritus in their diet.

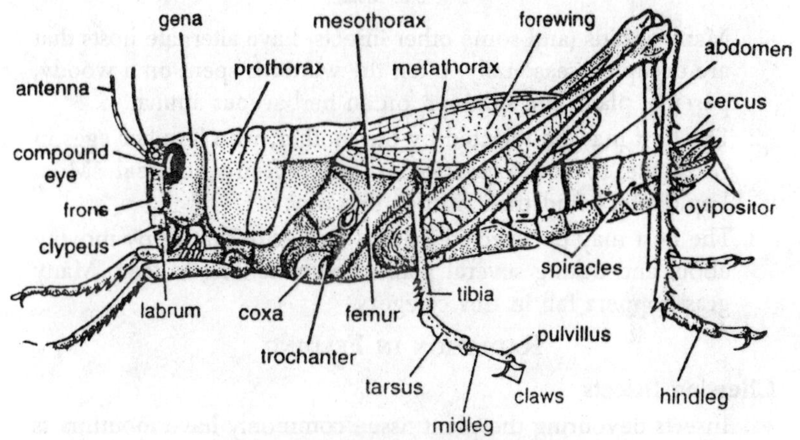

Fig. 3.2. Grasshopper (Lateral view).

Some insects are moderately discriminating in their tastes. Such insects are said to be *oligophagous.* The Colorado potato beetle feeds on plants of the genus solanum; the imported cabbageworm on various Brassicaceae; the monarch butterfly, on various kinds of milkweeds. We shall consider the chemical basis of host selection in a later section of this chapter. In behavioural terms, a phytophagous insect may reach its host in one of three ways:

1. The insect may live in an aggregation that extends through several generations, so that emerging young find themselves already settled on their host plant. Aphids are an example. However, aphids also have a Type I phase in their life cycle.

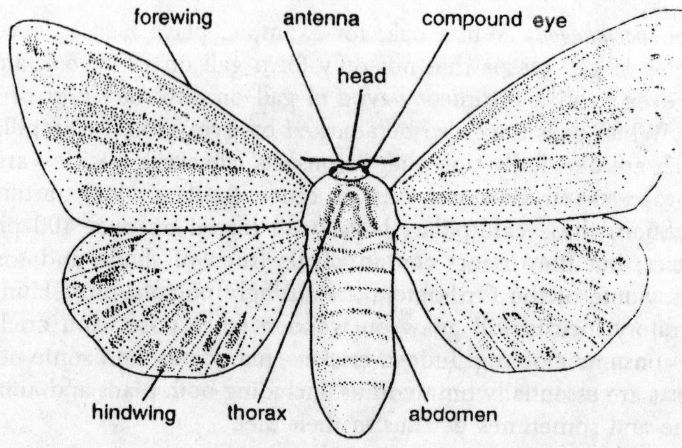

Fig. 3.3. Moth.

Many aphids (and some other insects) have alternate hosts that are occupied seasonally; often the winter is spent on a woody, perenial plant; the summer, on an herbaceous annual.

2. The host may be selected by the mother, who lays her eggs in response to some particular cue, often chemical. Most Lepidoptera find their host in this way.
3. The host may be selected by trial and error-that is, by moving about and tasting several plants before settling to feed. Many grasshoppers fall in this category.

Radiation in Feeding

Chewing Insects

Insects devouring the plant tissue commonly have mouthparts of generalized biting type, with stout, strongly musculated mandibles. Caterpillars may consume many times their own weight in plant tissue in the course of their development. Much fibrous tissue passes through the gut undigested and forms a major part of the large faecal pellets. Many insects begin feeding at the margin of the leaf, while others feed on either the upper or, more commonly, the lower surface. The leaf may be eaten all the way through, or the insect may scrape off the epidermis and parenchyma, leaving one layer of epidermis and its supporting veins intact. Such insects are said to be leaf skeletonizers. Others mine the interior of the leaf, feeding on the parenchyma and leaving both the upper and the lower epidermis intact. Such leaf miners tend to be small larvae that are pale in colour and flattened, and whose mouthparts project forward (prognathous) rather than downward (hypognathous); most of them are legless or nearly so. The patterns they form are often intricate and diagnostic of the species.

There are two general types. Some leaf miners chew out a broad patch, forming a blotchmine, while others move along a slender path, forming a linear mine, which is often quite tortuous. Linear mines tend to be very slender when the larva is small, but the broaden gradually, eventually ending in a small blotch. In India there are more than 1500 species of leaf-mining insects, belonging to orders of Hymenoptera (Coleoptera, Diptera and Lepidoptera).

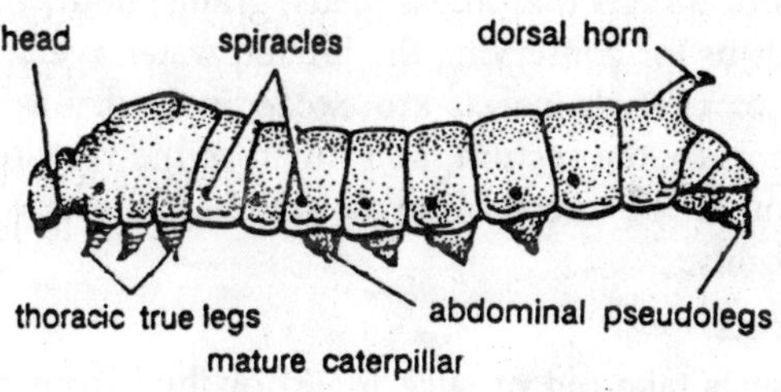

Fig. 3.4. Caterpillar of silk moth.

Plant insects belonging to Lepidoptera are usually *leaf tiers*, *leaf rollers*, and *leaf folders*. These insects use silk to hold leaves together in various ways to provide a retreat in which they molt or spend other inactive periods. Sometimes several such larve life together, forming a large, often unsightly mass of leaves and silk; one such species is appropriately called the uglynest caterpillar. Tent caterpillars build communal retreats primarily of silk, and they build silken trails that they follow to and from the nest during foraging trips. Borers in stems, trunks, and roots have some features in common with leaf miners: They tend to have larvae that are pale in colour, ***prognathous***, and more or less legless. Borers in trees have particularly powerful mandibles as well as proventriculus capable of grinding the hard particles to a usable size. Even so, much undigested material is passed off in the dry feces.

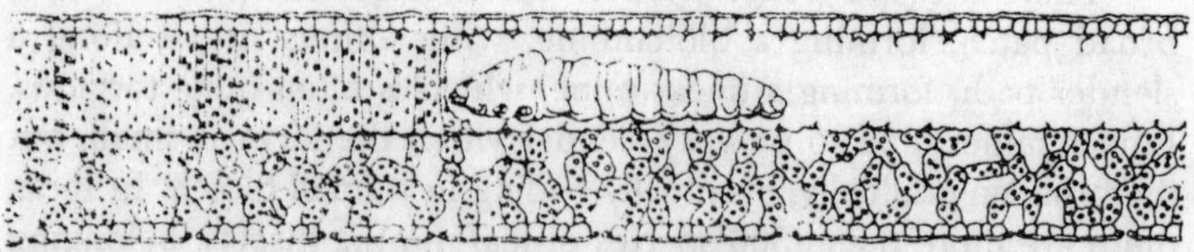

Fig. 3.5. Cross section of leaf being mined by a beetle larva.

Many borers have intestinal symbionts that assist in digestion, while others feed not on wood but on fungi that grow in the galleries. Larval insects of several orders live in fleshy fruits; others live in nuts and seeds, which like fruits provide a rich source of food, but unlike fruits provide a very dry environment. Thus, the many species of insects that infest seeds, grains, flour, and the like have mechanisms for conserving the limited water available in their food. The feces of such insects are exceedingly dry as a result of water extraction by the rectum, and much of the required water is obtained through the metabolism of starches, thereby show ater conserving ability.

Sucking Insects

Some insects take out or suck liquid or fluid from plants. The mouthparts of Hemiptera, are admirably adapted for piercing tissues and extracting fluids. Through muscular action, the four very delicate stylets are able to penetrate leaves, stems, and even the bark of trees. They are sufficiently flexible to pass between fibrous elements and, after considerable probing, reach the phloem or vascular bundles. Here the pressure of sap within the plant may induce a flow up the food channel between the stylets; or the flow may be assisted by the pharyngeal pump. At the same time saliva is being injected via the salivary channel in the stylets. Many sucking insects produce two kinds of saliva, from different parts of the salivary glands. One is a thin fluid that mixes with the sap and initiates digestion; the other, a more viscous substance that combines with fluids from accessory glands to form a sheath around the stylets. This sheath is formed of lipoproteins that gel on contact with air as result of the formation of hydrogen and disulphide bonds. Evidently the function of the sheath is to prevent the loss of sap and saliva.

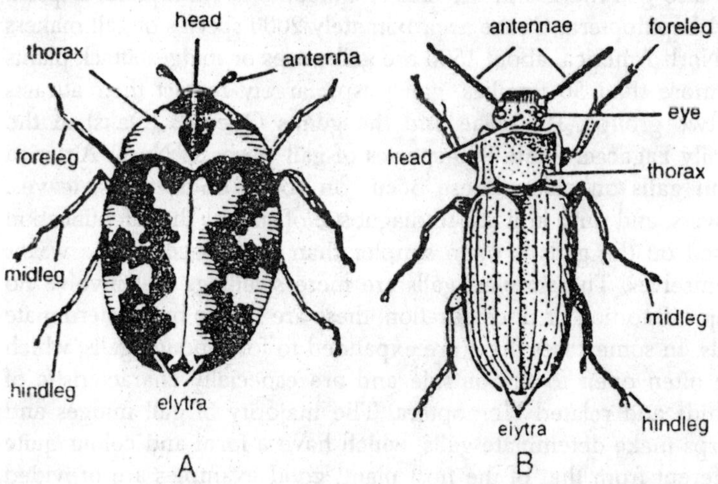

Fig. 3.6. Beetles. A–Carpet beetle (Anthrenus sp), B–Meal-worm beetle (Tenebris molitor).

Sucking insects must imbibe much fluid in order to obtain their requirements of protein, minerals and vitamins. Much of this fluid– chiefly water, carbohydrates, and some amino acids--is passed from the anus unmodified, and in fact in some Homoptera it bypass the midge by means of a filter chamber. The sweet, watery excrement of aphids, leafhoppers, and scale insects forms a stricky deposit on foliage, on the ground, and on the tops of automobiles parked under infested trees. This honeydew is fed on by bees, wasps, ants, and other insects, and it is also the medium on which a sooty fungus grows, often causing disfigurement of ornamentals. The honeydew of certain scale insects, and it is also the medium on which a sooty fungus grows, often causing disfigurement of ornamentals. Israelites use. These honeydews are called the *mauna from heaven*. Israelities use honeydew of certain scale insects of the Near East as food. These honeydews are called the *manna from heaven.*

Gall Insects

Sometimes plant develop abnormal growths on the buds, leaves, stems, or roots of plants, these are called *galls*. They result from the action of bacteria, fungi, nematodes, mites, or insects of several groups. Gall midges (Diptera, Cecidomyidae) and gall wasps

(Hymenoptera, Cynidae) are most frequently involved, but there are also gall makers in the orders Coleoptera, Hemiptera, Diptera and Lepidoptera. Of the approximately 2000 species of gall makers in Norh America, about 1500 are gall wasps or midges attack plants of more than 50 families, gall wasps largely restrict their attracts to two groups: Rosaceae and the genus Quercus (Oaks) in the family Fagacae. Some 800 species of gall wasps in North America form galls on oaks. These occur on roots, buds, twigs, leaves, flowers, and nuts, and are so diagnostic of species that identification based on the galls is often simpler than that based on the wasps themselves. The simplest galls are mere swellings that involve no major distortion or discolouration; these are said to be indeterminate galls. In some cases these are expanded to form pouch galls, which are often open to the outside and are especially characteristic of aphids and related Homoptera. The majority of gall midges and wasps make deteminate galls, which have a form and colour quite different from that of the host plant; good examples are provided by the willow cone gall and the oak apple gall.

The galls exist in such variety that many different terms have been employed to describe them, and guides have been written for their identification. Yet these elaborate structures consist entirely of tissue supplied by the host plant under stimulation by invading insects. Often they resemble abnormal fruits (cones on willows!), but rather than containing plant embryos, they provide insects with a rich source of food as well as protection from predators and from the elements. Galls may continue to grow as long as the stimulation presists, even though normal leaf or stem growth may have ceased for that season. Galls begin as small swellings at the point of oviposition by the female. In a sawfly gall of willow, studied by William Hovanitz, of the California Institute of Technology, the egg hatches in about five days, during which time the gall grows slowly. When the egg hatches and the larva begins to feed, growth of the gall continues. Dr. Hovantiz found that a fluid injected at the time of egg laying initiates gall formation. However, after about eight days the presence of the larva is required for continued growth. When the larva is required for continued growth. When the larva is removed from a gall, growth stops in about two days. A growth-promoting substance is secreted by both the mother, at the time of oviposition, and the larva secrete. The chemical similarity of those two secretions is still doubtfull.

Plant growth hormones are involved in gall formation, since abnormal growths produced by these hormones are similar to the cells and tissues in natural galls. Studies of needle galls of pinyon pine (*Pinus edulis*) by J. Wayne Brewer and his associates at Colorado State University have revealed levels of auxin and gibberellin in these galls many times higher than in normal needles of the same age. Extracts of the midge larvae that form the gall did not contain auxin and showed only traces of gibberellin, and it seems likely that the plants themselves produce these abnormal amounts of hormones under stimulation by the larvae. The substances secreted by the insects have so far defied full analysis. They must surely contain some components that differ from species to species, since the resulting galls on leaves of the same plant. It is possible that a fuller understanding of gall formation will shed light on processes of cell differentiation in all living things. What little we know of the gall-producing secretions of insects suggests that they contain adenine and other amino acids as well as nucleic acids.

Finding of Host Plant

How does a grasshopper "decide" which plants in the environment provide the most suitable food? How does an aphid or a butterfly locate an acceptable host plant for settling or for oviposition? These are questions of considerable importance, but only partial answers are available. The reaction's chains may involve several sensory modalities and highly specialized receptors. It is useful to recognize four stages in host finding:

1. Search for a suitable habitat;
2. Settling on a plant and the most suitable part of a plant (either for oviposition or prior to feeding);
3. Initiation of feeding ("tasting");
4. Feeding to satiation.

The insect may fail to receive appropriate stimuli or may actually be repelled, at any stage, necessitating a return to an earlier step in the sequence. Starvation or failure to lay eggs on a suitable substrate may result if the insect is unable to complete the series.

Cues for Search a Suitable Habitat

It has been repeatidely observed that apple maggot flies are attracted to yellow surfaces. Similarly, winged aphids can be collected in yellow-painted pans filled with water. In both cases yellow is probably not distinguishable from the green of the host plant, but yellow has higher reflectance properties than green and is thus a supernormal stimulus. Form as well as colour may be involved. Immature desert locusts are attracted to patterns of vertical stripes, evidently because these stimulate the grassy habitats they prefer. During dispersal, insects may become increasingly responsive to specific cues likely to guide them to a desirable habitat. These cues are evidently largely visual, but olfactory cues may also play a part, at least at short range.

Visual Cues

It is probable that visual cues often continue to play a role as the insect settles on a plant within a selected habitat. Female pipevine swallowtail butterflies (*Battus philenor*) alight on any leaves that resemble in shape those of their host plants, *Aristolochia*, but they lay their eggs only when, after drumming the leaves with their fore tarsi, they detect specific odor cues. According to Lawrence Gilbert, of the University of Texas, *Heliconius* butterflies locate their passion fruit vine hosts in tropical forests partly in response to their characteristic leaf shapes. These relatively long-lived butterflies return to the same roosting site each night and learn the location of the widely dispersed plants that provide them with nectar and with oviposition sites, returning to these regularly.

Learning

Learning probably plays little if any role in host finding in most insects. Most evidence suggests that olfaction is involved and that the insects respond innertly to sign stimuli in the form of odours of plant essential oils. In most cases these odours are attractants only within a range of few centimeters to a few meters. They are difficult to demonstrate experimentally, since insects often are responsive only for a short period following dispersal. However, there is a growing body of evidence as to the reality of such sign (or "token") stimuli. Sweet clover weevils are attracted to coumarin, an odorous constituent of Melilotus, their normal host plant. Adult females of the imported cabbageworm lay their eggs on cabbage, broccoli, mustard, and other members of the family Brassicaceae, all of which produce odorous mustard oils. Instances such as this have given rise to the statement that insects areoften "good botanists," capable of selecting plants that are taxonomically related. However, similar essential oils sometimes occur in quite different groups of plants. For example, methyl chavicol, anethole, and anisic aldehyde occur both in citrus and in members of the parsley family, and members of the black swallow tail butterfly group will oviposit and the larvae will feed on members of either group of plants– which may in fact be more closely related than is usually appreciated.

Olfactory cues

There are certain chemical substances that attract and stimulate attack are often called *kairomones.* These are also defined as interspecific messages that benefit the receiver rather than the sender. Thus, they stand in contrast to allomones, which benefit the sender and with pheromones, which are intraspecific messages. As we have seen, the olfactory receptors of insects reside mainly in the antennae, and there is evidence that insects are frequently able to distinguish between the odours of many plant species, even when the volatile substances are closely related chemically. Vincent Dethier, now at the University of Massachusetts, has shown that black swallowtail larvae do not confuse six essential oils, each of which characteristic of one of the food plants of the parsley family. When specific odours reach the antennae of caterpillars, the 16 olfactory receptors respond differentially: Nerve impulses in some increase and in others decrease, producing a pattern characteristic of each substance. These are "generalist receptors" differing from the specialist receptors for pheromones.

Response to a particular set of molecules is programmed in the central nervous system, and the behaviour released is best described as a chemokinesis or chemotaxis leading to reduced locomotion when the source is reached. While these remarks apply to many *monophagous* and *oligophagous* species, it is probable that many polyphagous species respond to more generalized cues and accept plants on the basis of the presence of phagostimulants or the absence of feeding detterents or toxins. *Polyphagous* species feeding below ground, such as white grubs and wireworms, may respond to respiration products plants, chiefly CO_2; but selective feeders such as cabbage and onion maggots may respond both to CO_2 and to volatile odours of the host. The identification of attractants may have great practical value. The oriental fruit fly, a major pest of many tropical fruits, is known to be strongly attracted to plants containing methyl eugenol. Oddly, it is mainly the males that are attracted this host plant odor, which is believed to serve as a "rendezvous stimulant," bringing the sexes together for mating.

Traps baited with methyl eugenol have been used for many years to monitor populations of oriental fruits flies. On the island of Rota, the fly has been eradicated by dropping fibreboard squares impregnated with kairomone mixed with insecticide.

Food Plant Acceptance

Insects accept or reject a food plant is finally accepted or rejected as a feeding or oviposition site through stimuli received on actual contact with the plant. These may be visual, relating to the shape or colour of the substrate, but are more often tactile, olfactory, or gustatory. A.J. Thorsteinson, of the University of Manitoba, Winnipeg, showed that female diamondback moths lay eggs more readily on rough than on smooth surfaces. When the rough surfaces are coated with mustard leaf juices, oviposition is further increased. In nature the moths lay eggs on various members of the mustard family, but to varying degrees, depending on the combined effects of tactile and olfactory cues. The initiation of feeding is commonly mediated by gustatory stimuli. In caterpillars, taste receptors are located primarily on the maxillary palpi. Removal of the palpi (along with the antennae) often causes these

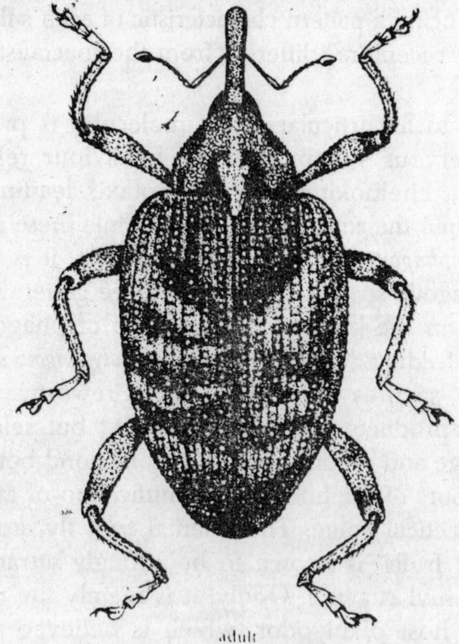

Fig. 3.7. Mango Seed Weevil.

insects to accept plants they would normally refuse. Tobacco hornworms feed only on solanaceous plants in nature, but when the palpi and antennae are removed, they will acccept such plants as dandelions and plantains.

Phagostimulants are the substances that induce feeding. Experimental evidence of the importance of phagostimulation has involved placing the substance to be tested on an abnormal plant or on agar or filter paper and recording the amount of feeding that occurs. As long ago as 1910, the Dutch entomologist E. Verschaffelt showed that imported cabbageworm larvae (*Pieris rapae*) would eat nonhost plants when these were smeared with the sap of cabbage plants or with sinigrin, a glycoside that is characteristic of members of the cabbage family. These substances serve both as attractants for ovipositing females and as stimulants for larval feeding.

Eastern tent caterpillars restrict their feeding mostly to Rosaceae, especially to wild cherry. According to Vincent Dethier, the leaves of wild cherry contain hydrocyanic acid (HCN) in sufficient quantity to poison cattle. Yet HCN, in combination with benzaldehyde, constitutes an odorous substances often called oil of bitter almonds, which is a mild attractant and a feeding stimulant for tent caterpillars. When the juice of wild cherry leaves is sprayed on filter paper, the larvae readily eat the paper; they also respond to emulsions of equal parts of HCN and benzaldehyde. Chemical sign stimuli with similar effects have been identified with respect to many insects.

Feeding by certain leaf beetles of the genus *Chrysolina* (Chrysomelidae) occurs only in the presence of hypericin, a substance present in the leaves of Klamath weed. As a result, the beetles have been used effectively in the biological control of these noxious weeds with little danger of them attacking desirable plants. The examples we have cited so far have involved secondary plant substances, which play no role in the basic metabolism of plants and have no apparent nutritive value for insects. In some cases these have evidently evolved as feeding deterrents. Mustard oils, for example, deter feeding by many polyphagous insects, although *Pieris* butterflies, diamondback moths, and some other oligophagous insects have evolved mechanisms not only for accepting these substances but also for using them as cues for host finding and feeding. Phagostimulants are, however, by no means always secondary plant substances.

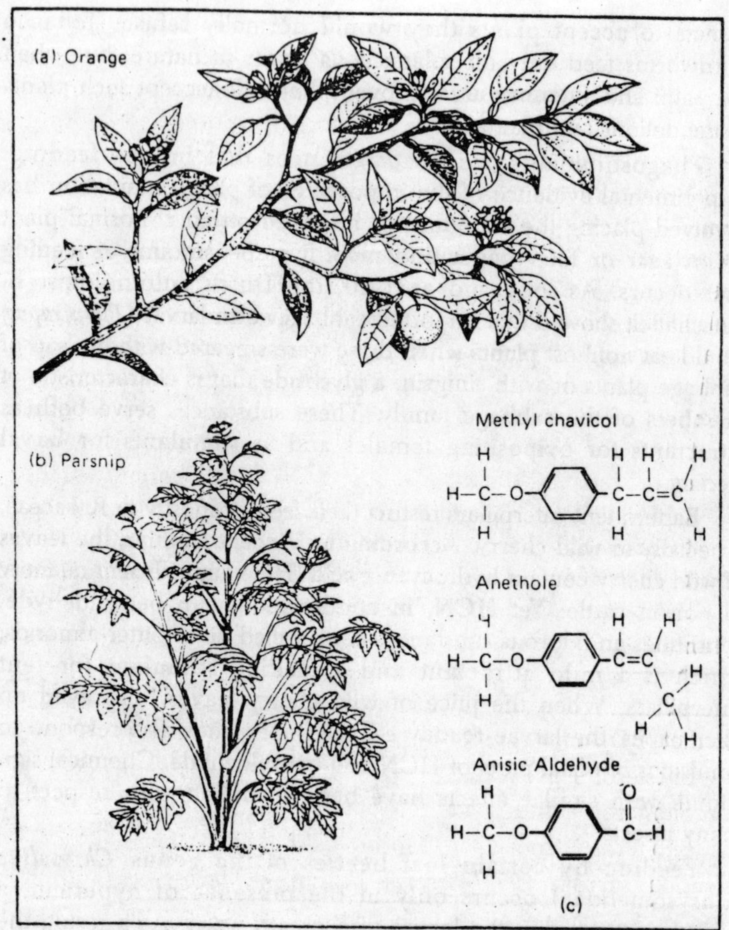

Fig. 3.8. Plants of the citrus family (a) and the parsnip family (b) produce the same three essential oils (c). These substances are phagostimulants for larvae of black swallowtail butterflies, which will feed on members of either plant family and will attempt to fed on filter paper soaked with these substances.

Nutrients, including minerals, amino acids, and especially sugars, often elicit feeding behaviour. European corn borers, for examples, show a preference for substances containing certain levels of sucrose. Most insects that have been studied have sucrose receptors but are capable of rejecting high concentrations of sucrose, which may be toxic. Often other plant substances are synergistic with sucrose. In diamondback moth caterpillars, for example, sinigrin acts synergistically with sucrose, and in Pieris ascorbic acid enhances the effects of sucrose. The silkworm of commerce, *Bombyx mori*, is one of the best studied of insects, and many experiments with artificial diets have been performed with these insects by Japanese workers. In this instance, substances such as morin and inositol synergistically increase the response to sucrose, although by themselves eliciting no feeding response at all. Ysuji Hamamura, of konan University, Japan, has identified substances involved as attractants and in the biting and swallowing responses of silkworms. Some of the results of his experiments demonstrating feeding stimulants are shown in Table 3.1, the number of fecal pellets being used as a measure of feeding. When all necessary substances know to elicit attraction, bitting, and swallowing are added to agar, along with known necessary growth substances, the larvae can be reared successfully in the total absence of the normal host, mulberry leaves. However, the resulting cocoon shells do not have the normal weight of silk, and development is retarded. It is evidently the diet of the first-instar larva that is critical. Thus silk producers do best to add mulberry leaves to the diet or to use them exclusively, at least until all dietary elements have been identified.

Table 3.1. Feeding Activity of Silkworms on Certain Diets

Diet	*No. of feces*
Basic diet (BD) only[a]	0
BD + Sitosterol + R1[b]	89
BD + Sitosterol + Inositol + R1	200
BD + Sitosterol + Inositol + Morin + R1	254
BD without sucrose + Sitosterol + Inositol + Morin + R1	60

[a] Basic diet includes all essential nutrients.

[b] R1 is a group of substances that include swallowing.

It is obvious from these and similar studies that even though plant tissue generally provides a suitable diet for many insects, many species require certain specific substances for normal development. Much progress has been made in the study of insect nutrition in the past two or three decades, and it will pay us to look at this important field of study at least briefly.

Dietary Acceptance

Many vertebrates also subsist largely or entirely on plant tissue, and it is interesting to compare the insects and vertebrates. As we have said, insects have limited ability to digest cellulose, although some have intestinal microorganisms that break down cellulose so that it is available to them. Grazing mammals, on the other hand, have complicated digestive processes that take advantage of bacterial fermentation to break down cellulose into digestible fatty acid. If we compare silkworms with cattle, for example, we find that both are about equally efficient at utilizing the protein in their diet; but cattle (with the aid of symbionts in their rumen) digest over 70% of the crude fiber in grass, while silkworms pass nearly all of the fibre of mulberry leaves undigested in their feces. Insects also differ from vertebrates in being unable to synthesize cholesterol and therefore requiring this or a similar sterol in their diet. Also, they donot require vitamins D and K, which vertebrates need for bone development. They do require several vitamins of the B group. They also require vitamin C. This is normally available in fresh plant tissue, but some insects are able to synthesize vitamin C. Vitamin A deficiency has been shown to result in reduced visual function in moths; apparently this vitamin is essential for the full development of visual pigments, as it is in vertebrates. Insects also require water, nitrogen, carbohydrates, amino acids, lipids, and minerals such as iron, phosphorus, zinc, magnesium, and sometimes sodium and others.

In recent review by J.M.Scriber, of the University of Wisconsin, and F. Slansky, Jr., of the University of Florida, it was pointed out that larvae of Lepidoptera show superior relative growth rates on leaves with a water content of from 60% to 90% and leaves of nitrogen between 2% and 6%. However, species adapted for feeding on tree leaves subsist on somewhat lower levels of water and nitrogen than do those feeding on forbs. Leaves show seasonal trends in the percentages of water and nitrogen, which may be accompanied by other changes in nutrients and in defensive mechanisms. Qualitative variations in nitrogen may be as important as variations in quantity. In young, moisture-rich tissues, nitrogen is available in amino acids and soluble proteins, and in nitrates, vitamins, and other substances. However, in older tissues it is largely in the form of insoluble proteins. Much work has been done on the amino acid requirements of insects. Apparently most insects require the ten "essential" amino acids also required by vertebrates. Various amino acids regarded as "non-essential" in the diet are commonly synthesized by insects. However, proline is required by silkworms, cystine by pale western cutworms, and others not among the ten essential amino acids by still other species.

Dietary deficiencies may restrict the laying down of sufficient food reserves, in the form of fat body, to carry the insect through the pupal and adult stages; or they may cause reduction in the size and functioning of the endocrine glands, resulting in abnormal development or the suppression of ovarian function. In extreme cases, they may of course cause starvation and death. Even accepted food plants may not always provide a perfect diet if foliage is old, for example, or deficient in water, or if the plant is growing in soil deficient in certain minerals. Rates of food intake by aphids are known to be affected by the levels of nutrients (sucrose, amino acids) in the plants. Varieties of peas more resistant to pea aphids, *Acyrthosiphon pisum*, are those containing lower concentrations of amino acids at stages of growth corresponding withthe period of aphid attack in the field.

Almost all adult butterflies and moths, require carbohydrates in the form of nectar; but many also obtain amino acids from nectar. Adults of some insects do not feed at all (giant silkworm moths, for example) subsisting for their short lives entirely on fat stored in the larval stage. Sucking insects, since they do not imbibe the fibrous parts of plants, are able to digest most of their intake; but as we have seen, they frequently discharge a large part of the water and sugars as liquid feces. Nutritional changes in plants, such as those occurring in senescence, may result in the production of winged morphs of aphids that disperse to other hosts. Diverse insects have symbiotic microorganism, either in the gut or in special organs called *mycetomes*, which play various role in nutrition. In the case of aphids and similar sucking insects, symbiotic bacteria are believed to supply nitrogen, which is not obtainable in adequate quantities in plant sap. In other instances symbionts are known to synthesize B-group vitamins, which cannot be obtained from foods such as dried grains. In termites and some other wood-eating insects, intestinal protozoa and bacteria play important roles in the breakdown of cellulose.

Artificial Diets

Artificial diets provide a tool for investigating the precise dietary requirements of insects. By omitting substances of altering the quantities of substances and determining the effect on development, number of eggs laid by the resulting females, and the like, one may determine the balance of nutrients required in nature. For precise nutritional studies, one must provide a diet in which the chemical nature of all ingredients is known. A basic diet must contain the following: water, carbohydrates, fatty acids, the 10 essential amino acids, cholesterol, choline, pantothenic acid, nicotinamide, thiamine, riboflavin, folic acid, pyridoxine, biotin, vitamin B12, vitamin/A, vitamin C, and several minerals. For chewing insects, it must also be (both nutritionally inert). Sucking insects require a liquid diet, which must be imbibed through a membrane. In every case the artificial diet must be sterilized and/ or supplied with a substances that inhibits microbial growth. Quantities of these ingredients must be adjusted appropriately, phagostimulants and additional nutrients added as required–the latter may depend on the ability of the species or its symbionts to synthesize the substance. Research on precise dietary requirements is a very active field at the present time. It was found, for example, that a diet satisfactory for pea aphids was not adequate for rearing green peach aphids. R.H. Dadd and T.E. Mittler, of the University of California at Berkeley, found that the addition to the diet of small amounts of iron, zinc, and maganese, as well as greater care to prevent loss of vitamin C, permitted rearing green peach aphids through many generations.

Discoveries of differences between species in their requirements for minerals, amino acids, vitamins, and other substances, provide the basis for a biochemical definition of an insect's food niche, a matter of much potential importance in the effort to control insects without resort to insecticides. The development of artificial diets has also been a boon to mass rearing of insects for experimental studies of many kinds, as well as for the rearing of predators and parasitoids of phytophagous insects for biological control. For mass rearing, it is cheaper and more convenient to use diets of underfined substances such as wheat germ or soyabean meal, which contain protein, fatty acids, minerals, and other essentials. To these water, sugars, and other more specific requirements must be added to provide a balanced diet. Artificial diets for many phytophagous insects are now available commercially.

The Advantages of Polyphagy and Monophagy

Polyphagous insects are able to take advantage of the nutrients in a variety of plant species, since they are not cued into specific attractants and phago- stimulants. Their food is available with limited searching, and they are in no danger of food shortage if a particular plant is decimated. (We live in a time when many plant species are regarded, and endangered, and some are already extinct, carrying with them any insects that are closely tied to them.) A polyphagous species may be able to extend its range widely, even to other continents if opportunities are provided. Some of our worst imported pests are polyphagous–the Japanese beetle and the gypsy moth, for example. So why be monophagous? Monophagy involves close adaptation to a single host species. The plant is easier to locate, since the sense organs and nervous system of the insect are programmed to respond to specific cues.

Mate finding may also be enhanced, since many phytophagous insects mate on the host plant. On its host, the insect it is able to resist or detoxify the substances the plant has evolved to deter its enemies, as has happened in the case of the monarch butterfly and other insects. Many of the plants utilized by monophagous or narrowly oligophagous insects are, in fact, avoided by generalist feeders because of their toxic or repellent properties. A good exmple is provided by ferns, which are generally avoided by grasshoppers and other generalists but exploited by a small group of specialists belonging to several orders of insects. There is, incidentally, little evidence that specialized feeders us their host plants with greater physiological efficiency than generalists.

It is probable that insects so adapted to specific hosts have limited capacity to reverse the process or to broaden their host acceptance. Their receptors may have evolved so as to perceive only certain molecules, and they may lack the capacity or the necessary symbionts to digest a different plant tissue. They may be able to detoxify (at some metabolic expense) a plant substance that would deter another insect, but that ability would not serve them well on another plant. It should be added that monophagous species can sometimes be put to use in the biological control of weeds; in this instance one must be sure that the insects lack the capacity to switch readily to another host. Clearly it is not correct to approach these questions from a purely entomological point of view. Plants undergo their own evolution, which is directed by many environmental influences. The development of secondary plant substances, of particular types of foliage or bark, of repellents or toxins–all of these require specific genetic events and many of them are energetically expensive, requiring physiological commitments that influence other events.

Under certain conditions it may be advantageous for a plant to protect itself against insect attack or against disease at a certain

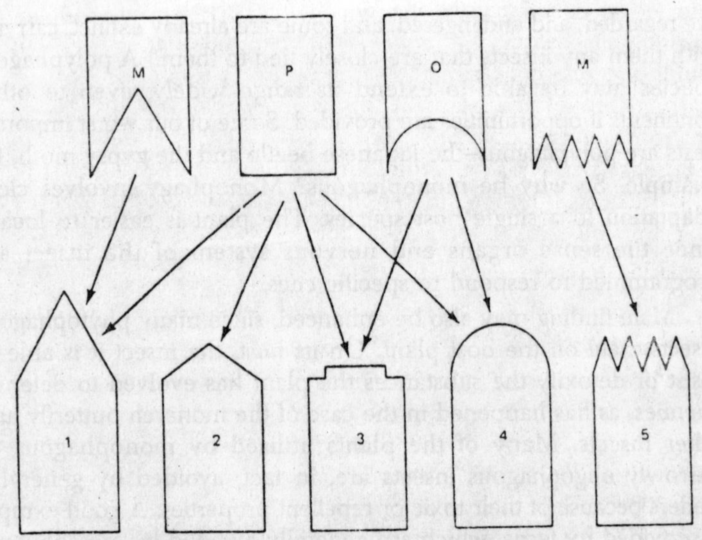

Fig. 3.9. Diagrammatic representation of the relationships between insects (top) and their host plants (bottom). M–monophagous; P–oligophagous. Plants 1-5 are assumed to produce chemical signals that impinge on receptors in the insects. Plant 5 produces substances to which only one monophagous species responds positively; to the remainder the substances are repellent.

cost in seed output. When this occurs, there are likely to be groups of insects that themselves find it advantageous to develop mechanisms for overcoming of the plant's defenses, even though it may have disadvantages from other point of view. In the many tens of millions of years in which insects and plants have been on earth, association of many kinds have evolved. As a general rule, most plants are attacked buy a somewhat limited number of hosts-that is, most are narrowly to broadly oligophagous. Natural selection, like politics, is often a matter of compromises that compound themselves to a degree of complexity that is often hard to fathom. Since polyphagy is prevalent among groups of insects that are usually considered more primitive-brisletails, cockroaches, crickets, grass-hoppers, earwigs, and the like–and most of the more advanced groups of insects show varying degrees of food specialization, it does appear that in the course of time the trend has been toward more intimate associations with specific food sources.

Transmission of Diseases

Host plants are also adversely effected by the inhabited insects. The effect of insects on their host plants often goes well beyond the actual consumption of plant tissue. This is especially true of sucking insects. As the styles reach the vascular tissue, salivary fluids are released and may have localized toxic effects that are carried throughout the plant. Insects may also pick up pathogenic microoraganisms from one plant to carry them to another. An insect whose feeding produces symptoms of disease is said to be *toxicogenic*, and the condition is spoken of as *phytotoxemia*. An insect that transmits diseases organisms is called a *vector*. A variety of insects, chiefly those with sucking mouthparts, may serve as vectors of plant diseases.

Phytotoxemias

A survey of literature reveals that little is known about the chemistry of the salivary secretions of sucking insects; they are believed to contain enzymes such as *amylase* but many also contain inhibitors of plant growth substances. Plugging or localized destruction of vascular tissues may also produce disease conditions. The reasons why certain insect-plant associations result in disease and others do not remain obscure. Several kinds of phytotoxemias are recognized, which are as given in the following paragraphs:

1. Localized lesions with development of more general symptoms. Two-lined spittle bugs (Cercopidae) may cause initial spotting followed by streaking and browning of leaf blades; tarnished plant bugs and other members of the family Miridae often produce disfiguring blotchs on leaves or fruits. In these instances vascular tissues carry the toxins some distance from the point of feeding.
2. Localized lesions at the feeding site, resulting in spoting or stippling, Leafhoppers and mealybugs have been indicated as causative agents of leaf spotting on citrus, pineapples, sunflowers, and other plants. Spots may be paler or darker in colour than the surrounding tissue, or in some cases reddish in colour. Size of the spot may depend on the time spent feeding at that site, since there is little diffusion of toxin from the point of insertion of the stylets and only localized damage to tissue.
3. Systemic conditions, including yellowing, wilting, reduction in growth, or killing of part or all of the plant. These conditions result from tansiocations throughout the plant of toxins produced by sucking insects. Aphids, leafhoppers, mealybugs, and other Homoptera have been indicated, and crops attacked include celery, sugar beets, corn and others. Psyllid yellows of potatoes is one of the best known of systemic toxemias.

4. Malformations of plants, including leaf curling production of witches' brooms, shortening of internodes, and other distortions. One of the most common example of this is a browning and curling of leaf edges produced by the feeding of leafhoppers (Cicadellidae) and often called "hopperburn." Several crop plants are subject to hopper burn, such as potatoes, melons, and lettuce. Gross malformations of plants are sometimes difficult to distinguish from certain microbial diseases or from true galls.

Plant Eating Insects in the Paleozoic Era

As plants evolved greater height and tree- like forms in the Devonian Period, a new habitat was created for terrestrial animals. The spores of these plants may have been an important items in the diet of insect scavengers. Measuring less than 200 μm in diameter, spores of the Lower Devonian plants could have been easily ingested when found singly or in windrows on the ground. At the tops of the sporophyte plants spores also could be found fresh and concentrated in exposed sacs or sporangia. An arboreal insect faces greater risk of desiccation than an insect that lives in moist litter, furthermore it must cling, while walking, to smooth and sometimes vertical surfaces, and it risks greater exposure to predators and parasitoids. Exposure to arboreal or aerial predators, however, was not a problem for Devonian insects because none existed. Nor were any other organisms in competition for food borne high on erect plants. It is likely, therefore, that some of the insect scavangers acquired the resistance to water loss and the tarsal modifications necessary to climb Devonian plants. The shift from scavenging to feeding on vegetative parts of plants also required physiological adjustments to the new diet.

Among the fossils preserved in Upper Carboniferous Period are insects already highly specialized for external plant feeding. The Carboniferous swamp forests would have provided an abundant food supply for phytophagous insects. The Paleopterous orders Palaeodictyoptera, Megasecoptera, and Diaphanopterodea had mouthparts prolonged in a break. The beak seems most suited for feeding on plants, but it is a matter of speculation whether the insects took tissue fluids by piercing and sucking, or probed for spores, pollen, and seeds in reproductive cones. The abrupt decline of the plant taxa that dominated the Carboniferous Period may explain the extinction of these breaked Paleoptera. Direct feeding on vascular phloem probably began with the evolutionary appearance of leaves. Feeding on stems probably did not take place until after the appearence of the arborescent gymnosperms known as the Cordaitales. The cambium and phloem of these plants were close to the surface and accessible by piercing mouthparts. The stems of other common plants had a thick cortex around the vascular tissue. The evolution of hemipterous piercing-sucking mouthparts, therefore, is correlated with the increasing availability of phloem tissue in the Carboniferous.

Numerous fossils of Hemiptera (Sub-order Homoptera) are known in the next period, the Permian. These are the oldest surviving insects that are exclusively phytophagous. When the number of species of breaked Paleoptera and Homoptera are compared with the number of other fossil insects, Carpenter (1970) estimates that nearly half the Paleozoic insects had piercing-sucking mouthparts. This is some measure of the amount of plant food available and the extent of its utilization by insects. The proportion of insects that are phytophagous has apparently remained approximately the same up to the present. The feeding habits of Paleozoic insects with chewing mouthparts are less easily associated with plant feeding. Some may have been predators or scavengers. The jumping Orthoptera were present in the Permian Period and probably were mostly phytophagous, as they are today.

The first possible evidence of insect damage to leaves was found in early Permian rocks in South Africa. Leaves of the ancient fern *Glossopteris* were discovered with marginal scallops resembling the notches made by edge-feeding, chewing insects today. Fleshy fruitlike or berrylike reproductive structures and nutlike seeds of gymnosperms have been found in the lower Permian. The fleshy fruits were probably eaten and dispersed by reptiles which then dominated the vertebrate scene. These fruits were probably also a surface of food for insects long before the origin of the fruits and nuts of angiosperms. Recall that the first endopterygote insects appear in the Permian. Among other advantages, larval insects were able to penetrate the tissues of plants and fleshy fruits for the first time. We do not know when mining and boring in living plants began. The earliest insects capable of such activity may have evolved repeatedly from scavengers that bored in dead wood or decaying vegetable matter, or from external phytophagous forms that extended their feeding into the plant from the surface. Today the endoptery–gotes are the most numerous of the phytophagous insects.

POLLINATING INSECTS

Flowering plants are the Angiosperms which are among the dominant land plants today. It is generally agreed that they owe the origin and much of their diversification directly to the behaviour of insects. The earliest flowering plants are known only from pollen grains. Like the fossil spores of the Paleozoioc, pollen is more readily preserved than plant fragments. The pollen of gymnospersm and that of early angiosperms are so much alike that no sharp distinction can be made. The first fossil grains with predominantly angiospermous features occur near the end of the Lower Cretaceous Period. Latter, at the close of the Cretaceous, angio-sperms pollen exceeds that of gymnosperms of fern spores. The evolution of angiosperms was so rapid that an astonishing 67 families are represented at this time. What are some of the events which led to the origin of this successful group? During the evolution of the gymnosperms, fertilization by swimming sperm was replaced by the growth of a pollen tube. Although the need for moisture or special fluids at fertilization was thus eliminated and drier regions became habitable, the pollen grains must lodge in contact with the ovule.

Sexual outcrossing among gymnosperms is assured because individual plants are unisexual, producing either pollen or ovules, but not both. Large quantities of wind-borne pollen are needed in order for a small fraction to land by chance in the correct sport for fertilization. The transfer of pollen from male to female structures is called pollination. Large pollen sacs, seeds, and other edible tissues probably attracted insects to the cones of Paleozoic gymnosperms. A pollen eating insects that moved only among male cones would not bring about pollination. Bisexual cones having both sexual organs, however, would be suited to benefit from such an insect. Attractive food would be combined with the receptive ovules. An insect would be able to transport pollen from the male organs to the female organs, increasing the likelihood of correct placement of pollen of the same species next to the ovules. The total amount of pollen needed would then by greatly reduced. Such an insects is called a *pollen vector*. The first pollen vectors may have included terrestrial or flying reptiles or early birds, but the most important were probably the flying insects. Among these, the Coleoptera are though to be the most significant. They were probably well diversified in the Mesozoic Era. Interest in them is heightened by the fact that a number of primitive angiosperms today are beetle-pollinated. However, many beetles consume the ovules of the plants that they pollinate. For this reason, certain flower structures such as carpels can be explained as defensive measures initially evolved against the powerful, chewing, jaws of these insects.

The coloured petals aid visual recognition and orientation are the other features to aid pollinators in locating flowers. Odours are emitted which are attractive to insects at a distance. The first floral odors may have initiated odours of fruits or decay that were attactive to scavenging beetles. Thus, the early flowers presumably had both pollen and ovules, sowy petals and odours, and their pollinators came mainly for pollen. The addition of nectaries, (glands that secrete nectar) probably came after beetles had established insect pollination as a regular part of angiosperm reproduction. Primitive beetle flowers generally lack floral nectar. Nectaries are lacking among all gymnosperms, but are present on new fronds of the fern *Pteridium*. A small amount of sugary fluid is secreted from the ovule of certain gymnosperms as a part of the pollination process, but this may or may not be significiant in the evolution of angiosperm nectaries. Nectar is an aqueous fluid rich in sugars. Recently Baker and Baker (1975) demonstrated the presence of other nutrients of value to pollinators amino acids, proteins and lipids. Other substances include ascorbic acid, possible serving as an antioxidant, and alkaloids, which might be toxic to certain unwanted flower visitors.

Nectaries are associated with the vascular phloem system of plants. The first nectaries of angiosperms may have been outside the flower. i.e., extrafloral nectaries, and they may have served a role different from the nectaries. Although nectaries were rare before the appearance of angiosperms, insects had access to a fluid of comparable composition beginning at least in the Permian, if not earlier. This is the honeydew excreted by the phloem-feeding Homoptera. The Homoptera living today are wasteful feeders, ejecting the phloem sap largely unaltered in quantity. Among the insects attracted to honeydew are natural enemies of phytophagous insects such as ants, predatory and parasitoid wasps, and lacewings, as well as other insects such as bees and moths. In some cases this is a regular or major part of their diet.

Insects like-fossils ants are known from the Cretaceous, they were as fond of honeydew as are their descendants. Ants and certain

Homoptera have evolved mutualistic associations in which the ants protect the Homoptera from natural enemies in return for honeydew. The host plants also benefit by feeding of ants on phytophagous insects. Even vertebrate herbivores are discouraged by ants on foliage. The first nectaries, therefore, may have been the plants' device for supplying imitation honeydew as an attractant for beneficial ants and other predaceous insects, without the injury of phloem-feeding Homoptera. The secretion of nectar inside the flower was an incentive to actively flying insects in need of carbohydrate fuel such as Lepidoptera, Diptera and Hymenoptera. Floral nectar differs in composition from both honeydew and extra-floral nectar in ways that suggests it is secreted expressly for the food needs of favoured pollinators. Advanced families of plants have higher concentrations of amino acids than primitive families, and butterfly-pollinated flowers have higher concentrations than beepollinated flowers. This is associated with the inability of most butterflies to ingest protein rich pollen, whereas bees obtain their amino acids from pollen. *Heliconius* butterflies, however, collect pollen on their galeae and digest it there, taking up free amino acids. Flies which feed on protein-rich dung are attracted to the nectars of fly-pollinated flowers that are high in amino acids.

Among the flower visiting insects, the most common pollinators are Coleoptera, Lepidoptera, Diptera and Hymenoptera. They have several features in common all are actively flying adults of neopteropus endopterygote insects. Their scarch for mates, oviposition sites, and plant or animal food is aided by a strong flight apparatus, highly developed senses, and, in some groups, learning ability. These same attributes aid pollinators as they search for and remember flowers. Individual insects which visit flowers of the same plant species during a single flight or longer period are said to be flower-constant. When all individuals of an insect species are restricted to visiting a single species of plant for food (nectar, pollen, other substances), the insect is said to be monotropic. If several, possibly related, plant species are visited the insect species is *oligotrophus* and if many are visited, the term *polytropic* applies. An individual bee may be flower-constant to each of a series of plant species during successive time periods and be a member of a polytrophic species. When visits are for pollen, the term *monoletic*, aligeletic, and polylletic are used. Flower constancy is beneficial to both plant and insects. It is advantageous for a plant to attract flower-constant visitors because they are the most effective cross-pollinators.

It is advantageous to insects to become temporary or permanent specialists because they reduce competition for food and forage more efficiently, learning to recognize a given flower and operate its floral mechanism. In a general way, the flower size, shape, position of reproductive parts, colour patterns, odour, nectar composition, and time of flowering can be matched with the sizes, anatomies, diets, sensory physiologies, rhythmic activity, and foraging behaviours of its pollinators. Even among related species of plants, different species may depend on quite different kinds of pollinators. This specificity attracts effective pollinators and tends to reduce losses of pollen and nectar to nonpollinating visitors. The reward given to pollinators is thereby more closely regulated to promote cross-pollination. Commonly nectar is situated deep within a floral tube so that casual visitors are unable to reach it. Elongation of the mouthparts into a sucking tube is consequently a frequent adaptation among specialized flower-visiting insects. The elongation is achieved in many ways and often independently. Curiously, the Hemiptera, with their long sucking beaks, never become regular flower visitors for nectar. The largest group of efficient pollinators are the bees, or Apoidea. About 20,000 species are known in the world. Bees evolved from visually hunting, predaceous wasps that frequently visit flowers for nectar. Both sexes of bees take nectar as flight fuel. Females eat pollen as a source of the protein used in producing their eggs. All females, except the cleptoparasites and social queens, also collect pollen and nectar for their larvae. Either this is stored in the nest as provisions, or, in certain social species, the pollen is fed directly to the larvae.

Table 3.2. Major Taxa of Terreestrial Phytophagous Insects[1]

External feeder on foliage, steams, roots, fruits, and/or seeds

Exposed feeders

- Isoptera (some)
- Dermaptera
- Plecoptera (some)[3]
- Orthoptera
    - Acrididae
    - Gryllidae
    - Gryllotalpidae
    - Tettigoniidae
- Phasmatodea (all)

Hemiptera (Heteroptera)

Coreidae

Largidae[4]

Lygaedae[4]

Miridae

Pentatomidae

Piesmatidae

Pyrrhocoridae[4]

Tingidae

Hemiptera (all Homoptera)

Thysanoptera (most)

Coleoptera (Adephaga)

Carabidae (some)

Coleoptera (Polyphaga)

Anthicidae(some)[3]

Anthribidae (some)[3]

Byrrhidae (some)[3]

Byturidae

Cantharidae (some)[3]

Cerambycidae (some)[3]

Chrysomelidae

Coccinellidae (Epilachna)

Curculionidae

Elateridae (some)[3]

Meloidae (some)[3]

Scarabaeodae (Melolonthinae, Rutelinae)3

Lepidoptera[2]

Bombycoidea

Geometroidea

Hesperioidea

Noctuoidea

Notodontoidea

Papilionoidea

Sphingoidea

Hymenoptera (Symphyta)[2]

Tenthredinoidea

Leaf rollers and *makers*[2]

Coleoptera

Attelabidae

Lepidoptera

Gelechiidae

Gracilariidae

Lasiocampidae

Pyralidae

Tortricidae (Torticinae)
Yponomeutidae
Hymenoptera (Symphyta)
Megalodontidae
Pamphilidae

Case bearers[2]

Coleoptera (Polyphaga)
Chrysomelidae (Clytrinae, Cryptocephalinae)
Lepidoptera
Coleophoridae
Incurvariidae
Psychidae
Tineidae

Open galls

Thysanoptera (some)
Hemiptera(Homoptera)
Aphididae (some)

Internal feeders of foliage, stems, and/or roots[2]

Borers

Coleoptera (Polyphaga)
Brentidae
Buprestidae
Cerambycidae (some)
Curculionidae
Languriidae
Platypodidae
Scolytidae
Diptera
Agromyzidae
Anthomyiidae
Chloropidae
Ephydridae
Lepidoptera
Cossidae
Hepialidae
Noctuidae
Pyralidae
Sesiidae
Tortricidae (Olethreutinae)
Hymenoptera (Symphyta)
Cephidae
Siricidae (some)
Syntexidae
Xiphydriidae

Leaf miners

Coleoptera

Buprestidae
Chrysomelidae
Curculionidae

Diptera

Agromyzidae
Anthomyiidae
Cecidomylidae
Chironomidae
Drosophilidae
Ephydridae
Lauxaniidae
Psilidae
Sciaridae
Syrphidae
Tephritidae

Lepidoptera

Coleophoridae
Cosmopterygidae
Cycnodiidae
Elashistidae (some)
Eriocraniidae
Gracilariidae
Heliodinidae
Heliozelidae
Incuraruiidae
Lyonetidae
Nepticulidae
Noctuidae
Opostegidae
Pyralidae
Tischeriidae
Tortricidae (a few Oleuthreutinae)
Yponomeutidae (Argyresthiinae)

Hymenoptera

Argidae
Tenthredinidae

Closed galls

Coleoptera
Buprestidae
Cerambycidae
Curcullionidae

Lepidoptera

Cosmopterygidae

Gelechiidae
Tortricidae (Olethreutinae)
Diptera
Agromyzidae
Cecidomyiidae
Hymenoptera
Cynipidae
Eurytomidae
Tenthredinidae
Flower feeders
Flower-tissue feeders[3]
Coleoptera (Polyphaga)
Anthribidae
Buptrestidae
Cantharidae
Cerambycidae
Chrysomelidae
Curculionidae
Elateridae
Meloidae
Melyridae
Nitidulidae
Scarabaeidae
Lepidoptera
Lycaenidae (some)[2]
Pollen feeders
Colembola (some)
Blattodea (some)
Dermaptera (some)
Plecoptera (some)
Hemiptera (Heteroptera)
Anthocoridae
Miridae
Thysanoptera (some)
Coleoptera (Polyphaga)
Cephaloidae[3]
Meloidae[3]
Mordellidae[3]
Nitidulidae
Oedemeridae[3]
Phalacridae[3]
Diptera (probably many)[3]
Anthomyiidae
Bibionidae

Bombyliidae
Calliphoridae
Mycetophilidae
Muscidae
Scatopsidae
Syrphidae
Tachinidae
Lepidoptera
Micropterygidae
Nymphalidae (Heliconius)[5]

Floer feeders

Hymenoptera
Apoidea[6]
Vespidae (Masarinae)[6]
Xyelidae[2]

Nectar feeders[3]

Neuroptera
Chrysopidae
Mecoptera
Panorpidae (Panorpa)
Diptera (many)
Lepidoptera (most)
Trichoptera (some)
Hymenoptera (most)[6]

Internal feeders in fruit and/or seeds on living plants[4]

Coleoptera (Polyphaga)
Bruchidae
Byturidae[2]
Chrysomelidae
Curculionidae
Nitidulidae
Scarabaeidae[3]

Diptera[2]
Cecidomyiidae (Contarinia)
Tephritidae (Ceratitis, Rhagoletis)

Lepidoptera[2]
Gelechiidae (Pectinophora)
Incurvariidae(Tegeticula)
Lycaenidae (Strymon)
Mpctiodae (Heliothis)
Nolidae (Celama)
Pyralidae (Ostrinia)
Tortricidae (Laspeyresia)

Hymenoptera[2]

Agaonidae (Blastophaga)

Eurytomidae (Bruchophagus)

---

[1] This list of taxa is intended to be representative, not exhaustive. The listing of a family without qualification does not necessarily mean that all species in the family have the same food habits. Unless otherwise indicated, both the immature and adult stages are believed to have the same food habits. Well known or exceptional genera are given in parenthesis.

[2] Feeding is mainly by larvae.

[3] Feeding is mainly by adults.

[4] Feeding is mainly on seeds.

[5] *Heliconus* adults ingest nutrients from pollen that have been dissolved in nectar.

[6] Adult females of Apoidea and Vespidae (Masarinae) also store pollen and nectar in their nests as food for their larvae.

Bee's Adaptations

The bees are associated with flower visitatiors have plumose or featherlike hairs; special pollen-transporting devices; modifications of the tongue, or glossa, for extracting nectar; a diet of pollen and nectar; and, in the honeybees, a highly developed system for communication. A few species of bees are *monotrophic*, but most bees are *oligotrophic* or *polytrophic*. Among the latter are honeybees which visit an enormous variety of flowers, including those designed to attract insects other than bees. During times of food scarcity, honeybees collect honeydew or fruit juices as nectar substitutes. When pollen is scarce, honeybees have observed to collect flour or even inert dust. Certain primitive bees, such as *Hylaeus* (Colletidae), are relatively hairless, like wasps. They eat pollen and later regurgitate it with nectar while preparing the nest provisions. But most bees have abundant, plumose hairs which retain pollen grains brushed on the body during flower visits. Female bees methodically groom them-selves and pack the pollen into special devices for transport to the nest. These are of two kinds: (1) pollen brushes, or *scopae*, of long, dense hairs on the hind legs of most bees or on the underside of the abdomen in *Megachilidae*, and (2) pollen baskets, or *corbiculae*, which are created by a circle of stiff hairs on the outer surface of the hind tibiae of honeybees, bumblebees, and their relatives.

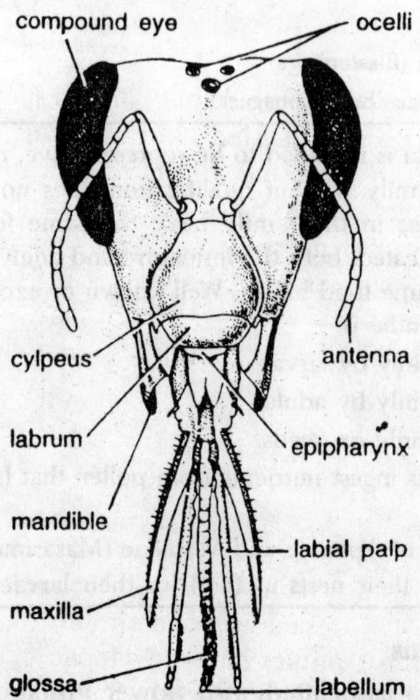

Fig. 3.10. Honeybee. Head and its appendages (Dorsal view).

Andrenidae, Colleticae, and Halictidae have short tongues suited to take nectar from exposed nectaries or flowers in which the bee can bodily enter. These are considered less specialized than the long-tongued Megachilidae, Anthophoridae, and Apidae, which can reach nectar hidden in the inner recesses of specialized flowers. The expansible crop carries nectar back to the nest and also functions on outbound flights as a fuel tank.

Bee Flowers

"*Bee flowers*" characteristically open at certain times during the day, emitting sweet or aromatic odours and presenting their pollen and nectar. Petals of bright blue, purple, yellow and other colours within the bee's range of colour vision are common. Recall that red is inivisible to bees; it is an uncommon colour of bee flowers. Patterns reflecting ultraviolet are seen by the bee but are invisible to us. Separate petals which create a broken outline are suited to detection as a mosaic image by the bee's compound eyes. The two-lipped form of certain bee flowers; such as those of legumes or mints, provides a landing platform. This places the bee in a position favouring access to the food and pollination. Distinctive stripes of spots serve as nectar guides which orient the bee to the food.

Two other large groups of pollinating insects shows conspicuous modification for flower visiting: certain Diptera and most Lepidoptera. In both orders, the special modifications are mainly elongation of the mouthparts to reach hidden nectar. The mouthparts of butterflies and moths are suited only to sucking fluids. Most species take nectar, but some moths do not feed at all as adults. The longest tongue of any insect is the 22.5-cm proboscis of the Madagascar hawkmoth. Xanthopan morgani predictor (Sphingidae). The moth is apparently the sole pollinator of the orchid *Angraecum sesquipedale*, a plant with nectar situated in an equally long tube and accessible only to this moth.

Mouthparts

In flies the mouthparts are variously modified for blood-sucking or sponging, the general ability seems to be retained to ingest nectar and small particles such as pollen grains. Certain species in each of several families of flies have developed exceptionally long mouthparts for probing deep flowers. Some species with long mouthparts are found in Bombyliidae, Apioceridae, Nemestrinidae, Acroceridae and Tabanidae. Shorter, but distinctly specialized, mouthparts are also seen on species of *Rhingia* (Syrphidae),

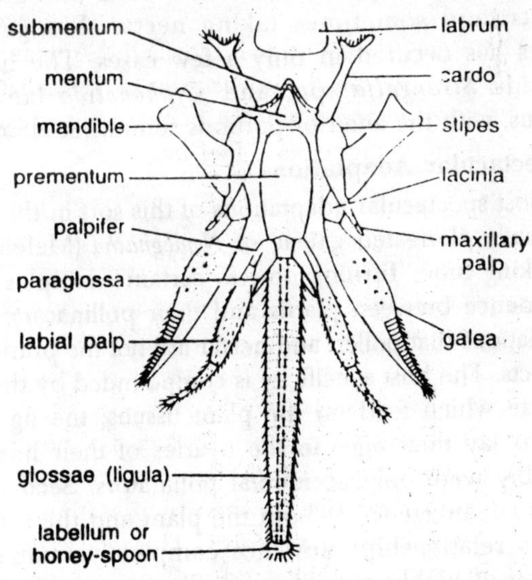

Fig. 3.11. Honeybee. Mouthparts.

Conopidae and Tachinidae. The long-tongued flies visit the same kinds of flowers that bees visit. The so-called "fly flowers" are mainly adapted to less specialized, short tongued insects which normally feed on fluids from dead animals, feces, or plant juices. The flowers depend more on odor than their appearance to attract these insects. The shallow flowers are often white or dull-coloured; the nectar is exposed; and the smell is often musty or rank.

Flower's Attractants

Flowers attractive to butterflies open during the day, have sweet odors, and often have nectar at the base of a deep tube. The flowers are erect and have a horizontal surface for landing. Red is visible to butterflies and is common colour of "*butterfly flowers*," such as carnations or the butterfly weed. *Asclepias tuberosa*. "Moth flowres" bloom in the evening or night; most of them have heavy, sweet scents; and they are often a highly visible pale or white colour. Hawkmoths (Sphingidae) characteristically hover in front of a flower and take nectar by extending their long probosces. Flowers visited by hawkmoths are horizontal or drooping, with the reproductive pats suited to contact the hovering moth. The Coleoptera which visit flowers are active fliers that frequent open, sunny places, in contrast to their terrestrial, cryptic relatives. Beetles tend to linger in flowers, feeding on pollen and flower parts with their powerful mouthparts, and sometimes taking nectar. Elongation of the mouthparts has occured in only a few cases. The heads of the cerambycids *Strangalia* spp. and *Cyphonotida* laevicollis are prognathous, with the anterior portions somewhat elongated.

Other Spectacular Adaptations

The most spectacular adaptations of this sort in the Coleoptera are the greatly elongated galeae of *Nemognatha* (Meloidae), which form a sucking tube. Turning now to certain examples of extreme interdependence between plants and their pollinators, we find in these associations that pollen and nectar are not the prime incentives for the insects. The host specificity is compounded by the specificity of the larvae which feed on the plant tissues: the fig wasps and yucca moths lay their eggs in the ovaries of their hosts. Initially they probably were only accidental pollinators. Seed production, however, is advantageous to both the plant and the seed-infesting insects. The relationships now are completely mutualistic. The largest group of highly specialized pollinators are the tiny wasps of the Family Agaonidae which pollinate figs. The genus *Ficus* (plant Family Moraceae), found in tropical regions, includes about 800 species. Virtually every species of wasp is confined to a single species of fig. The fig that we eat is actually an inflorenscence composed of an enlarged receptacle which encloses receptacle which encloses many small flowers.

Methods of Pollination

In general, the pollination of the caprifig, *Ficus carica*, by the wasp *Blastophaga psenes* takes place int the following manner. The fig has flowers of two kinds: pollen-producing male florets and female florets with short styles. The females wasp flies to a fig in the proper state of maturation and forcibly enters it through the narrow opening, or ostiole, which is guarded by scales. The restructed opening presumably excludes nonpollinating insects. In the process she loses her wings and antennal flagella. She penetrates deeply in the fig to reach the female florets. There she inserts her ovipositor through the short style of a floret to the ovarian region, where she lays an egg. She also is seen to remove some fig pollen with her front legs from special pouches on her body and brush it on adjacent stigmas, thus pollinating the female florets. After laying her eggs, she dies, still in the fig. Each wasp large feeds inside a floret, causing a ting gall. At maturity, the adult males are fightless and have reduced legs, eyes, and antennae. They chew their way out of their galls first and seek the galls containing adult female wasps. A hole is bored in the gall by the male, and he copulates with the female by means of an extensible abdomen. The male wasp then dies. The inseminated female emerges from her gall and seeks the pollen contained in the male florets nor the ostiole. She packs pollen into special cavities on her body, leaves the fig, flies to another fig in the proper state of development, enters, and the process is repeated. The exact cycle varies with the species of wasp and fig. When edible figs were introduced in California in the late 1800s, the Smyrna variety failed to produce fruit. An enterprising grower traveled to Turkey to learn the secrets of successful fig culture. After some difficulty, including an epidemic of plague, he returned with the knowledge that special pollination is necessary. The Smyrna variety has only female florets with long styles. The wasps and suitable pollen must be obtained from the caprifig, in which the normal cycle can be completed. Caprifigs fruits, containing pollen-laden females of *Blastophaga*, are hung in perforated bags among the limbs of Smyrna fig trees. The

Blastophaga emerge the enter the Smyrna figs. They are unable to lay eggs because their ovipositor is too short for the long-styled florets. The fig is nevertheless pollinated and normally ripens with seeds into an edible fruit. Another group of host-specific pollinators are the Yucca moths, *Tegeticula* and *Paralegeticula* sp. (Incurvariidae). All species of Yucca (plant Family Agavaceae) are American in origin, but they have been introduced elsewhere in the world. More than two dozen species of Yucca east of the Rockies and the Mojave Desert are pollinated by one moth species. *T. yuccasella.* In the west, Yucca brevifolia is pollinated by *T. paradoxa*, *Y. whipplei* by *T. maculata*; and *Y. schottii* by Parategeticula polleniferae, as well as Tegeticula yuccasella.

Nocturual insects like Eastern moths are active at night, the time at which flower scent is also strongest. The female *T. yuctasella* enters the white flower, climbs up the stamens to the anthers, and gathers the pollen in a ball. The pollen mass from one to four anthers is carried under her head, clasped by a prehensile elongation of the maxillary palpi and the bases of the forelegs. She then flies to another flower in the proper state of ovarian development. After inspecting the ovary, she bores in with her sclerotized, elongate ovipositor and lays an egg. Climbing the style, she packs some pollen on the stigma. This behaviour is often repeated after each egg is laid. On the average one egg is inserted into each of three compartments of the ovary. A few of the many seeds in pollinated flowers server as food for the moth larvae. Unpollinated flowers do not develop seeds. When fully fed, the larvae leave the seed pod and pupate in the ground. Emergence of the adults is timed to coincide with the flowering season.

An excellent example is provided by the male euglossine bees of the American Tropics, which visit orchids. Males take nectar as flight fuel from various flowers, but they visit orchids to obtain odours. The floral odours are created by species-specific blends of volatile compounds such as benzyl acetate, cineole, eugenol, methyl salicylate, and methyl cinnamate. Males of species of *Euglossa* and Eulaema are each attracted to certain odours. The males brush the odour-producing surfaces and apparently store the fragrance in special cavities in their large hind tibiae. The bee's use of the odour is not clear, but they seem dependent on an adequate supply. Possibly it is metabolized or converted to attractants for either or both males or females. During their contact with the orchid, the male bees become intoxicated. When disabled, they fall into a trap device. During their escape, a packet of pollen is attached to specific place on their body. At their next visit to an orchid of the same species, the packet is removed by the orchid thus achieving pollination.

Feeding in Modern Plant Eating Insects

Insects can be divided by their taxonomic groupings and by their general mode of feeding. The mouthparts and othe characteristics of the immature and adult insects should be kept in mind when considering each group and its food habits. Orthoptera and Phasmatodea have chewing mouth-parts in all stages and feed exterrnally on plants. The phytophagous species among the endopterygote orders Hymenoptera, Coleoptera, and larval Lepidoptera also chewing mouthparts. Some of these feed externally, and some bore bodily into the plant tissues. The rasping mouthparts of all stages of Thysanoptera are applied externally to disrupt cells and suck cell fluids. The piercing-sucking mouthparts of all stages of Hemiptera and the fruit-piercing moths take vascular or tissue fluids by penetrating the plant while the insect remains outside. Mouthparts of larval Diptera vary from the normal chewing type to hooklike structures which tear loose and ingest plant tissue and fluids. Larval flies are able to feed internally by boring. Adult Diptera, whether of the biting or sponging type, ingest only liquid food and particles in suspension. They feed externally and probably take fluids which are freely available without further damage to the plant. Nectar and pollen, as well as saps and juices from previously injured tissues or fruits, are eaten. The adult Lepidoptera have sucking types of mouthparts to suck the fluid of almost same kind.

External Feeders

On the other hand insects which eat the vegetative parts of plants can be broadly divided into external and internal feeders. During the life history of some species, especially small moths, the larvae may feed first inside then outside. External feeders may be freely visible to predators and parasitoids. Such *exposed feeders* are usually protectively coloured and patterened if they are large enough to be edible by vertebrates. The manner of feeding is characteristic of the species. Many caterpillars and sawfly larvae feed along the leaf edge. *Surface feeders* with chewing mouthparts may ingest whole pieces of leaf, leaving holes of notches. Others remove most of the photo-synthetic tissues, leaving a delicate skeletonized vascular network. The cellulose-rich feces are called *frass.* Exposed feeders simply drop frass from their feeding stations.

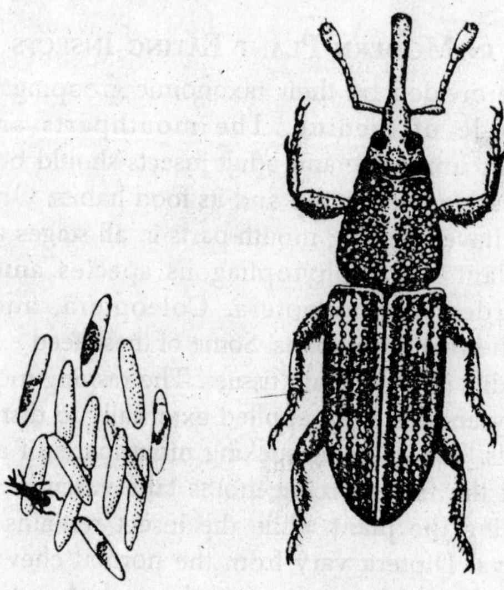

Fig. 3.12. Rice Weevil.

In Hamiptera the feeding injuries of leaf-feeding are often recognized as small, discoloured spots where the inner palisade and spongy mesophyll cells broken down and emptied. The insect's saliva sometimes has physiological effects on the leaf which are also injurious. Thysanoptera also remove the contents of the inner leaf cells, resulting in silvery air spaces inside the leaf. Damaged leaves and stems may wither, discolour, or fall prematurely. Some external feeders gain protection by feeding on roots or within enfolding leaves. Those that are able to spin silk may bind together leaves in a sheltering cluster. The *leaf rollers* are catepillars and sawfly larvae which roll or fold leaves to create a tube. The larvae hide in the tube, feeding at its edge or on nearby leaves. Some moth species pupate in the roll. Adult attebrid weevils cut the leaf bearing their egg, and the leaf forms a roll naturally. The weevil larva feeds inside. Parenthetically, we should note that the predaceous gryllacridid cricket *Camptonotus carolinensis* also rolls leaves and fastens them with an oral secretion.

Internal Feeders

Insects that feed inside the plant, completely surrounded by living tissue, are exclusively endopterygotes and usually the larval stages of the life history. *Leaf miners* are small larvae that eat some or all of the mesophyll tissues. between the outer layers of leaf blades or needles. Female insects select the host and lay their eggs on the leaf surface, or inserted in the mesophyll, or nearby on the plant. Usually mature leaves are mined. The larvae feed in a manner characteristic of the species, leaving a "signature" which is sometimes beautifully serpentine. Larvae of Diptera are cylindrical but soft-bodied and adaptable to the restricted mine. The other larval miners are highly flattened, usually colourless, legless, and

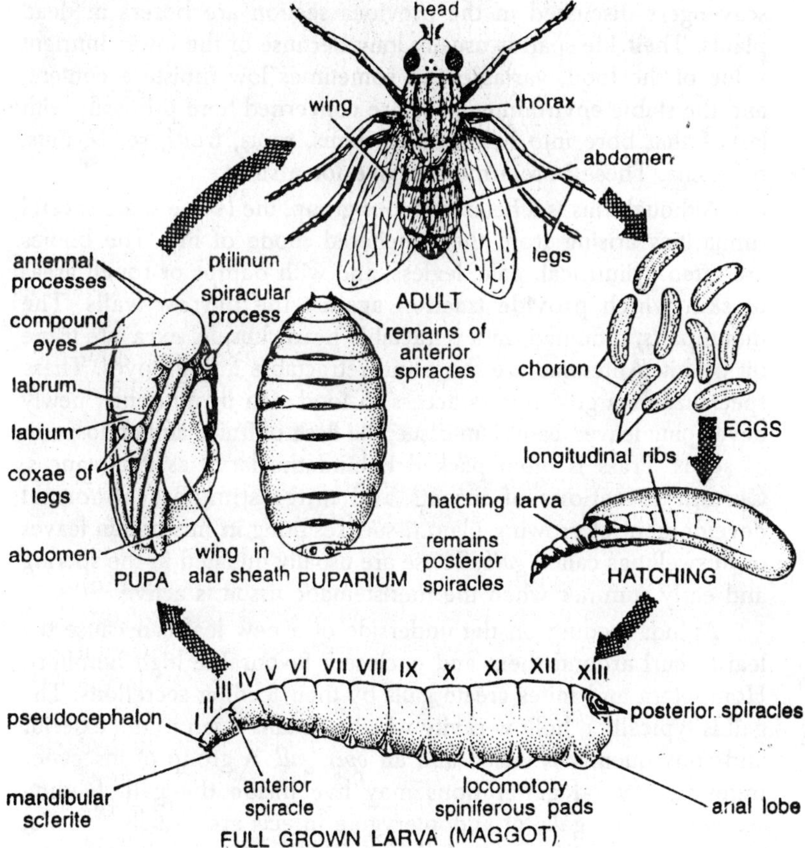

Fig. 3.13. Musca nebulo. Life cycle.

with a flattened, wedgelike head which slopes forward. No adult insects are leaf miners. Linear mines begin as a tiny channel and progressively widen as the larva grows in size and appetite. The direction of the mine may be altered when vascular bundles are encountered. Blotch mines are created when the larva feeds in various directions, eating both cascular and mesophyll tissues. The frass is often deposited in the mine or ejected through an opening. Lepidoptera may lay frass inside in a continuous line, or pack it at one end or plaster it randomly on the minefloor, while Diptera lay two rows. Hymenoptera scatter the frass about the cavity or in piles. Some miners pupate in the mine, while other species leave to form a cocoon elsewhere. Plant borers are insects that burrow in living or dead plant tissues other than leaves. The xylophagous scavengers discussed in the previous section are borers in dead plants. Their life-span is usually long because of the lower nutrient value of the food, variable and sometimes low moisture content, and the stable environment. We are concerned here primarily with larvae that bore into living buds, stems, roots, fruits, seeds, nuts, or grains. These borers may survive for a short.

Although this is a heterogeneous group, the larvae share several similarities arising from their sheltered mode of life. The bodies are often cylindrical, pale, legless, and with bumps or rough areas of skin which provide traction against the burrow walls. The mouthparts, mounted on a retractable head capsule, excavate tissue bit by bit. Antennae are short and retractable into grooves. These specializations give borers access to food rich in nutrients: newly developing leaves, cambium tissue, the flesh of fruits, and endosperm of seeds. Frass is often packed behind the larva as it advances. Certain secretions of insects and mites stimulate abnormal development of growing plant tissue, resulting in misshapen leaves or in swellings called *galls*. These are usually initiated in the spering and early summer when the meristematic tissue is active.

Aphids feeding on the underside of a new leaf can cause the leaf to curl around them and enclose a favourable high humidity. Homoptera and mites create galls by their feeding secretions. The gall is typically a hollow cavity which remains open to the exterior and consequently is called and an *open gall*. A group of insects or mites or several generations may live inside the gall, feeding secretions. The galls of endopterygote insects are usually, but not always, occupied by a single larva. Either the ovipositing female or the larvae or both seem to be responsible for the biochemical stimulants. The galls do not have a permanent opening and are called *closed gall.* The host plant responds with a growth of specific colour and design such that the insect can be identified by the gall it makes. The larva is surrounded by moisture and food and is largely protected from natural enemies. Some wasp parasitoids, however, are able to lay their eggs in or near the gall-making larvae so that several kinds of insects may emerge from a single gall.

Co-Evolution of Insects and Plants

Insects are very specific in their choice of food. Some exposed feeders accept a great variety of plants as food. The larvae of the gypsy moth, *Porthetria despair* (Lymantriidae), are known to feed on 458 species of plants in the United States. Such insects that accept a wide variety of plants for food are called *polyphagous,* Oligophagous insects feed on a few species of plants, often related to one another or having certain similar biochemical constituents. *Monophagous* insects feed on only one species. Insects with the more restricted range of hosts are usually the leaf miners, borers, and gall makers which are surrounded by plant tissues, or insects adapted to toxic plants. The early plant feeding Neoptera are assumed to have been polyphagous, exposed feeders, that are exemplified now by Orthoptera. The oligophagous and monophagous taxa evolved as specialists, with occassional reversions to polyphagous habits.

What factors favour the evolution of host specificity? To answer this question, we must consider in detail the coevolutionary nature of insect/plant relationships. Part of the explanation lies, on the one hand, in the defense strategies of plants against attack. We have seen that all parts of a plant may be eaten; roots, stem, leaves, sap, flower, fruits and seeds. Plant growth is affected by loss of food when photosynthetic tissue is eaten or phoulem sap is drained. Damage to roots and xylem deprives the plant of water minerals. Chemicals in insect's saliva or secreted by the ovipositor may have general or specific effects on the host's physiology. Injured cambium or meristematic tissue results in abnormal growth. These influences lead to stunting, lowered production of seeds, or death. Damage to flowers, fruits, and seeds directly affects the population growth of plants. Phytophagous insects also transmit plant diseases.

Co-evolution changes occur in insects and plants, the plant evolves a protective measure, then insects evolve effective means to overcome the plant's defense which in turn selects for new plant defenses, and so on. In the context of this interaction, a plant which is less damaged because of heritable characteristics is said to be resistant. Three basic *kins* of resistance have been identified by Painter (1958); (1) Resistant plants may be *nonpreferred* for oviposition, shelter, or food; e.g., some chemical or physical feature of the plant either is lacking and fails to attract insects or some feature is present which is repellent. (2) Resistant plants may adversely affect the biology of the insect. Physical or chemical properties of the plant may result in the insect's early death, abnormal development, decreased fecundity, or other deleterious conditions. For example, toxins, repellents, copious sap or pitch, or tissue which is nutritionally inadequate for insects could reduce or eliminate attack. This kind of resistance is called antibiosis. (3) Resistant plants may be *tolerant* and survive even when infested by insects at levels that kill or injure susceptible plants. Rapid replacement of lost parts, excess production of seeds, rapid wound healing, and detoxification of insect salivary toxins are some of the possible ways a plant might survive damage.

To prevent or reduce attack there are non-preference and antibiotic properties. These properties vary (as do other genetic traits) from plant species to species, geographically with the ranges of species, locally depending on ecology, and from individual to individual (see discussion of alkaloids in lupines). For example, the oleresins or pitch of conifers traps, expels, or is toxic to boring insects in needles or bark. The variation in physical and chemical composition of oleoresins give almost every tree a unique individuality. Angiosperms have so-called secondary plant substances such as essential oils, alkaloids, and glycosides whose primary function seems to be in chemical defense, although some may be otherwise integrated into the plant's metabolism. This toxic, biochemical shield has partially protected flowering plants against herbivorous, both insect and vertebrate. These chemicals, incidentally, also give us spices and flavourings.

The plant-feeding insects, on the other hand, compete for food. The resistance factors evolved by plants further limit the number of kinds of plants available to eat. Insects which are able to survive the antibiotic properties and develop a preference for a nonpreferred host will acquire not only food but also some relief from interspecific competition. Those able to bore into plant tissues must tolerate immersion in the host's chemical environment and physical constraints, but here also are food and relief from competition, plus escape from certain enemies and potection against desiccation, freezing, etc. Some insects store toxic plant substances and therby acquire a defense for themselves against vertebrate predators. Thus a number of advantages accrue to the monophagous or oligophagous insect that is adapted to the special condition of life associated with one or a few kinds of plant hosts. Some insects may be secondarily polyphagous because they are able to tolerate a variety of biochemical stresses also may permit them to resist or detoxify insecticides manufactured by humans.

The Defensive Mechanisms

The defensive mechanisms of plants and host specificity of insects have evolved in response to essentially two periods of contact; (1) the period of oviposition, during which the female seeks suitable host plants on which to lay her eggs; and (2) the period of feeding, during which the immature and sometimes adult insects eat the plant. Oviposition behaviour involves recognition and orientation to a host plant at some distance, the search for specific sites in the plant, and finally, the deposition of eggs, followed by dispersal. The behaviour is a series of complex events and involves many of insect's sensory receptors. Any heritable physical or chemical feature of the plant which reduces the numbers of eggs, and thus ultimately the number of feeding insects, will be selectively favoured. This is resistance of the nonpreference type and is achieved either by failing to attract ovipositing females or by providing some inhibitioin.

Feeding Behaviour

Feeding behaviour of insects on plants is similarly complex. Beck (1965) identified four steps; (1) host recognition and orientation; (2) initial biting or piercing of the plant; (3) maintenance of feeding; and (4) cessation of feeding, usually followed by dispersal. The behavioural response at act step depends on releasing stimuli provided by the plant and on the insect's response thresholds, which vary with its physiological state. Resistance of the non-preference type would be given a plant which lacked the appropriate releasing stimuli or which discouraged feeding at some step. Physical and chemical stimuli have been classified according to the response they elicit from insects. During orientation to a host plant at a distance, certain stimuli may act positively as *attractants* or negatively as repellents. When in close contact with the plant, positive stimuli may stop further locomotion, i.e., be *arrestants*, or act as repellents to hasten the insect's departure. At the initiation of feeding, positive stimuli are *suppressants.* Feeding is maintained by *stimulants* or terminated by *deterrents.*

Orientation Behaviour

The sensory apparatus and orientation behaviour of insects are finely tuned to the characteristics of the desired host plants. Secondary plant substances which are repellent to most insects are, in fact, often the feeding stimulants for the appropriate monophagous insects. Nutrients, including sugars, amino acids, phospholipids, and ascorbic acid, can also be stimulants to certain insects. The sensory receptors on the antennae and maxillae of caterpillars are chiefly involved in discrimination. When such receptors are experimentally removed, oligophagous caterpillars often accept as food a broader range of plant species. Although the female usually selects the host plant when she lays her eggs, the caterpillar must select the parts of the plant to eat, avoiding concentrations of toxins and finding the riches; food.

Since phytoxemias, especially when systemic, often resemble diseases caused by viruses, it is useful to recall that there are several fundamental differences:

Toxemias	*Virus diseases*
1. Toxin does not reproduce in plant	1. Virus reproduces in plant
2. Symptoms subside when insects are removed	2. Symptoms persist when insects are removed
3. Recovery common	3. Recovery uncommon
4. Degree of injury related to number of insects and length of time they feed	4. Degree of injury not necess-arily related to number of insects or length of time they feed
5. Disease not perpetuated by vegetative propagation or transmitted by grafting	5. Disease can be perpetuated by vegetative propagation and transmitted by grafting

Diseases Carrying Insects

Insects are involved in the transmission of bacterial, fungal, viral, and mycoplasmal diseases of plants. Some of these are all too well known; Dutch elm disease, a fungus disease transmitted by bark beetles; fire blight, a bacterial disease of fruit trees, transmitted by flies and other insects; and potato leaf roll virus, transmitted by aphids. Note that vectors need not be sucking insects, although these insects are especially well suited for inoculating plants, especially with viruses. Insect transmission of plant diseases is a large subject we can treat only briefly here.

Disease organisms may be spread in one of two ways:

1. *Transmission may be mechanical*; that is, the organisms are borne on the surface of the insect, usually the mouthparts, and in this way carried from plant to plant. When sucking insects ar involved, this type of transmission is spoken of as *stylet borne.* It is also spoken of as nonpersistent, since the microorganisms survive only a short time unless deposited on or in a plant.
2. *Transmission may be circulatory*; that is, the organisms are ingested by an insect, circulate in the body, and are later discharged in salivary fluids. In some cases the microorganisms multiply in the insect. This type of transmission is persistent, in the sense that it may occur over a considerable period of time. Transmission may be by inoculation, in the case of sucking insects, or by surface deposit of pathogens, which must then invade the tissue, often through a wound made by the insect. In some cases the relationship between insects and pathogens is obligate, which means that the pathogens are transmitted only in this way. In other cases the relationship is *faculative*, since the pathogens can be transmitted in other ways, for example, by wind, rain, or other organisms. We present here a single example of a disease caused by each major type of pathogen. Several others are listed in Table 3.3 Throughout the world several hundred plant diseases have been reported to be transmitted by insects and mites, and several thousand insect species have been incriminated as vectors. The brief overview given here will serve only as the briefest of introductions to this field.

Diseases Caused by Bacteria

Plant parthogenic bacteria are rod-shaped bacilli, usually non spore formers, which are able to enter plant tissue only through wounds or by inoculation. They have little capacity to live outside their plant host, and transmission is usually mechanical and non-

Table 3.3. Selected Examples of Plant Disease Transmitted by Insects

Disease	*Vectors*	*Disease organisms*	*Host plants*
Cotton boll rot	Various Hemiptera, such as stink bugs	Bacillus gossypina (bacteria)	Cotton
Fire blight	Various beetles, flies, other insects	Erwinia amylovora (bacteria)	Apples and other fruits
Soft rot	Beetles, flies, other insect	Erwinia carotovora (bacteria)	Cabbage, carrots, onions other vegetables
Stewart's wilt	Beetles, bugs, other insect	Erwinia stewartii (bacteria)	Corn
Ergot	Various flies, beetles	Claviceps purpurea (fungus)	Grasses, cereals
Brown rot	Fruit-feeding Coleptera and Lepidoptera	Monilinia species (fungus)	Apples,peaches, other fruits
Dutch elm disease	Bark beetles of two species	Ceratostomella ulmi (fungus)	Elms
Oak wilt	Beetles of several kinds (fungus)	Ceratocystis fagacearum Oaks	
Corn stunt	Leafhoppers of seveal kinds	Mycoplasma	Corn
Cucumber mosaic	Aphids of many species	Virus	Cucumbers, various weeds, and other plants
Potato leaf roll	Myzus persicae (green peach aphid)	Virus	Potatoes and other Solanaceae
Sugarcane mosaic	Aphids of several species	Virus	Sugarcane, sorghum millet, corn
Spotted wilt	Thrips	Virus	Tomatoes, other crops, weeds

persistent. Obligatory relationships between the bacteria and a particular insect host are unusual; more commonly a number of different insects and sometimes other agents are involved in transmission. However, the example we select does involve specific vectors.

Curcurbit wilt is caused by *Erwinia tracheiphila*, a bacillus that invades and blocks the vascular bundles of cucumbers and to a lesser extent those of cantoloupes, squash and pumpkins. The first symptoms are localized wilting; when stems are cut a milky ooze emerges. Eventually the entire plant may wilt and even die, a condition resembling dehydration resulting from drought. This disease occurs in many parts of the world and may be devastating if not controlled. Experiments have shown that when plants are protected from the attacks of two species of leaf beetles, the striped and spotted cucumber beetle, the plants are also protected from cucurbit wilt. Apparently these beetles carry the bacteria from infected to noninfected plants on the mouthparts or in their fecal pellets, the bacterial gaining entry via feeding wounds.

Diseases Caused by Fungus

Relationships between insects and fungi are common and often complex. The relationship may be intimate and mutually beneficial, as in the case of fungi cultivated by certain ants, terminates, and ambrosia beetles, or it may involve the casual transmission of spores from one plant to another. A great variety of rots, wilts, cankes, and root infections are produced by fungi that may be transmitted by insects. Unlike bacteria, fungi are often able to penetrate plant tissues directly, without requiring a wound. Transmission is usually mechanical, and sometimes several kinds of insects, or even physical agents such as wind or water, may serve as vectors. The example we select is, however, one involving a specific vector and a mutualistic relationship between insect and fungus.

Blue stain of conifers is caused by fungi of the genus *Ceratostomella.* These fungi cause discolouration of felled timber, reducing its value, and they also invade living trees via holes through the bark made by invading bark beetles. The relationship in beneficial to both beetles and fungi. The latter proliferate in the new host, weakening and eventually killing the tree. This renders the tree more suitable for development of the beetle larvae. When these complete their development, they fly to another tree while carrying the spores. The result is an expanding group of dead and dying trees. The mountain pine beetle of the Rockies and southern pine beetle of the Gulf states are especially notorious vectors. Both are members of the genus *Dendroctonus.* Bark beetles employ a complex chemical signaling system that enables them to attack trees *en masse.* The trees, in turn, have evolved chemical defenses.

Diseases Caused by Viruses

Symptoms of virus disease are diverse; they include blotching and mottling of leaves (termed *mosaic*), leaf curl, tumours, rosettes, distortions of flowers and fruits, yellowing, and necrosis. Symptoms are produced by destruction of tissue cells, often accompanied by abnormal growth of other cells resulting in disturbances in respiration and photosynthesis. Various inclusions are sometimes visible in prepared sections; these represent accumulations of virus particles. Like all viruses, they are unable to survive apart from host tissue. Transmission may, however, sometimes be mechanical andd nonpersistent, in which case either chewing or sucking insects may be involved. The more usual vectors are sucking insects such as aphids, leafhoppers and thrips. Transmission is frequently circulative and may involve relationship with a specific vector and multiplication within the vector. In some cases the virus is transmitted form one generation to the next transovarially.

Curly top of sugar beets is transmitted by the beet leafhopper, an insect occuring in the southwestern United States but undertaking seasonal migrations into the beet-growing regions of the Great Basin and the western Great Plains. Symptoms include leaf curl, stunting, and distortions of the roots. The only important mode of transmission is by the feeding of beet leafhoppers. The virus is retained within the blood of the leafhoppers through molts, but there is no multiplication within the body. The virus overwinters in various wild plants and in beets that have been left in the field. It is picked up by leafhoppers feeding in the spring and transmitted to seedlings.

Diseases Caused by Mycoplasmal

Before 1967, mycoplasmas were classed as viruses, since like viruses they pass through capable of retaining bacteria. However, they are now regarded as quite a different group of organisms having some features in common with bacteria: They can be cultured on agar media apart from the host, and they are susceptible to certain antibiotics, especially tetracycline, Mycoplasmas are pleomorphic, the cells undergoing changes in form through their life cycle. The most common disease symptoms are yellowing, stunting, and the development of "witches'brooms." At least 15 plant disease are now attributed to mycoplasmas, all of them transmitted by leafhoppers.

Aster yellows occurs in many partsof the world and infects not only asters but also plants of at least 40 families, including vegetable crops such as potatoes, carrots, lettuce and spinach. In addition to yellowing, plants may show dwarfing or various malformations. The common vector in North America is the aster leafhopper, *Macrosteles fascifrons.* Transmission is circulative, with multiplication occuring in the body of the vector. Leafhoppers are not infective until at least nine days after the mycoplasma has been ingested.

4

Carnivorous Insects

Carnivorous insects are those insects that kill or injure one or more other invertebrates before completing their life cycle. These are also called the *Entomophagous insects.* Most of these carnivorous insects feed on other insects and are said to be entomophagous. Snails, earthworms, millipedes, mites, and other terrestrial and freshwater invertebrates are also eaten by insects. Victims are called prey if killed directly, hosts if fed upon while still living. Carnivorous insects are divided into three major groups according to their mode of feeding: (1)predators kill and consume more than one prey organism to reach maturity; (2)parasitoids require only one host to reach maturity, but ultimately kill the host; and (3)parasites feed on one or more hosts, but do not normally kill the host. Parasitoids insects are intermediate between predators and parasites in the sense that they live at first parasitically at the host's expense, but ultimately kill the host. In entomological literature the word "parasite," alone and in combinations (e.g., hyperparasite), if often used loosely for an insect that lives at a host's expense, regardless of the host's fate.

From an ecological viewpoint, predators and parasitoids both act to eliminate individuals in the prey or host population, whereas true parasites permit the host to continue functioning in the community, though often at a subnormal level. This is an important distinction, especially when predators and parasitoids are used to control pests. Whenever possible in this text we have used the word "parasite" only for those insects that normally do not cause host mortality. However, we use the term "endoparasite" indiscriminantly for both parasites and parasitoids that are physically inside the body of the host. Although the three modes of feeding defined above can be easily distinguished among most insects, the spectrum of entomophagous habits is actually congtinuous. For example, within a species an individual of a normally predaceous insects such as a ladybird beetle, might complete its life cycle by eating only one large host and thus be technically, according to number of victims, a prasitoid. Alternatively, the host of a parasitoid might survive attack, or the host of a parasite might be fatally injured. These abnormal relationships make distinction somewhat arbitrary, but are instructive by indicating how one mode of feeding might evolve into another.

Although the insect may not be directly entomophagous, the net result is a reduction in the number of host offspring and an increase in the offspring of the insect that benefits Cleptoparasites, or "cuckoo" parasties, lay their eggs in the nests of other species, in the manner of cuckoo birds. The cleptoparasitic parent or the larva may kill the egg or larva of the host immediately, or the host larva dies of starvation after the cleptoparasite larva eats the

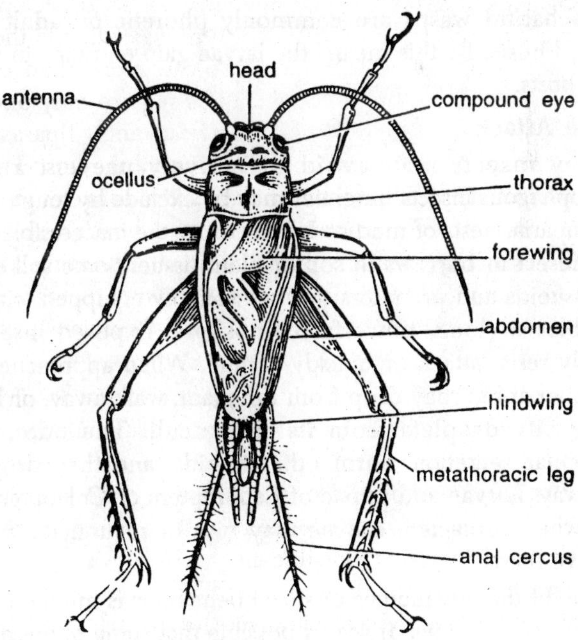

Fig. 4.1. Gryllus. (Dorsal view).

provision. In the former instance, the young cleptoparasites often posses enlarged mandibles suited for attack. A social parasite is a female that enters the nest of a social host and takes over the role of the queen.

The social parasite's offspring are fed by host workers at the expense of host offspring. The host is usually a closely related species. Parasites that spend much of their life their host's nests are known as inquilines. Slavery is practiced by certain ants. Workers pupae of another species are taken by slavemaking workers. The resulting adult slaves then become members of the colony any do most of the work. Nest robbing also involves social species. The robbing species enters the nest of another colony and removes the stored food. Phoresy is the transport of an insect on or physically insides the body of another insect. The insect that provides the transportation is usually not harmed. When the passenger is an adult female parasitoid or predator, however, the usual result is that the passenger remains abord until the host lays its eggs. The passenger then oviposits on the host's eggs. This habit occurs frequently in the wasp family Scelionidae. The first instar larvae of meloid and rhipiphorid beetles, stylopid parasites, and eacharitid wasps are commonly phoretic on adult of their favoured hosts. By this means the larvae gain entrance to the nests of the hosts.

Prey to Attack

Prey insects may avoid or actively prevent attack by entomophagous insects. Enemies may be exclude by tough cocoons, hard puparia, nests of mud or leaves, or by the inaccessible position of the insect in burrows in soil or plant tissue. Some still succumb to parasitoids and predators that are suitably equipped with strong mandibles or long, powerful ovipositors. Exposed insects may violently resist attack or quickly escape. When approached by an enemy, an aphid may drop from the plant, walk away, or kick and secrete oily droplets from its siphunculi. The odour of the siphuncular secretion alarm other aphids, and they drop off or walk away. Larvae and pupae of Lepidoptera or Coleoptera thrash about when contacted. Larvae may bits, or regurgitate or secrete defensive fluids.

One of the advantages of social behaviour is mutual defensive against insect enemies. It seems possible that some of the defensive strategies to avoid predation. By vertebrates are also effective against insect enemies. Certain aphids are distasteful and are avoided by insect predators after an initial contact. The gustatory senses and learning ability of the predator are therefore important to the success of this defense. The role of protective colouration and behaviour, however, is generally less effective because many insect predators, unlike vertebrate predators, do not depend on vision to find prey.

In some cases the insect hosts are able to respond defensively to endoparasite; by a reaction called *encapsulation.* Hemocytes collect around those foregion bodies that are to be engulfed by a single cell. A capsule forms as the inner layer of cells flatten over the object's surface and new are added outside. The inner layer secretes an envelope of connective tissue. The capsule may remain clear or become darkened by melanization. If the foreign body is an endoparasite, it is killed by the encapsulation. The process may take only a day to complete. Fourteen orders of insects are known to encapsulae foreign bodies, endoparasitic worms, and parasitoids. Among the parasitorids, larvae of certain species definitely stimulate encapsulation in hosts that are not the normal hosts; maggots of Tachinidae and some, but not all, wasp larvae of Ichneumonidae, Braconidae, Chaleidoidea (Encyrtidae, Eulophidae). Proctortrupoidea and Cynipoidea. Yet in their normal hosts, the parasitoids usually escape encapsulation. How do they succeed? There are certain species of wasps, attack only eggs. By rapidly completing their growth before the host's hemocytes develop, they avoid encapsulation altogether. Rapid destruction of a larval host also avoids encapsulation by simply killing the host quickly. If the parasitoid lingers for a while and allow the host larva to grow, a larger food supply becomes available, but at a greater risk of encapsulation.

Certain parasitoids insert their eggs precisely into an organ of the host where the endoparasite will be separated from the host's hemolymph by a layer of connective tissue. This prevents encapsulation. Others first invade the alimentary canal and are sheltered from the host's hemocytes. When the host is suitable larger, the parasitoid feeds voraciously and so debilitates the host that it is unable to react in time. Some parasitoids are initially invested in cellular membranes the cells of which become dissociated, enlarged, and circulate in the host's hemolymph.

Apparently the cells absorb nutrients and similarly act to reduce the hosts' ability to muster an encapsulation. It has been shown that the eggs and first instar larvae of the wasp Venturia coneseens (Ichneumonidae) are coated with a particulate layer that inhibits encapsulation by the hemocytes of its moth host, Anagasta Kuehniella (Pyralidae). The layer is first deposition on the egg by a specialized region of the wasp's oviduct and later is somehow transferred from the egg shell to the wasp larva once it is inside the moth larva. If the coating is disrupted, encapsulation ensues.

Evolution of Entomophagy

Among the earliest insect fossils in the Upper Carboniferous Period are Odonate. Today the dragonflies are exclusively predaceous, and we assume they were also in ancient times. Predatory habits have apparently evolved repeatedly thoughout the orders from scavenging and plant-feeding incestors. Parasitoids and entomophagous parasites did not appear until the holometabolous life cycle was evolved. Thus, we can conclude that predation is the oldest mode of the entomophagous habits. Other living orders present as fossils in the Paleozoic Era that were mostly predatory are the Neuroptera and Raphidioptera. Certain extinct orders. Have been suspected to be prdtaory. The Mischoptera of the Meganisoptera had stout, raptorial forelegs. Were these used to capture prey or to cling to vegetation? The giant Meganeura of the Meganisoptera had biting miles and large eyes. These insects resembled dragonflies and were doubtless predatory. Many fossils are merely wings without the body parts from which we could infer food habits.

The early of parasitoids and parasites is not well documented by fossils. Most parasitoids are Hymenoptera and Diptera. These orders must have evolved diverse forms during the Mesozoic Era following the first appearance of fossils of both orders in the Triassic Period. The parasitoids were probably part of this radiation. Fossil chalcidoid wasps are known from the Cretaceous Period. In contrast to most predators, parasitoids are host-specific. Presumably as a consequence of coevolutionary interactions with their hosts, parasitoid taxa also tend to have many more species than predatory taxa. The parasitic Strepsiptera are known from primitive forms preserved in Baltic amber. They may have existed no earlier than than Tertiary Period.

Table 4.1. Major Taxa of Entomophagous Insects

Predaceous as both immatures and adults

Collembola
- Isotomidae (*Istoma*)

Diplura
- Japygidae

Odonata (all)

Mantodea (all)

Dermaptera
- Arixeniina (probably)
- Forficulina (most)

Orthoptera
- Gryllacrididae
- Gryllidae (*Oecanthus*)
- Tettigoniidae (*Conocephalus*)

Psocoptera
- Caeciliidae (*Caecilius*)

Hemiptera (Heteroptera)
- Anthocoridae (*Anthocoris*)
- Belostmatidae
- Enicocephalidae
- Gelastocoridae
- Gerridae (Occasionally)
- Hebridae
- Largidae (*Euryapthalmus*)
- Lygaeidae (*Geocoris*)
- Miridae (some)
- Nabidae
- Naucoridae
- Nepidae
- Notonectidae
- Ochteridae
- Pentatomidae (Some)
- Phymatidae
- Pyrrhocordiae (*Dindymus*)
- Reduviidae (except Triatominae)

Saldidae (occasionally)
Veliidae (occasionally)

Thysanoptera
Aeolothripidae (*Aeolothrips*)
Phlaethripidae (*Leptothrips*)
Thripidae (*Scolothrips*)

Neuroptera (most)

Raphidioptera (all)

Coleptera (Adephaga)
Amphizoidae
Carabidae
Cicindellidae
Dytiscidae
Gyrinidae

Coleoptera (Polyphaga)
Anthicidae (*Anthicus*)
Anthribidae (*Brachytarsus*)
Cantharidae (many)
Cleridae
Coccinellidae (most)
Cucujidae (*Crptolestes*)
Elaterdae (some)
Histeridae
Lampyridae
Malachiidae
Melyridae (many)
Nitidulidae (*Cybocephalus*)
Nosodendridae
Orthoperdae
Phengodidae
Silphidae (*Xylodrepa*)
Staphylinidae (many)

Diptera (*Brachycera*)
Asilidae
Dolichopodidae
Empididae

Dipter (*Cyclorrhapha*)
    Anthomyiidae (Hylemya)
Hymenoptera (Apoerita)
    Chalcidoidea
    Chrysididae
    Formicidae
    Inchneumonodea
    Vespidae

Predaceous primarily as immatures

Ephemeroptera
    Siphlonuridae (*Isonychia*)
Plecoptera
    Setipalpia (most)
Megaloptera (all)
Neuroptera
    Chrysopidae (*Chrysopa*)
Coleoptera (Polyphaga)
    Dermestidae (*Thaumaglossa*)
    Drilidae
    Hydrophilidae (some)
    Lycidae (most)
    Meloidae
    Rhipiphoridae (most)
Diptera (*Nematocera*)
    Cecidomyiidae (*Aphidoletes*)
    Ceratopogonidae
    Culicidae (*Chaoborus*)
    Chironomidae (Tanypodiane)
    Culicidae (*Megarhinus*)
    Mycetophilidae (*Platyura*)
    Tipulidae (Hexatomiini)
Diptera (*Brachycera*)
    Bombyliidae (most)
    Mydaidae
    Rhagionidae
    Tabanidae (*Tabanus*)

Therevidae
Xylophagidae
Diptera (*Cyclorrhapha*)
Anthomyiidae (some)
Calliphoridae (*Stomorphina*)
Chamaemyiidae
Chlorophidae (*Siphonella*)
Drosophilidae (*Gitondes*)
Lonchaeidae (*Lonchaea*)
Otitidae (*Elassogaster*)
Phoridae (*Syneura*)
Sarcophagidae (*Sarcophaga*)
Sciomyzidae
Syrphidae (most)
Lepidoptera
Blastobasidae (*Holcocera*)
Cosmopteryidae (*Stathmopoda*)
Cyclotornidae
Heliodinidae
Lycaenidae (some)
Noctuidae (*Eublemma*)
Lyonetidae (*Ereunetis*)
Psychidae (Platoeceticus)
Pyralidae (*Laetilia*, *Macrotheca*)
Sesiidae (*Synathedon*)
Tineidae (*Dicymolomia*)
Tortricidae (some Tortrix)
Trichoptera
Hydropsychidae (*Hydropsyche*)
Polycentropodidae
Rhyacophilidae
Hymenoptera (Apocrita)
Proctotrupoidea
Sphecoldea (some)
Predaceous primarily as adults
Coleoptera (Polyphaga)

Cerambycidae (*Elytroleptus*)
Scarabaeidae (*Cremastocheilus*)
Mecoptera
Bittacidae (some)
Diptera
Anthomyiidae (some)
Blephariceridae
Calliphoridae (*Bengalia*)
Ceratopogonidae (some)
Hymenoptera (most Symphyta)
Hymenoptera (Apocrita)
Bethylidae
Dryinidae
Mutillidae
Tiphiidae

Parasitoids

Coleoptera
Carabidae (*Lebia, Brachinus*)
Colydiidae (*Deretaphrus*)
Meloidae
Rhipiceridae
Rhipiphoridae
Staphylinidae (some Aleocharinae)
Diptera
Acroceridae
Agromyzidae (*Cryptochaetum*)
Anthomyiidae
Conopidae
Nemestrinidae
Phoridae (*Megaselia*)
Pipunculidae
Pyrgotidae
Tachinidae
Lepidoptera
Cyclotornidae
Epipyropidae

Hymenoptera
- Bethyloidea
- Chalcidoldea
- Chrysidoldea
- Evanionidea
- Figitidae
- Ibalidae
- Ichneumonoidea
- Maglayroidea
- Pompiloidea
- Proctotrupoidea
- Scolioidea
- Sphecoidea (some)
- Trigonaloidea

Insect ectoparasites of insects

Diptera
- Ceratopogonidae (Some)
- Chironomidae (*Symbiocladius*)

Hymenoptera
- Scelionidae (*Rielia*)

True insect endoparasites of insects

Strepsiptera (all)

Hypermetamorphic taxa

Neuroptera
- Mantispidae

Coleoptera
- Carabidae (*Lebia*, *Brachinus*)
- Colydiidae (a few)
- Drilidae
- Meloidae
- Rhipiceridae (*Sandalus*)
- Rhipiphoridae
- Staphylinidae (Aleocharinae)

Strepsiptera

Diptera
- Acroceridae
- Bombyliidae

Calliphoridae (some)
Nemestrinidae
Tachinidae (some)
Lepidoptera
Cyclotornidae
Epipyroidae
Hymenoptera
Chalcidoidea (some)
Ichneumonoidea (some)
Proctotrupoidea
Trigonaloidea

This list of taxa is intended to be representative, not exhaustive. The listing of a family without qualification does not necessarily mean pull all species in the family have the same food habits. Well-known or exceptional genera are given in parenthesis.

Predaceous Insects

Predators are usually larger in body size than other entomophagous insects, and more than one and often many prey are eaten. The greater need for prey also make predators dependent on prey populations of higher density than other entomophagous insects. Prey are usually smaller than the predator but proportional to the predator's size. In other words, larger predators take larger prey and smaller predators take smaller prey. Wasps with stings and ants in groups can kill prey larger than themselves. Victims are rapidly subdued, and except when the predators are wasps that store prey for later consumption by their larvae, the prey are eaten immediately.

Monophagous predators feed on one specie for prey; *oligophagous predators* take several prey species; and *polyphagous predators* take many kinds of prey. The last type are the "generalists." They take individuals of prey species largely in proportion to their relative abundance. Thus polyhagous predators, by continually shifting to the most abundant prey, tend to stabilize populations of prey in the community. Furthermore, if mobile, they are able to thrive in disturbed communities where prey species vary in abundance as ecological succession proceeds. At the other extreme, monophagous predators are density-dependent on one species of prey, and may regulate the prey at lower levels than polyphagous predators.

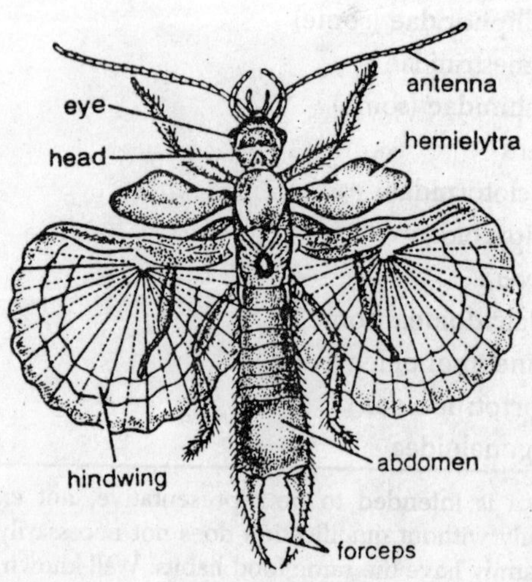

Fig. 4.2. Forficula.

Monophagous predators tend to be associated with undisturbed communities where their host maintains continuous, stable populations. Persual of Table 4.1 leads to the conclusion that a majority of predaceous taxa are carnivorous in all feeding stages. This does not mean they are always exclusively carnivorous; exact diets vary species by species. Adults of some of these visit flowers for nectar, as do certain adults of taxa that are predaceous only as immatures. Scavenging, honeydew, symbionts, and plant foods also may supplement a predator's diet.

Hagen and his associated (1970) have carefully examined the diets of lacewings of the genus Chrysopa (Chrysopidae). Adults of about half the species have larger mandibles and are predaceous like the larvae. Adults of other species have shorter mandibles and feed only on honeydew and pollen. The foreguts of the second group contain yeast symbionts and are supplied with larger tracheal trunks for increased respiration. By this means the essential amino acids are acquired even though the adult diet lacks animal prey. Predaceous adults of the first group lack the yeast and tracheal modifications. Females of both groups of Chrysopa require substantial amounts of food before eggs are produced. As a consequence, females lay eggs only near abundant food sources that will later supply the lacewing larvae. An advantage to the honeydew-feeding species is that prey of the larvae may include not only aphids but a variety of other insects attracted to honeydew.

Predators of both sexes must repeatedly find the subdue prey. They may be active during day and night. The eggs of predaceous insects are usually deposited by the females in close proximity to or at least in the vicinity of suitable prey. In this way the adult's well-developed compound eyes. Chemoreceptive organs, and ability to fly are used to search for prey-rich habitats within which the less well-equipped immatures can forage. Plants of a certain height, odours of honey-dew, odours of decay, and even the pheromones of prey are attractive to searching predators. The adult may eat prey or other attractive foods when they are found or only lay eggs near the prey.

FINDING PREY

Several kinds of strategies are used by insect predators in finding and capturing prey; (1) random searching, (2) hunting, (3) ambush, and (4) trapping.

Insects belonging to the first type roam in the appropriate microhabitat and seize prey after physical contact. Their orientation to objects in the microhabitat and their movements may increase the probability of encountering prey. For example, the predator may patrol leaf edges, veins, or stems and eat insect eggs or sedentary Homoptera. Random searchers may have either monophagous or polyphagous habits. The prey is accepted if proper incitants are detected by receptors on the forelegs, mouthparts, or antennae. The vitim is then devoured by use of mouthparts that are usually specially adapted for predation. After an initial contact or meal, further searching often involves more frequent turns so that the predator stays in an area of previous sucess. While seemingly inefficient, the random searchers are probably the most common type of insect predator and include species highly effective in regulating prey populations.

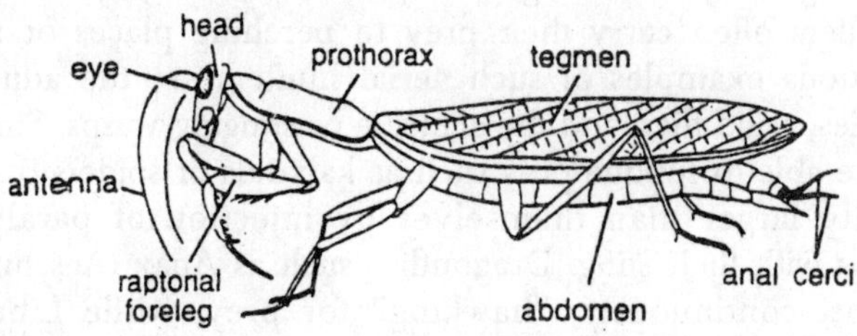

Fig. 4.3. Mantis.

Among Coleoptera predatory carabids are distinguished by long, sharply hooked mandibles that contrast with the broad, blunt jaws of plant-feeding relatives. Coccinellid beetles, syrphid larvae, and neuorpteran predators that feed on aphids also search at random. The mandibles of predaceous coccinellids may be incisors with one or two apical teeth and a basal tooth or they may be small with ducts for sucking prey juices. Neuopteran larvae have sickle-shaped jaws each formed by the mandible and maxilla locked together to create a tube through which body fluids of prey are sucked.

Most of Lepidoptera are *phytophagous.* Caterpillars of certain moths retain their close association with plants but attack other phytophagous insects such as coccids and leafhoppers as well as mites. Caterpillars of the lycanid butterflies may be phytophagous or partly or wholly predaceous on ant larvae and pupae or aphids. The relatively small size and high nutritive valve of insects eggs make them vulnerable to predation by insects of varied food habits. Scavenging, phytophagous, predatory, or parasitoid insects may consume eggs. Collectively they are called egg predators when more than one egg is consumed. These predators are of special ecological importance because the prey never function in the community. For example, the large clustered eggs of Orthoptera are subject to frequent attack by many predatory larvae; clerids, meloids, bombyliids, rhagionids, anthomyiids, calliphorids, otitids, phorids, sarcophagid and eurytomids. The adult female of the predator species finds the eggs, lay her own eggs, and the larvae feed essentially as random searchers.

The *hunting insects* differ from random searchers by utilizing sight or other stimuli to orient to prey at a distance. Visual hunters have enlarged compound eyes with overlapping fields of vision that permit distance perception. Mandibles may be sharply toothed, and the legs may be strong and spiny for seizing elusive prey. Strong fliers often carry their prey to perching places or nests. Conspicuous examples of such aerial hunters are the adults of dragonflies, asilid flies, and the aculeate or stinging wasps. Stinging wasps are able to subdue prey such as katydids or spiders that are physically larger than themselves by injection of paralyzing chemicals with their sting. Dragonflies such as Anax (Aeschnidae) fly almost continuously, "hawking" for prey, while Libellula (Libellulidae) remain perched until approaches, then quickly dash in pursuit. Asilids commonly perch and await flying prey, then return to the perch after the victim is caught. Adult cicindelid beetles pursue their prey by running fast on open ground.

Stimuli other than visual may be used by hunting predators. Although blind, the famous army ants (genus Eciton) of the New World Tropics are enormously successful predators. Odours and movement of prey are detected by chemical and tactile receptors while the colony is on one of its periodic raids. Prey many times larger than the individual worker ants are attacked by massive swarms. Cooperative foraging by "packs" of other kinds of ants also subdues large prey. The back swimmers (Notonectidae) are vorcavious acquatic predators. They flush and stalk prey by sight and are able to detect the vibrations of prey movements with receptors situated along the forelegs. The water striders (Veliidae) also perceive prey by sight and vibrations.

Insects that ambush prey conserve energy by simply waiting for prey to approach within striking distance. Reduviid and phymatid bugs often remain motionless on flowers awaiting flower visitors. Such predators are exposed to predation by vertebrates and have either concealing or warning colouration. Concealing colouration may also prevent detection by alert prey. Raptorial or clasping forelegs are frequently characteristic of insects that ambush, especially among the Hemiptera. Some reduviids aid prey capture by smearing sticky secretions or plant resins on their legs. The beaks of predatory bugs are stout and inject saliva that contains proteolytic enzymes. This is why their bites are painful to human in contrast to the mild bites of bugs that are parasites of humans. Some reduviids also have a potent toxin that quickly renders prey helpless. Once the tissues of the victims are digested, the resultant fluid is sucked by the bug.

The praying mantis is a familiar predator that strikes prey from ambush. The co-ordination of the depth-perceiving vision, mobile head and prothorax, and toothed, raptorial fore-legs has been carefully analyzed by Mittelstaedt (1962). The complete strike, timed by high-speed photography, takes 50 to 70 milliseconds. According to Roeder (1967), a fly or cockroach would require about 45 to 65 milliseconds to respond if startled by the first movement of the mantis; alas, too slow to escape the strike already in motion. The dragonfly naiad similarly grasps prey with quick strikes of its extensible labium. Cicindelid larvae seize prey that pass near the entrance to their burrows in soil. Only a few kinds of insects trap prey. The "ant lion" larvae of Myrmeleontidae (Neuroptera) excavate conical pits in fine, loose sand. The larva wails motionless and buried at the bottom until a small insect ventures into the pit. The ant lion fliks sand toward the victim to cause the unstable sand to slide down. Once the prey is seized, the ant lion extracts the body fluids with its sickle-shaped jaws. Larvae of the fly Vermileo. (Rhagionidae) also construct pits and trap prey. Wheeler (1930) aptly describes these insects as "demons of the dust."

Larvae of the glowworm, Arachnocampa luminosa (Mycetophilidae) live in moist caves. They spin slimy webs that dangle from the ceiling. They spin slimy webs that dangle from the ceiling. The larvae glow in the darkness, attracking small flies that become entangled in the sticky webs. Larval mycetophilids of the genus Platyura also secrete webs that entangle prey and contain a toxic fluid. Even though silk is secreted by various insects, webs are surprisingly rarely used to trap prey in the manner of spiders. The aquatic caddis fly larvae of the Hydropsychidae spin webs that filter small prey and other food from passing currents. Though they do not spin their own webs, one group of slender reduviid bugs, the emesines, wait in ambush at spider webs to attack trapped prey.

Parasitoid Insects

Parasitoid insects are intermediate between predators and parasites. The larvae of *holometabolous* insects, especially, Diptera and Hymenoptera, are parasitoids. Their unique attributes combine certain features of true parasites and predators. Parasitoids differ from predators in the following features: (1) only one host is required, (2) the host is larger than the parasitoid, (3) parasitoids are frequently host-specific, attacking one or several related host species, (4) a lower density of host population will sustain a parasitoid population, and (5) the victim is usually searched for and selected by the diurnally active, adult female. A species of parasitoid may be specific not only in the choice of host species, but also in the choice of the life stages of the host to be attacked. Eggs or young or both are most frequently attacked, but pupae and sometimes adult hosts are also eaten. Parasitoids may fed external as *ectoparasitoids* or internally as *endoparasitoids.* Exposed hosts are usually attacked by endoparasitoids, where hosts in protected situations such as leaf mines, falls, or nests, are attacked by either endoparasitoids, or ectoparasitoids. When the larva of a species characteristically develops in the ratio of one to a host, the species is termed a solitary parasitoid. When several larva of the same species normally develop in a single host, the species normally develop in a single host, the species is called a gregarious parasitoid. If the host is a phytophagous insect, the parasitoid is a primary parasitoid.

Parasitoids that attack other parasitoids are called *hyperparasitoids* or *secondary parasitoids. Multiple parasitoidism* occurs when two or more species of primary parasitoids attacks one host individual. Superparasitoidism occurs when a host is attacked by more larvae of the same species than can reach maturity in the one host. Adult parasitoids are free-living, and most of them are winged except for female velvet ants (Mutillidae) and certain other scolioid wasps. Energy for their activity and egg production is derived from food stored by the larva or by feeding as an adult. In the latter case, nectar and honeydew are frequently consumed. Some parasitoid wasps feed on the honeydew are frequently consumed. Some parasitoid wasps feed on the body fluids that are released from the host's body when it is punctured by the parasitoid's ovipositor. This is called host feeding. In some instances, the host is inside a cocoon or for some other reason can be reached only by the parasitoid's ovipositor. A tube of coagulated host's hemolymph, or possible of a secretion produced by the parasitoid, forms around the ovipositor. When the ovipositor is withdrawn, the parasitoid is able to suck the host's hemolymph as it wells up in the tube.

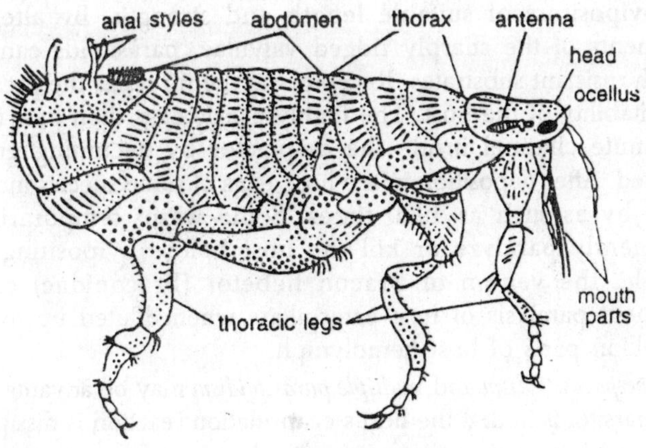

Fig. 4.4. Xenopsylla.

The adult parasitoid searches first for the habitat in which the appropriate hosts exist. Primary parasitoids are often attracted mainly by the plants that shelter favoured insects host. In this behaviour the parasitoids respond like phytophagous insects to shapes, colours, and odours of plants. Yet they do not feed on plant tissues. Once near the host, female Hymenoptera usually search for suitable individuals on which to ovipoist, using their tactile and olfactory senses. Their movements may be essentially random or somewhat systematic in response to stimuli detected at a distance. Female Diptera lacks the well-developed antennae of wasps and the needlelike ovipositor for precisely inserting eggs in hosts. Consequently, the flies commonly depend on first instar larvae to actually locate hosts after the eggs are laid in the proper habitat. Insects that have two or more successive larval instars specialized for different modes of life are hypermetamorphic. Among the hypermetamorphic parasitoids, the first instar larva is specialized for active host finding and the subsequent instars specialized for feeding. In Hymenoptera the ovipositor plays an important role in selecting the host, sometimes in paralyzing it, and in delivering the egg to the host.

Dethier (1947) demonstrated that chemoreceptors on the ovipositor tip responded to a variety of chlorids, aliphatic alcohols, and hydrochloric acid in much the same way as taste receptors on the anterior appendages. These receptors function in the final discrimination of hosts before the egg is laid. Hosts inside tough cocoons or sclerotized puparia, in leaf mines, or even deep in burrows in solid wood can be reached by parasitoids equipped with ovipositors of suitable length and strength. By alternate movements of the sharply ridged valvulae, parasitoids can drill through resistant substrates. Receptors at the tip sense the presence and suitability of the host. The highly elastic eggs then pass down the minute channel inside the ovipositor by becoming greatly elongated. After deposition in the host, eggs of some species increase in size by as such as 1000 times. Some wasps temporarily or permanently paralyze or kill the host before ovipositing. For example, the venom of Bracon hebetor (Braconidae) causes permanent paralysis of host caterpillars when diluted up to 1 in 200 million parts of host hemolymph.

Superparasitoidism and *multiple parasitoidism* may be advantageous to the parsitoids in that the host's ecapsulation reaction is dissipated. The resulting competition among larvae also may be dis-.

advantageous. First instar larvae of certain wasps are specialized to compete successfully in such situations. Some have large sickle-shaped mandibles and kill competitors; others release inhibitory toxins or, if larger, deprive smaller competitors of oxygen. Competition is also commonly reduced by the behaviour of the female wasp. For example, Trichogramma (Trichogrammatidae, Chalcidoidea) reject hosts on which are females have walked. Some wasps seem to be able, by means of sensilla on their ovipositors, to select healthy host individuals from among those already parasitized. Endoparasitoids are immersed in the body tissues and fluids of their hosts.

Many synchronize their development with that of the host by responding to the host's hormones. Some obtain food in the normal manner via the mouth, but others, especially those in eggs, are greatly simplified and absorb food through the outer integument. Endoparasitoid wasp larvae commonly exchange gases through their outer integument with or without a tracheal system and rarely with open spiracles. Respiration may be aided by caudal filaments, anal vesicals, or an averted hindgut. Endoparasitoid fly larvae usually have at least a metapneustic tracheal system. Some perforate the host's trachea or integument and obtain atmospheric air. Others take advantage of the host's encapsulation reaction and mold a respiratory "funnel" or tube of host tissue connected to a trachea or the integument.

In parasitoids, Hymenoptera the sex is determined in this order by a haplodiploid mechanism. The ratio of sexes is often unbalanced in favor of females. A peculiarity of Aphelinidae (Chalcidoidea) is that the female may select the host according to the sex of the egg to be laid. Furthermore, the sex of the larva may influence its feeding behaviour; males may be regular hyperparasitoids. Some wasps are exclusively parthenogenetic, in contrast to other entomophagous insects that are rarely so. The unusual phenomenon of polyembryony and parasites; it occurs in some endoparasitoid Hymenoptera and a few Strepsiptera. As many as 1500 embryos in a single caterpillar have been counted after oviposition by one female of Litomastic (Encyrtidae).

Parasites

A parasite is the organism that lives at the cost of other. True parasitic relationships between one insect and another are surprisingly rare. As we have seen in the previous sections, the usual result of attack is the death of the victim. Entomophagous parasites may be divided into ectoparasites and endoparasites. Adult biting midges of the genus Forcipomyia (Ceratopogonidae) and several other genera are ectoparasites that take blood from both vertebrates and insects. When feeding on the latter, the flies puncture the wing veins or intersegmental membrances. The flies visit the host only during feeding.

The small wasps of the family Scelionidae lay their eggs among the freshly laid eggs of host insects. Females of several genera have been observed attached phoretically to the bodies of female hosts. After the host's eggs are laid, the wasp immediately oviposits. Young winged females of the French Rielia manticida (Scelionidae) search for and attach themselves to mantids of both sexes. Female hosts are more frequently selected. Once attached by their mandibles, the wasps shed their wings and await the host's oviposition, which may not take place for several months. In the interim, the wasp feeds as an ectoparasites on the mantid's hemolymph. When the mantid's ootheca has been deposited and is still soft, the wasp leaves and mantid to lay its own eggs. The wasp is said then to attempt to return to its host's body. An unusual instance of ectoparasitic behaviour is provided by the acquatic larvae of the midge Symbiocladius (Chironomidae). The larvae attach themselves behind the wing buds of mayfly naiads and feed.

Strepsiptera are the only entomophagous insects that are true endoparasites. Numbering about 300 species, they probably evolved in the early Tertiary Period from some group, as yet unknown, of parasitoid Coleoptera. Fossils have been found in Baltic amber. Hosts include insects in the orders Thysanura, Blattodea, Mantodea, Orthoptera, Hemiptera, and Diptera, but most of them are aculeate Hymenoptera. The infested hosts are said to be "stylopized," because the name of a common genus is Stylops.

On bright warm days in February and early March, adult bees of the genus Andrena emerge from their burrows in soil and visit flowers of the buttercup, Ranunculus. As many as 16 per cent are parasitized. The visible evidences are the puparia that protrude from the body, commonly between the fourth and fifth terga of the abdomen. The male Stylops is winged and less than 3 mm long. On emergence from the puparium, it immediately begins a rapid, vibrating flight in search of a female. The latter is reduced to virtually a sac of reproductive organs inside the puparium. Females release a sex attractant while the host flies from flower to flower. Males by upwind, tracking the odour, until the female is located on the dorsum of a feeding bee. The male Stylops lands and inserts the aedeagus by puncturing the female's puparium. Once inseminated the female no longer releases the attractant. Males probably die the same day as they emerge and mate but once. Except during copulation their intense flight activity never ceases and they rarely feed. The tarsi even lack claws for clinging to objects.

Linsley and MacSwain were able to capture the rare males by putting bees with virgin female parasites in cages and placing the cages among flowers. The first-instar larvae are triungulins, and they develop from the fertilized eggs while still inside the female's body. During the next 30 to 40 days, the female dies and the larvae move into a median brood passage in preparation for their exit. A large female Stylops probably produces 9000 to 10,000 triungulins. On warm day while the host is visiting flowers, the active larvae emerge through the ruptured puparium. Parasitized hosts move more rapidly among flowers than normal hosts. The triungulins are brushed singly or several at a time onto the flowers. In contrast to the triungulins of Meloidae, Rhipiphoridae, and possibly other Strepsiptera, those of Stylops pacifica do not readily attach themselves to new hosts, but are ingested with nectar by bees.

The bee returns to its nest and prepares a ball of pollen mixed with regurgitated nectar. In this way the triungulin is deposited in a nest cell with the bee's egg. The triungulin penetrates the egg, transforming as it does into the second instar, and becomes an endoparasite. As the host feeds and grows, the Stylops larva feeds on the host's nonvital tissues and passes through an unknown number of instars. When the bee pupates, the Stylops protrudes the anterior portion of its body through the intertergal membrance and also pupates. The last larval skin is not shed, but forms a tanned puparium.

Strepsiptera have no known natural enemies. Major causes of mortality are the initial losses of triungulins that fail to find a female host of the proper species, and superparasitism. Linsley and MacSwain noted that usually only the larger female bees with a solitary female parasite lived long enough for the triungulins to escape. Smaller male hosts, those with several Stylops, or those with Stylops plus nematode parasites, usually died before the cycle was completed. Bees that survive parasitism are variously affected. The genitalia of both sexes may be reduced. Secondary sexual charactes, especially in the female, may be shifted toward the opposite sex. Stylopized females may have more malelike yellow colour and reduced pollen-collecting hairs on the legs.

5

SOCIAL LIFE

Living together of insects develop different types of inter-relationship of which the social life is the highest development of intra specific inter relationship. From simple animal aggregations there may evolve complex animal societies composed of specialized types of individuals such as the colonies of bees, ants and termites. Although ants (like termites) live solely in organized societies, bees and wasps may have either a solitary or a social way of life. Some times the boundary between the two is hazy. Various simple forms of family life are shown by a few mining bees of the genus Halictus, as described earlier, and by a few vespid wasps. Vespid wasps of the genus *Stenogaster* include both solitary species and species whose members live in united families. Among the latter, the females reared by the mother remain in the nest for a short time giving help. After their departure, they are replaced by younger sisters. Frequently such offspring, instead of establishing their own nests at a distance, attach them directly to the maternal structure, forming a home that contains a loose federation of families.

HONEYBEE AS A SOCIAL INSECT

Honeybee is a social insects and exhibits a clear cut phenomenon of Division of Labour. For the first two three days of their life, honeybees take care of cleaning tasks and the regulation of warmth. By means of community co-operation, they maintain a continuous brood temperature of about 95°F to 97°F (35°C to 36°C). Ventilation and cooling are accomplished by mean of whirring the wings, in case of need even by fetching water and pouring it over the combs. On the other hand, heating result from vibratory activity of the

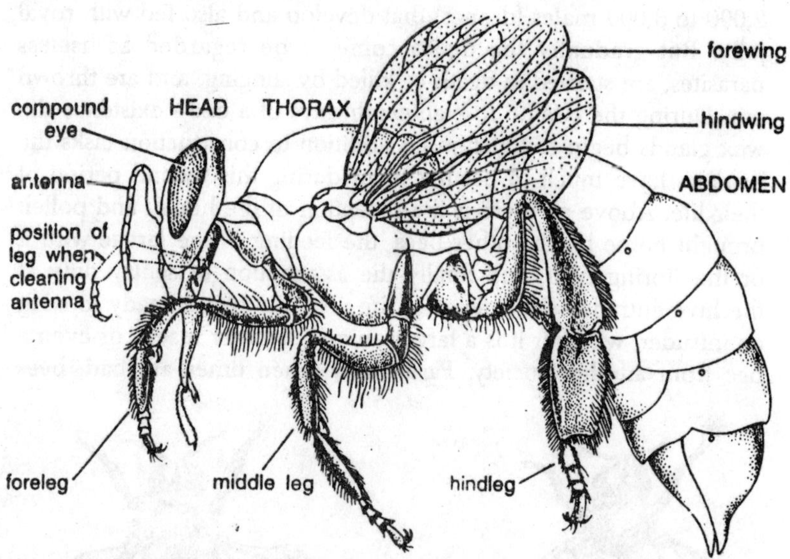

Fig. 5.1. Worker honeybee (Lateral view).

flight muscles, from which the wings have been uncoupled, and thus from burning the carbohydrate that has been derived from the honey. From the third to the sixth day of their adult life the bees use pollen and honey to feed the medium-sized to full- grown larvae, which are four to six days old, Both of these foods are stored, carefully separated, in different groups of cells in the combs. From the seventh day onward, for about a week, two large glands develop in the bee's head. These are vitally important to the colony in that they secrete an extremely growth - promoting, predominantly protein - containing fluid, the royal jelly. This flows from the worker's mouth and is fed to the young larvae and the queen. Unlike them, she is given royal jelly continuously, for to ripen about 100 eggs every hour of the day and night her body must have an enormous metabolic turnover. The "royal court," whose members feed and lick their queen, consists of individuals of various ages.

The queen does nothing but deposits an egg at the bottom of each cell prepared for it. Every day she lays over 2,000 eggs, with a total weight more than her own. At any one time the bees in a colony have perhaps 10,000 larvae to feed, and each of these may receive several thousand feeding visits during the six days it takes to mature. In a strong hive with 10,000 to 80,000 individuals, the

2,000 to 3,000 males (drones) that develop and also fed with royal jelly. But gradually the males come to be regarded as useless parasites, are starved to death or killed by stinging, and are thrown out. During the twelfth to eighteenth days of a bee's existence the wax glands begin to function; in addition to construction tasks the workers have much to occupy them during this second period of their life. Above all there is the reception of the honey and pollen brought home by the other bees, the feeding of the larvae with it or the storing of it, and finally the assumption of sentry duty at the hive entrance. In defense of the nest, the bee is ready to sting an intruder, whether it is a larger animal, another insect, or even a bee from another society. Particularly when times are bad, bees

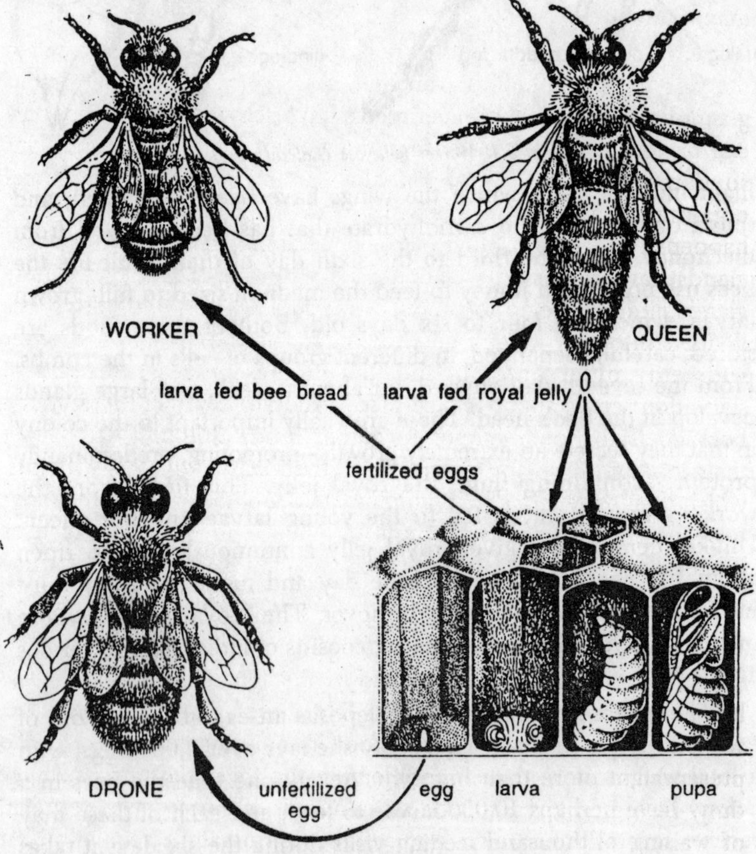

Fig. 5.2. Honeybee. Life cycle.

turn into honey thieves. If they succeed in overpowering the guard at the gate, they fetch their companions and raid the nest.

Among the sting less honeybees (meliponines) there is even one species whose societies specialize in organized predatory invasions of other nests. There they pillage the honey stores and also steal wax for use as building material. The third phase of a bee's life, beginning at an age of about three weeks, is one of active service in the field. The worker prepares herself by undertaking several preliminary orientation flights in which she impresses upon herself the location of the nest and its immediate surroundings. Now she begins her food collections, and when she finds sources of food learns how to pass on information about them to her comrades. After perhaps ten days of this, the bee's life is at an end; the average age is 30 to 35 days, with a probable maximum of 55 days.

One might be tempted to regard the division of labour within the bee society as merely an automatic consequence of various glandular functions. But experiments have shown that such conduct can be repressed or can be forced ahead of schedule if conditions so demand. In other words, the requirements of the social organization are the moving force of all behaviour. Thus, if a hive happens to lose its normal building workers, old bees whose wax glands have regressed with experience a redevelopment of these glands and will begin to produce wax again. Or suppose a hive is divided into two nests, one with only young bees and the other with only older ones. In the first of these the catastrophe of starvation seems about to take place. And yet many of these young bees go flying out betimes to gather food, although this is not their job, and they do this in spite of their active royal-jelly glands, which then degenerate prematurely. On the other hand, in the older bee's nest, these glands continue to function in a number of bees longer than normally, simply because, for lack of young nurse bees, this has become a vital necessity.

The Workers

In other social bees and wasps, the female progeny do not leave the home, but continue to live there. They participate, as workers, in the care of the brood. in building activities, in the preservation of hygienic conditions in the dwelling, and in sentry duty. Domestic tasks include cleaning out of empty cells, removal of walling off of foreign bodies, and temperature regulation an renewal of air by fanning the wings in the vicinity of the nest entrance and air vents. This last is done in concert and mostly silently, except by bumblebees. Early in the morning a single bumblebee, dubbed a "trumpeter," takes a position immediately beneath the roof of the nest and ventilates it, with a loud buzzing.

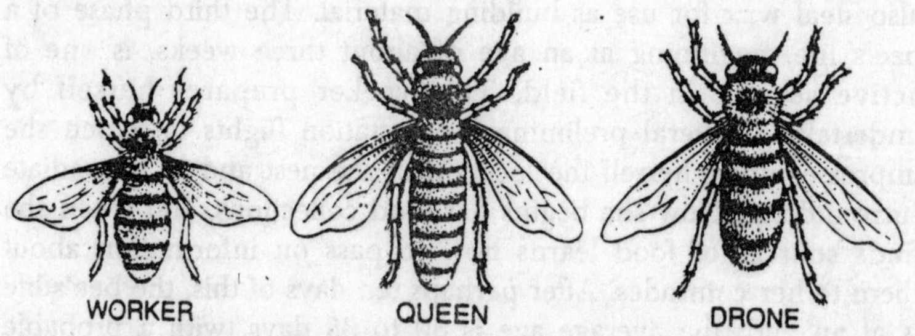

Fig. 5.3. Castes of honeybee.

Another chore of the workers is to receive food and building material fetched home by their foraging sisters. The food is eaten by the workers and fed to the larvae right away or is first stored in special provision cells. The building material is used at once. An especially important job is the care of the brood and, for honeybees and ants, care of the nest mother or queen. With out proper attention the eggs may die. The larvae, which live in groups within the structure, have to be shifted again and again from chamber to chamber, according to the temperature and the humidity. Even newly matured insects may need to have the cells opened for them when they emerge, or may have to be freed from the pupal skin, as happens with ants.

Aside from these chores, nursing is confined to distributing food adapted to the age of the brood. In general, wasps and ants proffer food prepared from captured insects, spiders, or even fresh meat; the wasps provide it in the form of little balls saturated with saliva, the ants usually as droplets of predigested juice. Bees, on the contrary, feed the brood only with pollen and honey. Just as young songbirds, apparently by means of their strikingly bright-coloured and widely opened gullets, keep alive and augment the urge of the old birds to continue feeding them, the larva of wasps and many ants (but not of bees again and again offer a drop of secretion on their extended mouthparts. This is freedily licked off by the older insects and thus indirectly stimulates them to feed the larvae. Such exchange of food, called trophallaxis, may also induce a swarm to rear as many offspring as possible. Moreover, many ant larvae do not produce this delicacy from the mouth, but instead exude it from peglike processes or even from the entire body surface. In the vespids of the genus *Belonogaster* the assisting daughters are just like the mother. Since their urge toward nest construction and search for food awakens only with sexual maturation, they are occupied during their first week solely with cleaning operations and with distribution of food in the nest. Then they begin to lay eggs, gather food and enlarge the nest as needed. Nevertheless the family still needs the original mother, for if she were lost the daughters would finally devour their own brood, and the entire society would gradually die off. But otherwise the family gets bigger and bigger, and males, too, are developed. For four to five days the males are fed in the nest with stored up honey by their sisters or else, pilfering on the sly, lead a parasitic life, as is also the case with various species of the wasps *Polistes.* If the members of the community become so numerous that for technical reasons the structure no longer can be enlarged sufficiently, some of the wasps depart and found a new home. But they do this in a most indelicate manner. They not only bite off material from the old home for use in the new one, but even pillage it of the larvae, feeding them to the new brood, with the result that old nest soon is destroyed. *Polistes* wasps are of a rather predatory nature anyway, and occasionally they plunder one another's nests.

In other wasps, such as *Ropalidia* and some species of *Polistes* the worker caste begins to show some of the differences that among the rest of the social bees and wasps make it a clearly distinct form. The offspring of the nest mother are smaller females and have reduced fertility or lack it entirely. This results from a restricted supply of food. Except in honeybees and ants, the later offspring, as the family grows in size and the amount of feeding becomes larger, develop into bigger and bigger individuals. With wasps such as Polistes and occasionally even with bumblebees (*Bombus*), the later-emerging females scarcely can be distinguished from the mother any longer, but with the other wasps and hornets (*Vespa*) they always are clearly smaller than the queen. The tasks of the workers in the nest frequently are determined by their age-that is, by the degree of maturity of definite internal organs- and thus the life cycle of the individual unfolds in a predetermined sequence. Such a sequence of duties has been best studied in honeybees and, with some exceptions, follows the schedule described here.

Ant Workers

Ants are also social insects. In an ant society, though brood care and division of labour are more highly organized, the type of work seems to be less definitely regulated than it is with bees, and its individual facets are only slightly dependent on the age of the individuals, or not at all. So the society is much more adaptable. In place off males that merely are reduced in size and more or less sterile, there are workers with distinctly different body structures specially suited to the tasks at hand. The workers of a given ant species may be more or less uniform in size and shape (monomorphic), may be variable (polymorphic), or may appear as two distinct types (dimorphic). These last, known usually as "*workers*" and "*soldiers*" may represent the extremes of an originally continuous chain of variations, the intervening links of which no longer exist.

In genera with a single type of worker, only one variant, usually a smaller one, has remained, with ants low in the evolutionary

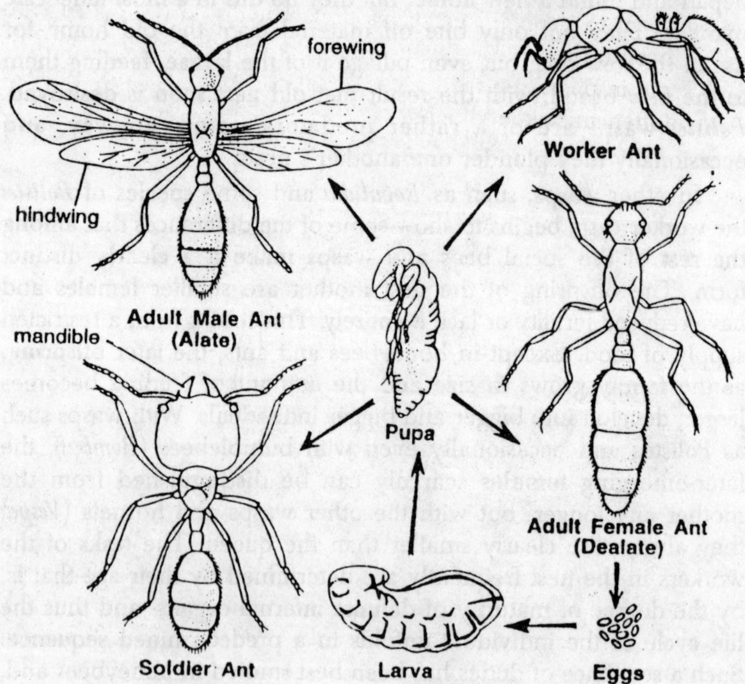

Fig. 5.4. Ants. Castes and life cycle.

scale such monomorphism may be a primary manifestation. Where a species has different types of worker, the large ones are designated as "*soldiers,*" yet by no means are they always the defenders or the warriors of the society. Big heads and mandibles of extra ordinary shape may be based upon other biological pre-requisites, for instance, they may be necessary for mincing up seeds or animal prey. In some species the nest entrance is closed off by the head of a "porter" stationed there; in certain genera the porter's head is so constructed that the door is not only barricaded but is concealed besides. Various worker forms, ranging from about 0.08to 0.6 inch in length, are found among the leaf cutting ants (*Atta*). As in other ant societies, the different tasks appear to be parceled our to the types best fitted for them. The dwarfs take care of the mushroom garden, the middle-sized individuals gather leaves and work them over in the nest, and the big workers guard and defend the nest.

Different tasks in the nests of ants all can be carried out by any worker, and usually a single worker executes a particular job only for a limited length of time. Nevertheless, the small individuals seem to devote themselves preferentially to the actual care of the brood. Both the eggs and the larvae are licked over and over and put in piles, arranged according to their age, and the pupal coccons (often mistakenly called "*ant eggs*") are also separated and heaped

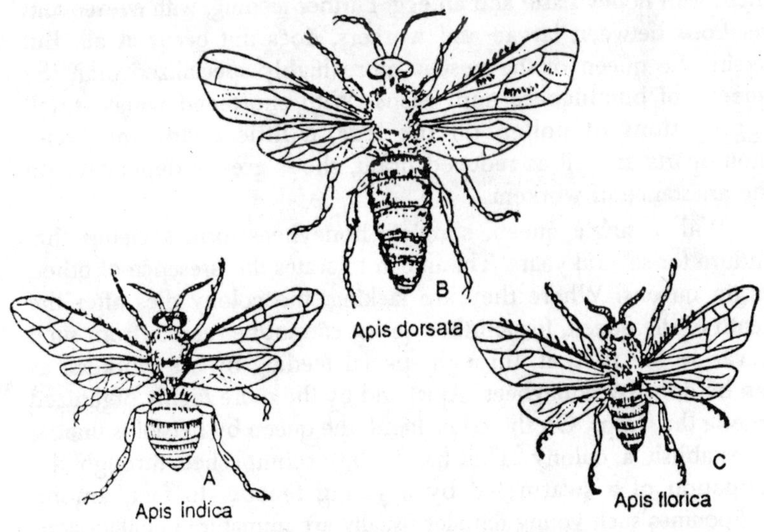

Fig. 5.5. Workers of three species of honeybee.

up in different chambers. Ants mostly feed their larvae, as they do their queen, their fellows, and their guests, from mouth to mouth with drops of nutritive juice from a "crop" or "social stomach" that, like the honey stomach of bees, lies in the fore part of the abdomen. But more primitive species, such as many ponerines, proffer chopped-up insects; and a few genera offer balls of minced tissue as food, in some cases from a special oral pocket.

Forms of Social Life

After this consideration of the worker caste and its duties, we return to the forms taken by family and civic life. In the tropics the life span of families of vespid wasps and bumblebees (Bombus) does not depend on portion of the hive together with one or several queen-or else, as in other wasps, simply by individual fertilized females. These *Polybia* wasps, moreover, have the strange peculiarity of flying soundlessly, and are particularly prone to attack intruders within a circumference of about two years of their nest. The bleeding puncture made by their sting is very painful and makes the affected part of the body whiten at once, yet causes only a slight swelling. The frequently large societies of the "stingless" honeybees (*Melipona, Trigona*), which do have a sting, though an ineffective one, seem more primitive than the societies of *Polybia.* For these honey-bees close off their brood cells, as do the solitary bees, after providing them with honey paste and an egg. Further feeding, with consequent relations between larvae and workers, does not occur at all. But again, the queen of the nest is more highly specialized than the queens of bumblebees and wasps. With shortened wings, small aggregations of pollen on the legs, a little head, and weak mouthparts as well as reduced brain, she is greatly dependent on the assistance of workers.

With a single queen, stingless honeybees form societies that endure for several years. The queen tolerates the presence of other, virgin queens. Where they are lacking, the colony dies after the death of the queen, for another queen cannot be reared, from such larvae as are present, through special feeding by the workers, as can be done by honeybees (Apis) and by the more highly organized among the wasps. On the other hand, the queen by herself is unable to establish a colony. This has to be accomplished through the formation of a swarm led by a young female. In fact, among meliponines such young females usually are immature initially; later, the great swelling of their abdomen makes flying out impossible.

SWARMING

Swarming is the mating of individuals away from their place to residence. After the sexual forms-males and females (queens) have been reared, swarming occurs. Since in many species, including the honeybees, these forms are larger than the workers, sometimes far larger they have to be reared in special large cells. One might imagine that honeybees would construct their cells of a size proportional to their own body mass. But where do honeybee workers get the proper measure for their fatter brothers, the future males (or drones), which they have never seen before in their lives? What signal suddenly causes these bees to build larger cells in the comb, and impels the queen to deposit eggs that are unfertilized and that therefore produce males? Truly they have great power over the life of their folk, these workers. Ultimately, too, they determine the number of the young queens. These are reared with special food in the few big, pendant, peg-shaped queen cells, which strangely enough bear on their surface the size–cornered pattern

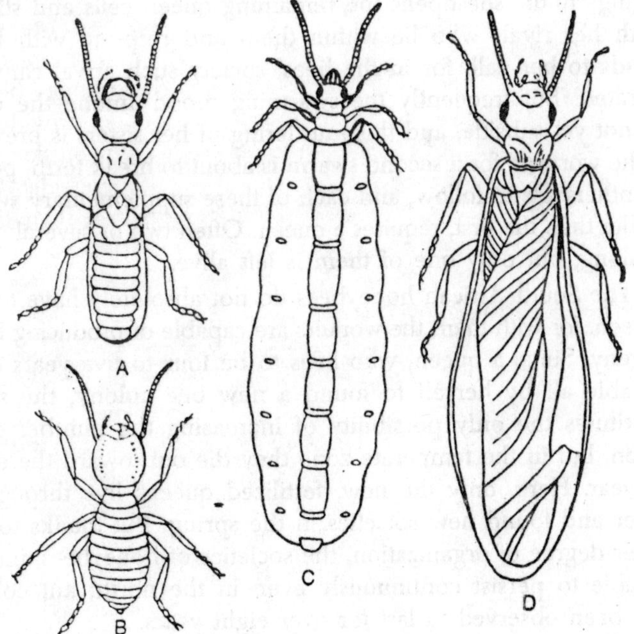

Fig. 5.6. A–Queen of termites without wings after nuptial; B–Wingless sterile workers; C–Mature queen; D–The winged male before nuptial.

of the comb. Inside the cells the young larvae hang head down ward. When they are ready for pupation, their cradles are closed over with waxen covers. The worker caste is then gripped by that feverish, pervasive agitation, the swarming mood. This leds to the departure, as if on order, of about half the population, after each of the individuals has filled up on honey from the plentiful stores in the nest. In the course of only a few minutes, they fly off from the exit, perhaps some 30,000 of them, together with the old queen. Not being an especially good flier, she soon settles some where on a branch, and the whole billowing swarm of bees gathers around her like a bunch of grapes. From this place, usually a few hours later, they make off to a new nesting place that has been explored mean while by scouts. Or perhaps the swarm is packaged up by a beekeeper and given a home in his apiary.

It is precisely this situation that has enabled man to culture bees a practice known to the Egyptians over 5,000 years ago. Among the ancestral stock that stays behind when the swarm leaves, there emerges after about a week the first of the young queens. Excitedly singing "tu-tu" she opens the remaining queen cells and stings to death her rivals who lie within them and respond with hollow sounds to her call, for in the bees' society such arival cannot be tolerated. But frequently the swarming mood among the colony has not yet subside, and this murdering of her sisters is prevented by the workers, for a second swarm is about to break forth, perhaps still others are to follow, and each of these supplementary swarms, smaller than the first, requires a queen. Often two or several queens fly along, but only one of them is left alive.

The South African honeybees do not absolutely have to have a queen, for with them the workers are capable of producing female progeny. Since a queen, who lives to be four to five years old, is not able all by herself to found a new bee colony, the swarm constitutes the only possibility of increasing the number of the season, but in the temperate zone they die out toward the end of the year. Here, only the new, fertilized queens live through the winter and found new societies in the spring. But thanks to their higher degree of organization, the societies of honeybees and ants are able to persist continuously even in the north; ant colonies have been observed to last for over eight years.

Many wasps and bumblebees have simpler and more similar forms of social life, and an example will describe their normal course. A bumblebees queen, after over wintering, finds a site that appeals to her and prepares it for nesting. After plastering a spot on the ground with wax, she mixes a lump of pollen and honey, builds a circular wall of wax on top of it, lays some 7 to 16 eggs inside, and closes the cell with a wax roof. From the same material she builds a storage vessel for pollen and honey to one side, placing it where she can reach it without leaving the waxen cell. Sitting on the cell, she keeps the eggs inside war, and they soon hatch into larvae. These are fed from time to time be the mother, who bites a hole through the lid of wax and then closes it afterward. The larvae also eat the mass of pollen and honey on top of which they were born in the cell, which the mother enlarges, as they grow bigger, until finally they spin cocoons that hang together; in these they pupate. The queen then attaches more waxen egg chambers here and there to the old cell, using in part material from it, so that the lighter-coloured cocoons now lie bare. After the first bumblebee workers emerge, these cocoons are used as storage vessels. The workers help the mother, and the more numerous they become the less she works herself, until she perhaps has become nothing but a producer of eggs. Thus the number of queen in two ways embodies the transition from the solitary to the social bees. First, she gathers food like the solitary bees and closes off the cell after laying eggs, but on the other hand constantly feeds the young through openings. Second, in the beginning she carries out all the tasks, but later merely lays eggs like a queen bee.

Somewhere in the midsummer, when the family has become numerous and food is plentiful, new queens are reared. Bigger and bigger workers develop, and some lay unfertilized eggs. If there are excess eggs, the bumblebees frequently feed on them. The unfertilized eggs always develop into males, as is everywhere the case among social Hymenoptera. Thus males may be descended from workers. But more often they are produced by the queen, who is able voluntarily to prevent access of semen to the egg she is laying. This semen, which suffices for the entire duration of her life, was received from the male with whom she mated in the preceding year. The emerging males, fed by their sisters, remain in the nest for a few days until their hairy covering, at first a whitish-gray, has gained its full colour. Then they fly off, visit flowers and mate. In the tropics bumblebee societies may last for years; then, instead of a single queen, a whole group of them is active in the nest. In the far north, there frequently is another variant. During the very short period of the year when flight is possible, the mother's offspring may not develop into workers but instead directly into males and new queens, as is customary in the solitary bees.

The family life of many vespid wasps corresponds in great measure with that of the bumblebees, apart from the wholly different manner of constructing the nest and feeding the young. In both groups, the mature insects imbibe nectar and sweet of fermenting plant substances, as well as honeydew. But the vespids, with the apparent exception of the nectarines, feed their with a paste of chewed-up insects. In the spring two or several vespid queens occasionally found a nest in common. In *Polistes* wasps, this urges for sociability may even extend to where a female will leave her own nest and her still-young brood in the lurch and will attempt to join a distinct family. Whether she succeeds is another matter.

For the majority of species drive away intruders, even those of the same line of descent; the cause of this is nothing else than that very possessiveness toward the brood that to some degree has been lost by the worker caste in correlation with their degenerate sexuality, a loss that has made possible the selfless cooperation of the individuals in a society. Through this loss a certain loosening of the rigidly established instinct for the care of the brood also entered and brought with it space for other, new tasks. The most highly developed social organizations among the wasps are those of certain species of *Polybia*. They are said to surpass in this respect even the social organizations of honeybees, which are very similar. With one and the same *Polybia* species the establishment of new colonies can take place in very different manners according to the circumstances. It can take place by swarming, as in honeybees-that is, through the departure in flight of a bee population ad thereby preserving the species. Lack of food or any kind of disturbance or menace an also provoke bees into swarming; such migratory swarms then represent merely a change of dwelling place and not a fractionation of the population.

The honeybee Apis mellifica, originally from the Old World, has been Spread by man over the entire earth. Until recently its races were not products of intentional breeding, but were primal geographic forms. Today, at all event, crosses can be produced by artificial insemination of queens. Of the remaining three *Apis* species (dorsata, flora and indica), all of which live in Asia, only the last has been domesticated.

In ants a quite different kind of swarming is seen among ants. Excited workers appear before the best entrance and get more and more numerous. Finally, there appear the winged, sexually mature individuals, which usually are reared in large numbers only years after the society is founded. First come the males, and then the females. They rise into the air, perhaps like clouds of smoke, later to cover the earth with a layer of ants. This is no swarming in the sense of the bees, for these are mating flights.

In many genera the males and females, which as a rule are equipped with huge wings, are almost equal in size, but in other genera they are very different. One may even see a tiny male riding about on his partner's abdomen as if on a dirigible. Verging on the incredible are the differences in size between sexually mature forms and works, especially in legionary ants (dorylines). Here, a male may measure 1.2 inches, and the smallest worker a mere 0.08 inch; or a worker 0.12 to 0.16 inch long may have a queen more than 2 inches long.

Role of Males

In hymenopterous colony the males are of no use. After mating, they die. The fertilized females then found their purely female societies, as queens with a final cohort of many myriads of workers. There are ant socities with over a million individuals, and one population my hold sway over an area of hundred, even thousands, of square yards. Certain *Farmica* ants pile on top of their nests heaps of earth more than a yard high and perhaps four yards in diameter. Such digging by ants is of great benefit to man in loosening and aerating the soil.

Life Span

On the average the workers live to be about three years old. Thus they have much more time than bees, of instance, to gather impressions and gain experience, and their adaptability is decidedly greater than one might credit insects with. The queens may live for as long as three decades, not as rulers, it is true, but merely as producers of eggs, for among ants even more exclusively than among bees it is the colony that does the deciding. In the simplest case, a fertilized female founds a colony by enclosing herself in a hole she has dug or in some other hiding place, after shedding her wings, which is facilitated by a performed line of rupture. Here the ant remains closed off from the world at large until the emergence of her first workers, which may take a year. She ingests nothing but moisture, living on her bodily reserves. A good part of which come the breaking down of the thoracic flight musculature.

This source of nourishment also enables the queen to rear her first larvae on glandular secretions in spite of the long period without food. The queen may eat some of the eggs she lays, but they of course, are products of her own body. When the larvae are fully grown, the mother helps them in the spinning of their cocoons by providing them with bits of dirt, which they incorporate in the structures. Later, she helps the workers emerge from the workers emerge from the cocoons. At this point her instincts for the care of the brood usually disappear, since subsequent brood care is taken over by the workers; only in the smaller societies of certain ponerines and *Leptothorax* does the queen continue to help.

Societies of a few of the primitive poneriness may be no larger than ten individuals. In founding them the queen does not enclose herself, but goes on the hunt, fetching food for her brood in the manner of the solitary wasps. Frequently several females work together in establishing the nest, but when the first brood of workers appears, the mothers battle one another until only a single one survives. But several or even many of them are tolerated when spatially more extensive societies are concerned. The young queens then settle down in outbuildings after their mating flight. This procedure is also followed by species in which queens have lost their capacity for digging or for some other reason are unable to found a society by themselves. By remaining within the existing society, they have ant helpers at their immediate disposal.

In the genus *Carebara* a young queen solves this problem by taking along with her on her mating flight a number of tiny workers from the nest; these cling fast fast to her en route. Or a female of Formica sanguinea, a species that is particularly many-sided in its methods of establishing colonies and its ways of life, first joins forces for founding a common nest with a female from the company of one of her slave species; then, as soon as the latter's first workers appear, she kills her. When females that are not self-sufficient but are dependent on worker.

Ants establish branch settlements of a flourishing colony, there frequently are schisms and estrangement between the original population and it offshoots. On the other hand, the adoption of young queens is by no means limited to populations of their own species, but strangers of the same genus are accepted, resulting in mixed communities. These occur frequently, especially in the genus *Formica*, and also in *Lasius* and others. They are only temporary, however, is the queen of the original population dies, and often enough she is beheaded or murdered by the adopted stranger or the latter's workers. In this case a pure stock of the adopted species gradually arises. Communities that have lasted a long time frequently consist of several mixtures and a confusing diversity of interrelationships. For instance, a female of a third species may establish her nest in a *Formica* colony that already consists of two species; and, in addition, the workers may sally forth to capture slaves, bring home pupae of still another species, and let the ants that hatch from them likewise become servants of the state. Such mutual associations of different species stretch through countless possibility, including mere neighbourly activity, the mutual benefits of symbiosis, hospitality, and larceny. They range in one direction to slavery (in which only the ruling ants produce males and new colonies, while the helping ants are allowed to produce only workers), and in the other direction to parasitism.

Slavery

The dependence on slaves varies from one society to another. The custom of capturing slaves seems to have developed from the dependence of those queens that are able to establish a nest only with the help of workers. The pupae of foreign societies are stolen, and a permanent supply of the enslaved working a population that merges from them is assured to the society by forays repeated annually. But occasionally the slaveholders cradicate all the potential sources of slaves in the area and have to return gradually to self-sufficiency. Slave hunts may be extended over such great distances that the soldiers establish guarded camps at intervals of a day's march from home. The saberlike mandibles of the Amazon ants (Polyergus) make them fine fighters, but also make them so incapable as workers at home that they are fully dependent on their slaves, which usually far outnumber them. On the other hand, many *Formica* community and the greater its own resources, the fewer slaves it maintains. The species of *Formica* that steal slaves, although they are less capable soldiers than *Polyergus*, thus remain good workers.

The Amazon ants must depend on their slaves even to feed them, for their salves even to feed them, for their dagger like jaws, tailored for piercing the head of adversaries, do not permit them to chew their food. In addition, these implements are no good for caring for the brood or for construction, and apparently this formal specialization is accompanied by a degeneration of the corresponding instincts. Thus, the true masters in the Polyergus community are the slaves, which happen to be Formica species. These would by no means exchange their life of slavery for one in the maternal nest, for among the warrior-folk they are better protected, quit apart from the fact that ants feel at home in the place where they were born. And more than ample food is at hand, for during the summer the daily slave-hunting expeditions yield an abundance of captured larvae and pupae that can be eaten.

The predatory attacks of *Amazon* ants on *Formica* communities are carried out with refined tactics adapted to prevailing conditions, and are uncannily abrupt. After the return of scouts sent out to ascertain the route to be followed for attack and the proper initial disposition of the forces, the soldiers assemble before their nest, excitedly drumming each other's heads with their antennae. As if responding to an order, part of the population, numbering from many hundred to perhaps 2,000 warriors, sets out in a close column, marching directly at full speed and perhaps straight through heavy grass to the gathering place near the selected nest. When the attackers have consolidated here, they carry out an assault so suddenly and with such concentrated force that the Formica community is caught by surprise and in a few minutes is sacked. Nevertheless, in the first confusion of battle, many Formica ants succeed in fleeing to a place of safety, carrying part of the brood in their mandibles. A *Polyergus* colony has been observed to undertake 44 expeditions in 33 days, collecting a booty estimated at 40,000 slave pupae. In addition to various species of *Formica* and *Polyergus*, the myrmicine genera *Strongy lognathus* and *Harpagoxenus* take slaves.

Stronglylognathus ants have saber-like mandibles, but unlike the Amazons are capable of feeding themselves. And they have different tactics or battle methods. Thus certain *Stronglylognathus* communities maintain throughout the duration of their attack, a constant liaison between their own nest and the *Tetramorium* nest they have assaulted; their aim is to intimidate the inhabitants of the latter, namely to render their more or less defenseless by pinching them suddenly from behind with the mandibles. But if battle is joined, nonetheless, fights break out between the robbers slaves, participating in this expedition, and the similar workers of the besieged community, from which it is possible that salves themselves have descended, having been captured in an earlier raid. Consolingly, however, such an occasion may end in reconciliation, with attacker and attacked giving up the battle and joining forces. The methods of slaveholders vary. Colonies of *Harpagoxenus sublmevis* drive *leptothorax* societies out of their own nest, set themselves up in it, and obtain the indispensable helping ants from the Leptothorax pupae left behind. For these slaveholders, which even try to make workers out of already-hatched queens by biting their wings off, are unable rear their own brood.

Composite Nests

Composite nests also are shared by very different groups of ants and even by ants and termites. Frequently societies of this sort are brought together by such things as the presence of

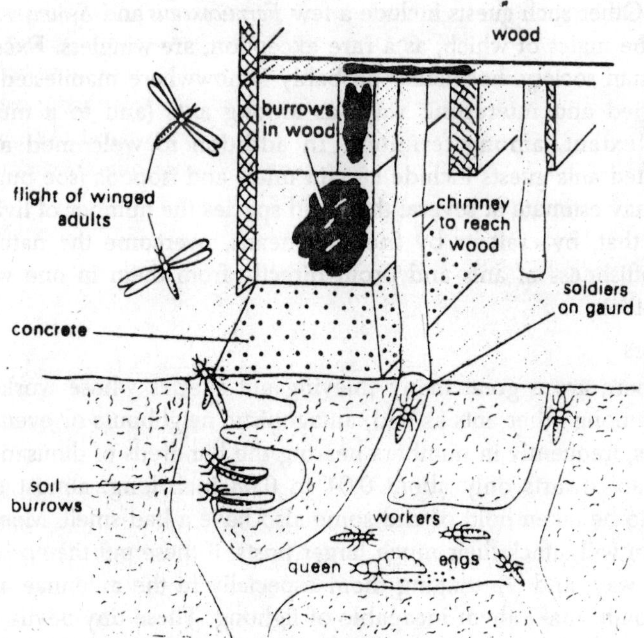

Fig. 5.7. Diagram showing the termite colony.

favourable dwelling places–for instance, hollow branches or the hours made by termites. Each of the species, more or less forced by circumstances into a neighbourly existence, has its own dwelling, but with a common entrance to the nest. This may result in various forms of symbiosis. A defenseless species may live with a species capable of defending itself and may bear a deceptive resemblance to it. The combined defenses of two or more populations living in a given habitat may enhance the safety of each. In the outer parts of the ant gardens of a *Componotus* colony, for example, a population of *Cremastogaster* may establish itself. Under moderate menace only the latter takes the appropriate measures of defese, but when seriously threatening attack occur, the fundamental owners appear from the depths of their spherical nest.

Guests

A few genera of tiny ants live as *obligatory lodgers*; they build their little cities and dwellings in the middle of the nests of much larger species, being tolerated a friends of the latter and fed by them on solicitation. Certain *Leptothorax* species are ant guests of this sort. They often ride on the back of their host ants, licking them. Other such guests include a few *Formicoxenus* and *Symmyrmica* ants, the males of which, as a rare exception, are wingless. Except in human society, hospitality probably is nowwhere manifested in so varied and interesting ways as among ants (and to a much lesser extent, among termites). In addition to welcomed and uninvited ants guests include certain mites and isopods (soe bugs). One may estimate at several thousand species the number of living forms that, by craft or by friendly means, overcome the natural standoffishness of ants and profit directly from them in one way or another.

Thieves

There are a great many thieving ant species whose workers settle among other ants as bad, antagonistic neighbours or even as lodgers, frequently in numbers nearing the hundreds of thousands. They are dwarfs only about 0.04 to 0.08 inch long, almost too small to be taken hold of and some also have a bad smell. Masses of them will attack their much larger hosts, if these put themselves in the way, and by stinging them especially in the antennae and the mouth, make them incapable of fighting. These tiny devils eat the host's brood and get into the hosts' chambers by making fine passageways that course like a network throughout the structures of the host ants; the hosts cannot possibly pursue them through these little tubes. *Soienopsis* and many other genera are thief ants of this sort, and they dwell also among termites. Since the smallest workers are best adapted to this away of life, these species have developed from originally larger forms, as is still shown by the sexual individuals, above all by the queens, which frequently are veritable giants in comparison with their tiny workers. Thus, the workers of *Carebara*, which live with termites, seem like lice hanging to the legs of their queen. Less harmful thieves are those highway men that settle in the vicinity of another nest with the intention of plundering the ants that are bringing foodstuffs home. Both robbery and quarrels over boundaries can lead to open war or perhaps to feuds lasting for years between different stocks.

Contest of a more ritual nature also occur among ants. *Camponotus* ants, for example, run headlong against one another, with the abdomen audibly beating the ground. But ants in general engage at once in mortal combat. The principal weapons are the mandibles and, among the more primitive ants (ponerines, myrmicines), the toxic sting. In lieu of stinging. *Formica* curls the abdomen forward between the legs and sprays out its poison, formic acid. Certain species disseminate repugnant odours (for example, many *Lasius*), and dolichoderines plaster the enemy's antennae with a sticky, ill-smelling secretion from the rear end. And finally, the soldiers of many species are able to blockade the nest entrance with their head.

Harvesting and Hunting

Ants usually feed upon plant juices, nectar and honeydew, as well as on insects (ants not excepted), worms, and other small animals. Insect-eating ants can be highly useful to man. A large community of wood ants of the genus *Formica*, for example, destroys about 100,000 insects daily. In specializing on certain types of food, some ants have developed unusual organizations and ways of life. Ants that feed on honeydew, a sweet excretion of aphids, are good example. These ants in their own nest raise aphids, feeding them leaves or roots and even building special chambers-stalls, so to speak- for such "milk cows." The ants comprehend to an astounding degree the essentials of aphid care, a capacity that has probably developed in great part from the instincts for the care of their own brood.

Walking Honey Jars

Certain ants (such as Mymecocystu; Prenolepic, Leptomyrmex, Melophorus, a few Componotus and Plagiolepis) of warm and dry Countries store juice in the bodies of their own workers. A number of such workers are fed by their sisters until their expandable abdomen inflates into a gigantic, amber-coloured, shimmering sphere, upon which the normally contiguous segmental sclerites are widely separated from one another. These "walking honey jars" can creep forward only slowly and with difficulty, and usually remain hanging calmly in groups on the ceiling of a cavity in the nest; particularly when food is scare they dole out their store from the mouth drop by drop to their enterating sibs.

Man gathers and eats the bodies of such honey ants, or at least their contents, rating them as a delicacy that surpasses the honey of bees. Incidentally, the honeydew of aphids is not the sole constituent, for these ants frequently also gather nectar and exudations from plant galls, in America particularly those from oaks. Quite different specialists are the harvest ants, especially *Messor*, *Pheidole*, *Holeomyrmex*, and *Pogonomyrmex*, the last of which strings energetically. These collect plant seeds, preferentially those of grains and grasses, and accumulate them in their nests. The seeds, later to be broken up and eaten, are laid out in the sum to dry; if they nevertheless sprout, the sprout is often bitten off. Certain species throw such unwanted seeds out of the nest, and surrounding the nest entrance there soon grows a dense fringe of grass, a phenomenon that has led to the erroneous designation of these ants as 'farmer ants." The harvest ants are not only eaters of seeds, but also hunters of insects; in addition, a hard-stinging American species (Solenopsis geminata) eats fruit and cultures aphids.

Ant Farming

In contrast, the well-known leafcutting ants (Atta) of tropical and subtropical America are one-sided specialists that grow "vegetables." There vegetables are the protein-containing bodies that grow on the mycelia of certain fungi. And the majority of these fungal species will flourish only in the gardens of certain leafcutting ants. The culture medium in which the fungi grow is prepared by these ants and consists of a fermenting mass of chewed-up leaves carefully manured by the ant's own excrement. The amount of plants that an *Atta* community will cut up in a very short time and carry into the fungal chambers is so great that the result is frequently as destructive as a catastrophe of nature. The nest workings, which may be many yards below ground and spread over a wide area, may perhaps have a volume of several hundred cubic yards.

Rocking and reeling, the ants hustle nervously along their routes, stumbling past their fellows coming the other way, who draw aside for them, and, falling suddenly into the aperture of a nest entrance, are swallowed up by the earth, one after the other, or even two at a time, over and over again. On her flight into new county, a young queen, as founder of a colony-to-be, brings with her not instruct for tending a fungus garden but, in a special pouch of her head in back of the mouth, a tiny was of fungal mycelia from the maternal home. In the new little brood chamber the ant puts this down, deposits five or six eggs on it, and manures the fungus by pushing little bits of it beneath her anus and then returning them to the pile. The wad grows, and yet the young queen takes none of it; instead she eats some of her own eggs and also feeds her first hatching larvae exclusively on these.

The tiny workers that after a few weeks have developed from these larvae are the first to consume the "vegetables" that have grown on the fungus plant, which now is about 0.8 inch in diameter. The mother and brood still partake only of eggs. The workers also manure the precious fungus, open the brood chamber perhaps some ten days later, and gather and chew up pieces of leaves. This now permits the vitally important vegetable to develop adequately.

There are certain species of insects which are not leaf cutters, but culture their fungus on insect feces, especially that of caterpillars, or on all sorts of decaying plant tissues. In great contract to these vegetarians are two groups of warlike carnivores. First, there are the primitive Australian ponerines of the genus *Myrmecia*, known as bulldog ants and notorious for their stings and furious attack. Like certain other ponerines (*Odontomachus*, *Harpegnathos*), they are able to leap upon their adversary by snapping themselves backward from the ground with their powerful mandibles. Otherwise, ability to jump has been seem among Hymenoptera only in a few chalcids, and here is done with the legs. In their organized predatory raids, the bulldog ants hunt mostly termites. The second group of carnivores are the driver, or migratory, ants (dorylines). They also prey on termites, as well as on other ants. In tropical America the driver ants are represented especially by the genus *Eciton*, and in

Africa by *Anomma.* Their sexual individuals are gigantic, the males resembling wasps more than ants, the females wingless from the beginning. The workers, remarkably unequal in size, have saber-shaped mandibles. The workers, remarkably unequal in size, have saber-shaped mandibles.

Army March

The driver ants travel in well-organized formations that differ according to the species of ant. Some columns may be more than 100 yards long and several yards wide. At times, the workers with the most formidable mandibles march on each side of a column. Biting, stinging, murdering, and pillaging, these armies move over the countryside, some travelling more through open territory, others through woods and thickets. Whatever is unable to flee from them is torn to pieces and carried forward in little bits. The ground, vegetation and trees, holes caves, nests, even houses are in a short time completely cleared of every living thing, and even larger animals and people take flight. Clamoring antbirds accompany these horrifying processions and eat their fill of the insects before them. Many driver ants migrate below ground, and others under the protection of leaves and fallen debris. In case of need they may even build uninterrupted tunnels above ground over open stretches. The driver ants build no nest, but in a sheltered place crowd together in a heap, leaving cavities that constitute the brood chambers. These satisfy all requirements for the care of the brood. In the nomadic life of these ants, resting intervals approximately twenty days long alternate with slightly shorter migratory periods, a rule that is determined by the rhythm of egg-ripening in the queen, for she hardly could take part in the march with her abdomen swollen with ripe eggs.

In a circle of as much as a mile around the resting places, everything edible is soon eradicated, yet the signal for a new migration apparently does not stem from a shortage of food, but rather from the 25,000 or so larvae produced from the previous egg-laying period. In fact, this signal seems to emanate from their secretions, which are licked up by the attending workers and stimulate them. Because of their social effects, these secretions are called "*social hormones.*" Certain other insects go along as undisturbed guests on the expeditions of migratory ants. Rove beetles (Staphylinidae) frequently accompany the ants. That they are allowed to do so may seem odd, for they rarely resemble the hosts in appearance. But with the exception of the males, the ants have no compound eyes, and at most have ocelli; the majority are blind and gauge their environment exclusively by the senses of smell, test and touch. Evidently these guests meet the specifications covered by these senses deceptively well. Along with such adaptation, the guests have developed to an astonishing degree the ability to respond satisfactorily to antennal "fingering," the so called "antennal language," despite the difference in the structure of the host's and guest's antennae. When the migratory ants raid termite houses, frequently some of the accompanying guest get into the termite chambers, perhaps lose touch with the ants when these withdraw, and are left behind, usually to starve to death or to killed. But certain of these guest are able to make themselves acceptable to the new hosts and live henceforth as guests of the termites.

Insect Communication

The ability of insects to live together in communities rests on the means of communication among them. Such communication is based first of all on the species and nest, which are perceived through the antennal sense organs, Sentries at a nest entrance admit no individuals contaminated with the odour of strange nest, and populations refuse to accept them. But if strangers somehow succeed in taking on the community odour, they are adopted without further ado. Some ants regularly use this "fifth column" system to gain entrance to a nest, and certain others are rendered inconspicuous naturally by the possession of deceptive body odour. Mother and children (or workers) recognize one another by smell, even after long separation, at least up to a certain age; after recognition may fade, for older workers often change. Odours serve not only for recognition, but also for marking pathways. By their odours, for example, honeybees become direction signs for their companions returning homeward. In front of the hive entrance, bees stand and do a waggle dance, during which fragrance is emitted from an exsertile, membranous organs between the last two segments. The odour is disseminated of the winds by means of continual fanning movements of the wings. In the same way bees that have discovered a source of food call in their fellow. Or the fragrance of the flowers, clinging to a bee that returns home successful, can induce her nest-fellows likewise to scour the vicinity in search of such blossoms. The simple urge for imitation may play an especially large role in ant communities; it seems that there the older workers, leading the way in all actions, set an example for the younger ones.

When a large prize is being transported by several individuals, for instance, there is a constant participation of new individuals, that attracted by what is going on, hasten up and crowd into the work. Thus, there is a constant succession of more or less compulsory relief of those already on duty, and the object passes through many "hands" before it is finally fetched home. Such community transport takes much longer than that engaged in by individual ants, parincipally because the different temperaments and intentions of the participants are likely to be a mutual hindrance. Quite similarly, any moods or conditions of excitement may be disseminated, via imitation, throughout the population, yet ants also have at their disposal concrete means of expressing themselves. As one instance, there is capability of many species of producing high-pitched chirping notes by rubbing two abdominal segments together. And, of course, there is the "antennal language," which is rooted in the sense of touch. Definite signs and signals are given by antennal movements, which include stroking, tapping, and drumming; with these may be included similar motions of the foreleges, as well as blows with the head. Added precision frequently is given the communications not only by the varying intensity of the movements, but also by the production or transmission of odorous materials. Within the community, the most frequent signs of this sort no doubt are connected with the mutual solicitation of food.

Language of Bees

The honeybees (*Apis mellifica*), and quite similarly a few advanced wasps, have developed an incredibly highly differentiated sign language. For if a honeybee has found a source of food in the immediate vicinity of the nest, say up to 50 or even 100 yards distant, once home she executes a small circle on a comb, first to the left and then to the right, and so on. This is done in the midst of her sisters, both busy, and idle, and they carried away by the display of so much zeal in their immediate vicinity, are infected by her spirit and join the leader in her circular dance. Thereafter the bees fly out and round about the hive until, by means of the particular flower odour on the leader of the dance, they have located the source of honey or pollen she has announced so ingratiatingly by her dancing. But this is only part of the story. For if the flowers are farther away, then the discoverer, still dancing alternately to the left and to the right, changes from a single circle to a broad figure eight, the longer middle portion of which is a straight line. Every time the dancing bee runs over this stretch, she waggles her

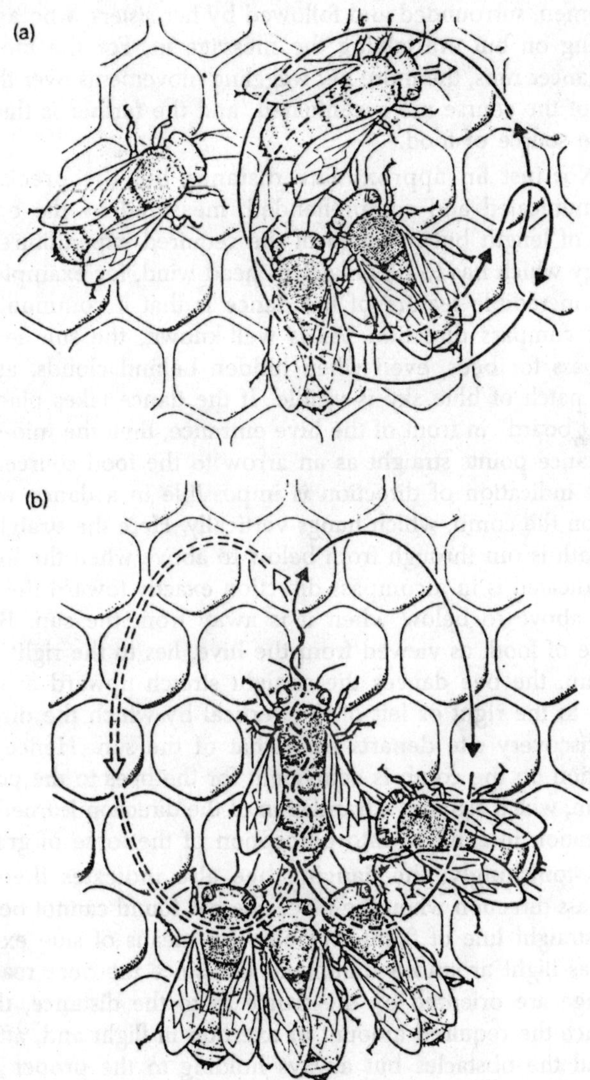

***Fig. 5.8.** The dance of the honeybee tells the location of food sources in relation to the hive. (a) The 'round' dance, which says that food is nearby, but does not show the direction to it. (b) The waggle dance, which says both the distance to and direction of a food source. The distance is told by the rate of the dance; the direction by the angle of the centre part of the dance, during which the dancer wags her abdomen: the angle of this part of the dance relative to the vertical is the angle of the food source relative to the sun.*

abdomen, surrounded and followed by her sisters, who are not just looking on but are talking the message in. For the more slowly the dancer runs, the more her waggling movements over the middle part of the course are accentuated, and the further is the distance to the source of food.

Not just an approximate distance but the precise one is communicated and comprehended, measured by the bees no in units of length but in terms of the required expenditure of flight energy which has to be greater in head wind, for example. Yet the most incredible feature of the dance is that it communicates the exact compass direction. As is well-known, the sun serves as a compass for bees, even when hidden behind clouds, as long as clear patch of blue sky is visible. If the dance takes place on the "flight board" in front of the hive entrance, then the mid-stretch of the dance points straight as an arrow to the food source. But this direct indication of direction is impossible in a dance within the nest on the comb, which hangs vertically. Here the straight part of the path is run through from below to above when the location to be indicated is in a compass direction exactly toward the sun, and from above to below when it is away from the sun. But if the source of food, as viewed from the hive, lies to the right or left of the sun, the bee dances the straight stretch upward at the same angle to the right or left of the vertical by which the direction of the discovery site departs from that of the sun. Hence, vertical direction on the comb is equivalent for the bees to the position of the sun, which signifies a translation of the direction learned through orientation in sunlight into perception of the force of gravity.

Astonishingly, the dancing bee also indicates then precise compass direction when the place she has found cannot be reached in a straight line of flight, but only by means of side excursions, such as flight around a mountain. But since the bees reading her message are oriented by the dance as to the distance, they now produce the required amount of exertion in flight and, after flying around the obstacles but always holding to the proper compass direction, they arrive at the goal. The dance expresses, in addition to distance and direction, the productivity of the source of food, which is communicated by the degree of liveliness.

After a rich discovery the dance is a correspondingly excited one, and even an arousing one. The dance is also used by swarming bees in searching for a new home. Each bee returning from the search dances on the hanging mass of the swarm. She communicates not only the location of the place she has found but also how good it is. The discoverers of the best situation prove to be the most infectious dancers and induce more and of their comrades to fly out and inspect the place. These return, dance with equal elan, and soon the other scouts, which are behaving less convincingly, are ignored and either give up or simply join with the rest. Thus finally all are doing the same dance, the decision is made, the swarm breaks forth, and, led the many dancers, they fly off to the new home. In this way, the swarms is likely to select the best hollow tree even though it may be miles away.

Winter Swarming

Bees swarm densely not only when seeking a new home, also when passing the winter in the hive. A single inactive bee, like all resting insects, has inappreciable body warmth of its own and is therefore at the mercy of the environmental temperature. When the temperature drops to about 46°F (8°C), the bee is incapable of moving and soon dies. Nevertheless, together in the hive, bees are able to survive a rigorous winter. They cluster in a thick spherical mass around the empty inner combs, and individual bees inside the cluster produce heat by vibrating their wing muscles. In this way they are able to maintain a temperature in the center of the swarm of 68°F to 86°F (20°C to 30°C) even when it is as cold as- 4°F to -22°F (20°C to 30°C) outside. The temperature of the outer layer of the swarm is about 45°F to 46°F (7°C to 8°C). When raising a winter brood, the raise the cluster temperature to 95°F (35°C). Ants spend the cold season of the year in deeper chambers or in special winter nests. A few house ants–such as the notorious yellow pharaoh's ants (Manomorium pharaonis), which have been spread all over the world and which establish their colinies everywhere, even in the handle of a table knife-live in comfort in the heated houses of humans and avoid the problem of overwintering altogether.

Social Parasites

Among certain bumblebees, wasps, and ants, individuals social species have become parasites. They differ from their hosts in that they have lost their worker casts, and in part, their own instincts for brood care. Whereas a few *Polistes* and *Vespa*, which resemble their wasp hosts almost to the hair, live quite peaceably in the latters nests and lay their eggs in the comb, the female parasites of bumblebees and ants are compelled to gain possession of foreign workers for

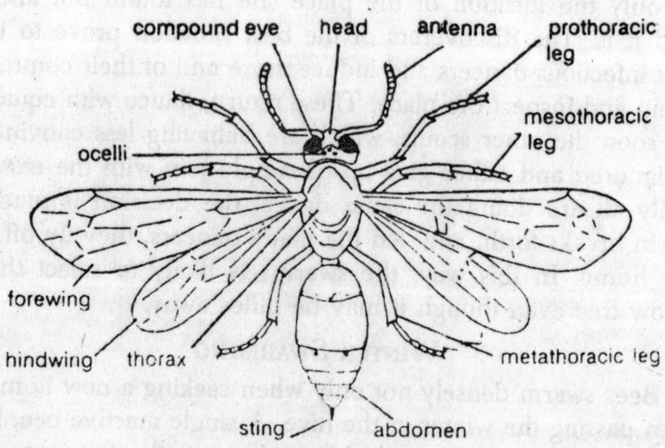

Fig. 5.9. Wasp (Vespa). Female.

rearing their own brood. With some exceptions, this done usually with naked force, but among certain ants with treacherous guile.

A parasitic bumblebee (*Psithyrus*) forces her way into the nest of her host (*Bombus*), intimidates the workers with her bites slaughters the queen, and destroys eggs and larvae. In some cases, *Psithyrus* will spare the life of the bumblebee queen, which then no longer pays any attention to her own progeny. In either the bumblebee colony perishes. In rare instances a genuine bumblebee (*Bombus*) adopts the despotic method of a *Psithyrus*, kills as relative, and with the latter's workers found her own community. But Psithyrus found no new community, merely rearing a few males and females. In ants, social parasitism may transpire in a significantly more peculiar manner.

Parasitic queens frequently have arisen from former slaveholders, whose own workers, instincts for brood care, and even ability to feed themselves independently become superfluous and gradually were lost altogether. Some of these of these ants are pitifully reduced creatures, unable to rear their offspring, yet are endowed with the refined cunning needed to compel others to carry out these tasks. The female of a *Bothriomyrmex* for instance, climbs onto the back of the much larger queen of a *Tapinoma* community, bites her head off, and now usurps her place in the society, to which she gained access thanks to her similar body odour. Other parasitic females are not in so simple a position; first they must patiently make friends in the outlying quarters with the workers, and attempt to accustom these to themselves, until their own body odour has become imbued sufficiently with the family smell.

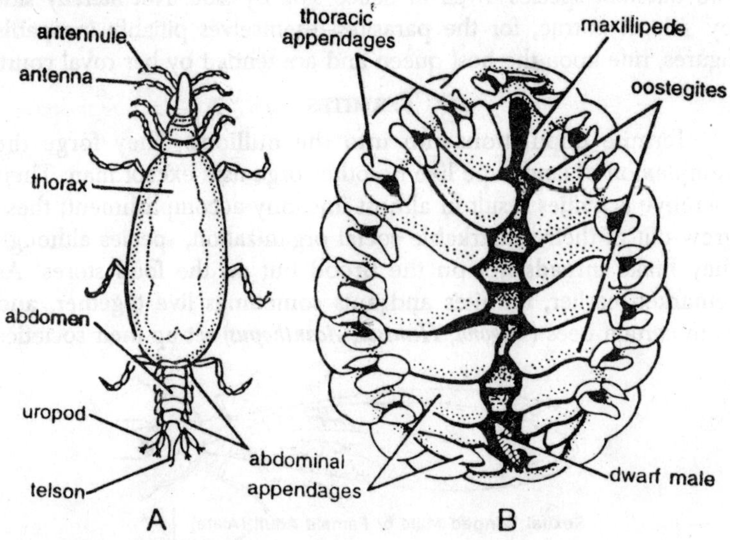

Fig. 5.10. Parasitic Isopoda. A–Pranzia larva of Gnathia, B–Female Bopyrus.

Still another procedure is used by the female of *Epimyrmex stumperi*; she takes a firm seat, from which she cannot be dislodged, on the back of a worker of the selected *Leptothorax* colony, and for a long time brushes the worker as well as herself with her bristly forelegs. Thus the crafty *Epimyrmex* perfumes herself with the *Leptothorax* odour. And then finally, being unrecognized in the darkness of the passageways and chambers of the nest, she comes upon the queen, rolls her over, and bites a firm hold in the throat of her victim, who is much larger but does not defend herself. There the Epimyrmex hangs for days, until the queen dies or is completely exhausted.

Workers are in the vicinity no doubt, but none offers help against the perfumed strangler; no one knows of the drama that is being played out silently in the middle of the community. Finally their own queen, perhaps still alive though paralyzed, is not heeded any further and must slowly starve to death, while the monster sets out after another queen, if there is one, but leaves her unmolested if she is young and has not been fertilized. What a horrible triumph of bare natural instincts. How friendly, in contrast, seems the relationship of *Teleutomyrmex schneideri* with their hosts ants *Tetramorium caespitum.* For their these little invaders are tolerated, and hence, as a rare exception in ant communities, the queens of two different species dwell in peace side by side. Not literally side by side, it is true, for the parasites, themselves pitiably incapable figures, ride upon the host queen and are tended by her royal court.

Termites

Termite populations run into the millions. They forge the complexion of landscape like no other organism except man. Their secretive activities result in almost uncanny accomplishment; these grow out of their remarkable social organization, species although they make inroads not on the brood but on the food stores. As remarked earlier, termites and ants sometimes live together, and even certain bees (*Trigona*, *Hemisia*, *Acanthopus*) set up their societies

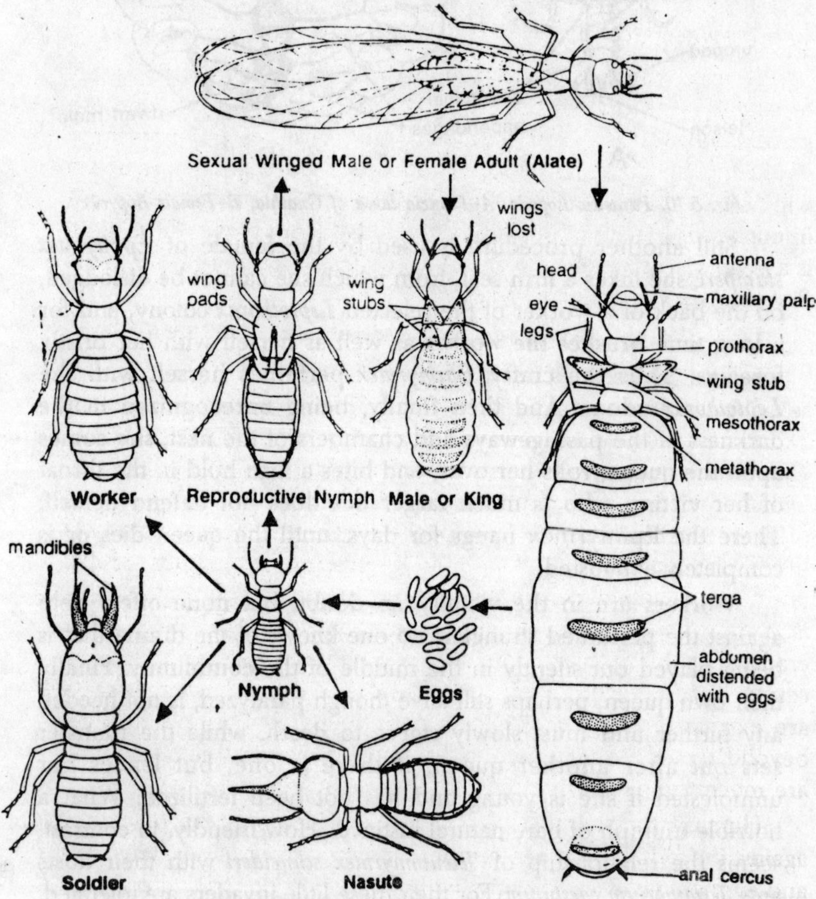

Fig. 5.11. Termites. Castes and life cycle.

unmolested in the arboreal nests of termites. The same is done by a few parrots; and in Africa the great Nile lizard (*Varanus noiloticus*) lays its eggs in a hold it has clawed out of a termite hill, where the constant temperature and humidity are favourable for brooding.

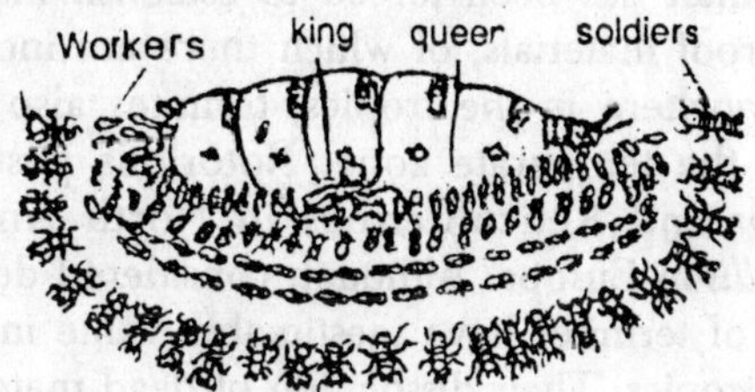

Fig. 5.12. A scene of the royal chamber inside a termitarium.

Termite guests include members of the most varied insect orders, as well as mites and spiders. Many insect guests that are greatly desired and well-tended by the termites have strongly swollen, often grossly misshapen abdomens (physogastry), as in the wingless flies *Termitoxenia*, which live in the fungus gardens, and in numerous beetles of the family Staphylinidae, among which there is a viviparous species from Brazil (Corotoca-melantho) that deposits first-stage larvae instead of eggs.

Broader Relationships

Indeed all insect-eating organisms are active as enemies of the sexual individuals, which swarm in huge numbers, and during such swarming even man traps termites as food. But the pursuit termites within their structures requires specialists. Among vertebrates these include the armadillos, anteaters, and a few others usually adept at digging. Among birds the woodpeckers take precedence. But none of these is an destructive as bulldog ants, migratory ants, and a few others. These decimate the termites, emptyping an entire house at a time and then frequently talking possession of it, or even eradicating the termites from whole localities. In general termites cannot cope with these enemies. But the so-called "nasute" forms are exceptions. The sticky excretions rom their "nose" gum up beyond repair the antennae and mouth of the attacking ants thus are much more effective than implements for biting.

Humans that inhabit the warm Countries must be on guard against the termite horders, for these insects devour almost anything and everything, even (possibly with the help of their solvent saliva) metal and ivory. Coming from the ground, they penetrate into the wooden parts of the houses, hollowing everything out yet leaving the outer layers deceptively untouched, and in the same way get into furniture and utensils of all sorts. In addition to nonliving materials, they attack trees and other plants. As proof of their destructiveness, man has been forced to establish institutions for testing termite-proof materials, of which there are incredibly few. A problem everywhere in the tropics, termites also threaten to become one in the temperate zone. Notorious pests there are *Kalotermes minor* and Reticulitermes in North America, and *Kalotermes flavicollis* in Europe. Although considered destructive by man the armies of termites have inestimable value in the natural economy of the tropics. Their dissolution of dead materials, above all fallen timber, and their loosening up of the soil enable new generations of plants to grow. On the other hand, in certain districts the secretions give off by termites during the activities of building can harden the ground to such an extent that it becomes unsuitable for cultivation.

Classification

About 2,000 tropical and subtropical species of termites are known. They range from about 0.08 to 0.9 inch long (large queens up to 4 inches). Of the three to five families of termites, the kalotermitids and the mastotermitids seem the more primitive; the latter are represented only the single Australian species *Mastotermes darwiniensis*, which has been designated the most primitive of all. Rich in species and more advanced in their organization are the termitids and rhinotermitids. Only among them is there, strictly speaking, a true worker caste, namely fully developed workers and soldiers no longer capable of transformation. Those of the other groups are merely nymphs of different stages. Such nymphs can be developed into workers, soldiers, or sexual individuals by special feeding applied at a definite period of their development. At first glance, insect societies may seem strangely similar, or at least comparable, in some ways to human societies. But it is well to keep in mind the infinite distance that separates the two. Our social existence is based on the family as a unit, has been developed by human intellect, and depends upon the free will of the participants. Insect societies, on the other hand recognize no unit other than the population as a whole, are governed by instinct, and have compulsory membership. There is really no need to make comparisons; the life of social insects is marvelous enough in itself.

6

DIVERSITY OF INSECTS

This is a tremendous variety of insects that we see today. From the limited fossil record it would appear that the earliest insects were wingless, *thysanuran* like forms which abounded in the Silurian and Devonian periods. The major advance made by their descendants was the evolution of wings, facilitating dispersal and, therefore, colonization of new habitats. During the Carboniferous and Permian periods there was a massive adaptive radiation of winged forms, and at this time most of the modern orders had their beginnings. Although members of many of these orders retained a life history similar to that of their wingless ancestors, in which the change from juvenile to adult form was gradual (the hemimetabolous or exopterygote orders), in other orders a life history evolved in which the juvenile and adult phases are separated by a pupal stage (the holometabolous or endopterygote orders). The great advantage of having a pupal stage (although this was not its original significance) is that the juvenile and adult stages can become very different from each other in their habits, there by avoiding competition for the same resources. The evolution of wings and development of pupal stage have had such a profound effect on the success of insects that they will be discussed as separate topics in some detail below.

EVOLUTION OF WINGED INSECTS

Origin and Evolution of Wings

The origin of insect wings has been one of the most debated subjects in entomology for more than a century and a half, and even today the question remains far from being answered. Most authors agree, in view of the basic similarity of structure of the wings of insects, both fossil and extant, that wings are of monophtyletic origin; that is, wings arose in a single group of ancestral apterygotes. Where disagreement occurs is with regard to (1) the position (s) on the body at which wing precursors (pro-wings) developed (and, related to this, how many pairs of pro-wings originally existed); (2) the nature of the original functions of pro-wings; (3) the nature of the selection pressures that led to the formation of wings from pro-wings and (4) the nature of the ancestral insects, that is, whether they were terrestrial or aquatic and whether the pro-wings arose in larval or adult forms. Oken (1811) apparently made the first suggestion as to the origin of wings, namely that they were derived from gills, Woodworth (1906) having noted that gills are soft, flexible structures perhaps not easily converted (in an evolutionary sense) into rigid wings, modified the Gill Theory by suggesting that wings were more likely formed from accessory gill structures, the movable gill plates, which protect the gills and cause water to circulate around them. The gill plates, by their very functions, would already possess the necessary rigidity and strength. This proposal receives, support from embryology, which has shown abdominal segmental gills of larval Ephemeroptera to be homologous with legs not wings.

Wigglesworth (1973, 1976) has recently resurrected, and attempted to extend, the gill theory by proposing that in terrestrial apterygotes the homologues of the gill plates are the coxal styli, and it was form the thoracic coxal styli that wings evolved. Wigglesworth, a firm believer in a monophyletic origin for the arthropods, attempts to strengthen his argument by pointing out the enormous diversity in structure and function that has evolved from the primitive exite (supposedly homologous with the coxal stylus) and endite of the ancestral biramous limb. *Kukalova-Peck* (1978) states that the homology of the wings and styli as proposed by Wigglesworth is not acceptable and points out that wings are always located above the thoracic, spiracles, whereas legs (even if ancestrally biramous) always articulate with the thorax below the spiracles. However, in support of Wigglesworth's (1973, 1976) proposal, it should be noted that primitively wings are moved by muscles attached to the coxae and are tracheated by branches of the leg trachea.

The Spiracular Flap Theory of Bocharova. Messner (1971), cited in Kukalova-Peck, 1978) suggests that wings evolved from integumental folds which initially protected the spiracles. The folds enlarged, became hinged, and began to function as a ventilating mechanism for the tracheal system. Improved efficiency of ventilation would be achieved by a flattening of the flaps and an increase in their surface area. At the same time the flapping motion would also assist with locomotion over the substrate and, it is proposed, leg muscles might be used to coordinate leg and flap movements. Further selection pressure would lead to the development of wings from the flaps.

Bradley's Fin Theory (1942), proposed that wings arose in group of terrestrial insects that became amphibious, that is, spent much of their time in water, returning to land for mating and dispersal. These insects originally possessed nonarticulated pro-wings that had, perhaps, a protective function. As the insects spent increasing amounts of time in water the pro-wings took on new functions as fins in propulsion and attitudinal control. In other words, they would become articulated, strengthened by the development of veins and would develop suitable musculature. Selection pressure would also lead to enlargement of the fins to the point at which they would be also capable of moving an insect through air. The most popular theory for the origin or wings is undoubtedly the Paranotal Theory, first proposed by Woodward (1876, cited in Hamilton, 1971), and supported by Crampton (1916), Sharov (1966), Hamilton (1971), Wootton (1976), and others, Essentially the theory proposes that wings arose from rigid, lateral outgrowths (paranota) of the thoracic terga which became enlarged and, eventually, articulated with the thorax. Although large paranotal lobes are found on the prothorax of many fossil insects, fully developed and articulated wings occur only on the meso- and metathoracic segments.

Presumably, whereas three pairs of paranotal lobes were ideal for attitudinal control (see below,) only two pairs of flapping wings were necessary to provide the most mechanically efficient system for flight. Indeed, as insects have evolved there has been a trend toward the reduction of the number of functional wings to one pair. This freed the prothorax of other functions such as protection of the membranous neck and serving as a base for attachment of the muscles that control head movement. Various suggestions have been made to account for the development of paranotal lobes. For example, Alexander and Brown (1963) proposed that the lobes functioned originally as organs of epigamic display or as covers for pheromone-producing glands. Most authors believe, however, that the paranota arose to protect the insect, especially, perhaps, its legs or spiracles. Enlargement and articulation of the paranotal lobes were associated with movement of the insect through the air. Packard (1898) suggested that wings arose in surface-dwelling, jumping insects and severed as gliding planes that would increase the length of the jump. However, the almost synchronous evolution of insect wings and tall plants supports the idea that wings evolved in insects living on plant foliage. Wigglesworth (1963a,b) proposed that wings arose in small aerial insects where light cuticular expansions would facilitate takeoff and dispersal. The appearance later of, muscles for moving these structures would help the insect to land the right way up. Hinton (1963a), on the other hand, argued that they evolved in somewhat larger insects and the original function of the paranotal lobes was to provide attitudinal control in falling insects.

There is an obvious selective advantage for insects that can land "on their feet," over those which cannot, in the escape from predators. As the paranotal lobes increased in size, they would become secondarily important in enabling the insect to glide for a greater distance. Flower (1964) has examined the hypotheses of both Wigglesworth and Hinton. Flower's calculations show that small projections (rudimentary paranotal lobes) would have no significant advantage for very small insects in terms of aerial dispersal. However such, structures would confer great advantages in attitudinal control and, later, glide performance for insects 1-2 cm in length. The final step would be transition from gliding to flapping flight, that is the development of a hinge so that the wings became articulated with the body. Some authorities believe that the hinge evolved initially in order that the projections could be folded along the side of the body, thereby enabling the insect to crawl into narrow spaces and thus avoid capture.

Movement of the wings during flight was therefore a secondary adaptation. Perhaps a more likely explanation is that the earliest movements of wings during flight were simply to improve attitudinal control. Only later did the movements become sufficiently strong as to make the insect more or less independent of air currents for its distribution. In this hypothesis the earliest flying insects would rest with their wings spread at right angles to the body, as do modern dragonflies and mayflies. The final major step in wing evolution was the development of wing flexing, that is, ability to draw the wings when at rest over the back. This ability would be strongly selected for, as it would confer considerable advantage on insects that possessed it, enabling them to hide in vegetation, in crevices, under stones, etc., thereby avoiding predators and desiccation. An implicit part of the paranotal theory is that this ability evolved in the adult stage. The paranotal theory is based on three pieces of evidence: the occurrence of rigid tergal outgrowths (wing pads) of modern larval exopterygotes (ontogeny recepitulating phylogeny), the occurrence in fossil Paleodictyoptera of rigid pronotal expansions whose venation is homologous with that of wings, and the assumed homology of wing pads and lateral abdominal expansions, both of which have rigid connections with the terga and, internally, are in direct communication with the hemolymph. This evidence has been severely criticized by Kukalova-Peck (1978).

In a thought-provoking paper Kukalova-Peck argues that the fossil record supports none of this evidence. Rather, it indicates just the opposite sequence of events, namely, that the primitive arrangement was one of freely movable pro-wings on all thoracic and abdominal segments of juvenile insects, and it was from this arrangement that the fixed wing-pad condition of modern exopterygote larvae evolved. According to Kukalova-Peck, numerous fossilized juvenile insects have been found which possessed articulated thoracic prowings. However, with few exceptions even in the earliest fossil insects, both juvenile and adult, the abdominal pro-wings are already fused with the terga and frequently reduced in size. Some juvenile Protorthoptera with articulated abdominal prowings have been described and in Ephemeroptera the abdominal pro-wings are retained as movable gill plates. In proposing her ideas for the origin and evolution of wings, Kukalova-Peck emphasizes that these events probably occurred in "semiaquatic" insects living in swampy areas and feeding on primitive terrestrial plants, algae, rotting vegetation, or in some instances, other small animals. It was in such insects that pro-wings developed.

The pro-wings developed on all thoracic and abdominal segments were present in all instars, and at the outset were hinged to the sides of the body (not the terga). With regard the selection pressures that led to the origin of pro-wings, Kukalova-Peck used ideas expressed by earlier authors. She suggests that pro-wings may have functioned initially as spiracular flaps to prevent entry of water into the tracheal system when the insects became submerged or to prevent loss of water via the tracheal system as the insects climbed vegetation in search of food. Alternatively, they may have been plates that protected the gills and/or created respiratory currents over them, or tactile organs comparable to (but not homologous with) the coxal styli of thysanurans. Initially, the pro-wings were saclike and internally confluent with the hemocoel. Improved mechanical strength and efficiency would be gained, however, by flattening and by restricting hemolymph flow to specific channels (vein formation). Kukalova-Peck speculates that eventually the prowings of the thorax and abdomen became structurally and functionally distinct, with the former growing large enough to assist in forward motion, probably in water.

The new function of underwater rowing would create selection pressure leading to increased size and strength of pro-wings, improved muscular coordination, and better articulation of the pro-wings, making rotation possible. These improvements would also improve attitudinal control, gliding ability, and therefore survivial and dispersal for the insects if they jumped or fell off vegetation when on land. The final phase would be the development of prowings of sufficient size and mobility that flight become possible. Probably more interesting than her views on wings origin are Kukalova-Peck's suggestions as to the evolution of fused wings pads and wings flexing. As noted above, the earliest flying insects had wings that stuck out at right angles to the body. Thus as they developed the insects would be subjected to two selection pressures.

One exerted in the adult stage, would be toward improvement of flying ability; the other, which acted on juvenile instars, would promote changes that enabled them to escape or hide more easily under vegetation, etc. In other words, it would lead to a streamlining of body shape in juveniles. In most Paleoptera streamlining was achieved through the evolution of wings, which in early instars were carved so that the tips ware directed backward. At each molt, the curvature of the wings became less until the, "straight-out" position of the fully developed wings was achieved. Two other groups of paleopteran insects became more streamlined as juveniles through the evolution of a wing-flexing mechanism, a feature which was also advantageous to, and was therefore retained in, the adult stage. The first of these groups, the order Diaphanoptera, remained primitive in other respects and is included therefore in the infraclass Paleoptera. The second group, whose wing-flexing mechanism was different from that of Diaphanoptera, contained the ancestors of the Neoptera. The greatest selection pressure would be exerted on the older juvenile instars, which could neither fly nor hide easily.

In Kukalova-Peck's scheme, the older juvenile instars were eventually replaced by a single metamorphic instars in which the increasing change of form between juvenile and adult could be accomplished. To further aid streamlining and, in the final juvenile instars, to protect the increasingly more delicate wings developing within, the wings of juveniles became firmly fused with the terga and more sclerotized, that is, wing pads. This state is comparable to that in modern exopterygote (hemimetabolous) insects. Further reduction of adult structures to the point at which they exist until metamorphosis as undifferentiated embryonic tissues (imaginal discs) beneath the juvenile integument led to the endopterygote (holometabolous) condition, that is, the evolution of the pupal stage. In concluding this discussion on the origin of wings, the differences and similarities between the paranotal theory and Kukalova-Peck's proposal may be summarized. In the paranotal theory wings arose in truly terrestrial insects and environmental water played no part either in their origin or evolution. Kukalova-Peck's theory is that wings arose in semiaquatic insects, and both their origin and evolution were water-related. Both theories suggest that the original function of pro-wings was not a movement-related but one of protection.

The paranotal theory suggests that the legs may have been the structures protected, hence the development of pro-wings only on thoracic segments. Kukalova-Peck proposes that spiracles or gills were protected by the pro-wings which thus arose on all thoracic and abdominal segments. A major difference between the two theories concerns the site of origin and the onset of articulation of the pro-wings. The paranotal theory proposes that pro-wings arose as rigid outgrowths of the thoracic terga, and articulation came later; Kukalova-Peck argues that pro-wings developed between the spiracles and terga and were hinged from the outset. According to the paranotal theory, enlargement of the pro-wings was associated with improved gliding ability in insect living on terrestrial plants; the evolution of articulated pro-wings led to improved attitudinal control. In Kukalova-Peck's theory, the increase in size of the already articulated pro-wings was related to their use as oars for rowing an insect through water. Both theories are in agreement that the further increase in size and improved articulation of the wings was associated with their use in actively propelling insects through air.

Although both theories agree that the selection pressure leading to the evolution of wing flexing was the need to escape from predators or to avoid desiccation, they differ as to a life stage on which this pressure was exerted. The paranotal theory implies that wings flexing evolved in adult insects. Kukalova-Peck argues that the need for concealment arose in juvenile stages, and it was in these that the wing-flexing mechanism developed, though its advantage would also apply to adults. Thus, in her view, the well-sclerotized, nonarticulated wing pads of modern exopterygotes are secondarily derived.

Table 6.1. The Major Groups of Pterygota.

Infraclass	*Orders*	*Divisions within Neoptera* *Martynov's scheme*	*Hamilton's scheme*
	Paleodictyoptera		
	Megasecoptera		
	Diaphanoptera		
Paleoptera	Protodonata		
	Protephemeroptera		
	Odonata (dragonflies, damselflies)		
	Ephemeroptera (Mayflies)		
	Protoblattoidea		
	Protorthoptera		
	Dictyoptera (cockroaches, mantids)		
	Isoptera (termites)		
	Orthoptera (grasshoppers, locusts, crickets)		
	Miomoptera		Pliconeoptera
	Protelytroptera	Polyneopetra	
	Dermaptera (earwings)		
	Grylloblattodea (grylloblattids)		
	Phasmida (stick and leaf insects)		
	Embioptera (web spinners)		
	Hadentomoida		
	Caloneurodea		
	Protoperlaria		
	Plecoptera (stoneflies)		
	Zoraptera (zorapterans)		
Neoptera	Glosselytrodea		
	Psocoptera (booklice)		
	Phthiraptera (biting and sucking lice)	Paraneoptera	
	Hemiptera (bugs)		

Thysanoptera (thips)	
Megaloptera (dobsonflies, alderflies)	Planoneoptera
Raphidioptera (snakeflies)	
Neuroptera (lacewings, mantispids)	
Mecoptera (scorpion flies)	
Lepidoptera (butterflies, moths)	
Trichoptera (caddis flies)	Oligoneoptera
Diptera (true flies)	
Siphonaptera (fleas)	
Hymenoptera (bees, wasps, ants ichneumons)	
Coleoptera (beetles)	
Strepsiptera (stylopoids)	

Regardless of their origin, the wings of the earliest flying insects were presumably well-sclerotized, heavy structures with ill-defined veins. Slight traces of Fluting (the formation of alternating concave and convex veins for added strength) may have been apparent. The wings (and flight efficiency) were improved by a reduction in sclerotization, giving rise to the condition seen in Paleoptera. Only the basal part of a wing and the integument adjacent to tracheae within the wing remained sclerotized forming definite basal plates and veins, respectively. Fluting of the wing was accentuated in the Paleoptera, and the distal portion of the wing was additionally strengthened by the formation of nontracheated intercalary veins and numerous crossveins. This is known as the Archedictyon arrangement. Primitively, and still seen in Ephemeroptera, the base of each wing articulated at three points with the tergum, the three axillary sclerites running in a straight line along the body.

In the evolution of Neoptera the axillary sclerites altered their alignment; specifically, the second axillary sclerites moved away from the tergum and became articulated with the pleural wing process so that each wing articulated with the tergum at only two points. This alteration of alignment made *Wind Flexing* possible. It will be seen from Figure 2.2B that the two axes along which folding occurs (the basal and vannal folds) cross in the vicinity of the first axillary sclerite. Contraction of the wing flexor muscle which runs from the pleuron to the third axillary sclerite pulls the wing dorsally and posteriorly over the body. A second important consequence of the altered articulation of the wing was a further improvement in flight efficiency. In Ephemeroptera and, presumably most or all fossil Palepoptera the wing beat is essentially a simple up-and-down motion: in Neoptera each wing twists as it flaps and its tip traces a figure-eight path. In other words, the wing "rows" through the air, pushing against the air with its undersurface during the downstroke yet cutting through the air with its leading edge on the upstroke. To carry out this rowing motion effectively necessitated the loss of most of the wing fluting. Only the costal area needs to be rigid as this leads the wing in its stroke, and fluting is retained here. Another evolutionary trend, again leading to improved flight, has been toward a reduction in wing weight, permitting both easier wing twisting and an increased rate of wing beating. Concomitant with his reduction in weight has been a fusion or loss of some major veins and the loss of crossveins.

The extent and nature of fusion or loss of veins follow certain well-definable pattern which, together with other structural features, are important characters on which conclusions about the evolutionary relationships of neopteran insects can be based. Hamilton (1972c) recognizes three primary venational types, in addition to the archedictyon. The Polyneurous arrangement, which has crossveins more widely separated and parallel compared with those of the archedictyon, is characteristic of Protorthoptera, many fossil and extant orthopteroid orders, and the ancestors of the hemipteroids and endopterygotes. In the *Costaneurous* pattern crossveins remain only between the costal and subcostal veins (and in Plecoptera between the medical and plical veins). In addition to the wings of Plecoptera, those of Isoptera, Meglaoptera, Neuroptera, and the fossil order Protelytroptera show the costaneurous condition. All hemipteroids, most endopterygotes, and members of the orthopteroid orders Dermaptera and Embioptera have wings that almost entirely lack crossveins, a condition described as *Oligoneurous*. In some insects the basic pattern for the order to which they belong is secondarily modified, especially by the addition of crossveins.

Phylogenetic Relationships of the Pterygota

There are about 26 orders of living insects and 12 containing only fossil forms, the number varying according to the authority consulted. Elucidation of the relationships of these groups have been hampered by a lack of fossil evidence from the Devonian and Lower Carboniferous periods during which a great adaptive radiation of insects occurred. By the Permian periods, from which many more fossils are available, almost all the modern orders had been established. Misidentification of fossils by early paleontologists led to their drawing incorrect conclusions about the phylogeny of certain groups and the construction of rather confusing nomenclature. For example, Eugereon, a Lower Permian fossil with sucking mouthparts, was placed in the order Protohemiptera. It is now known that this insect is, in fact, a member of the order Paleodictyoptera and is not related to the modern order Hemiptera, as was originally concluded. Likewise the Protohymenoptera, whose wing venation superficially resembles that of Hymenoptera, were thought originally to be ancestral to the Hymenoptera. It is now appreciated that these fossils are paleopteran insects, most of which belong to the order Megasecoptera. To aid subsequent discussion of the evolutionary relationships within the Pterygota the various orders referred to in the text are listed in Table 6.1.

It is generally assumed that the Paleoptera and Neoptera had a common ancestor in the Devonian period [see, for example, Ross (1965) and sharov (1966), although there is no fossil record of this ancestor. By the Upper Carboniferous period, when conditions became suitable for fossilization, a number of paleopteran and neopteran orders had evolved. Among the Paleoptera at least two major evolutionary lines evidently appeared. The more primitive of these, the ephemeropteroid series, includes the Protephemeroptera (Upper Carboniferous) and Ephemeroptera (Permian-Recent). The fossil *Triplosoba* looks very much like the modern mayflies except that it has two identical pairs of wings. It is therefore placed in a separate order, Protephemeroptera, and considered to be slightly off the line of evolution leading to the Ephemeroptera. The second group, the odonatoid series, contains three, possibly four, fossil orders and the extant Odonata.

The Paleodictyoptera (Upper Carboniferous Permian) and Megasecoptera (Upper Carboniferous-Permian) were closely related orders, the former possibly ancestral to the latter, whose members were large and heavily built. The Paleodictyoptera were phytophagous or saprophagous and usually equipped with chewing mouthparts. However, one group, which includes Eugereon, may have had a sucking proboscis and fed on plant juices. The Megasecoptera were rather like the modern dragonflies in their habits: that is, they were carnivorous, catching their prey on the wing. Venational and other morphological features indicate that the Paleodictyoptera and Megasecoptera were allied to the Protodonata (Upper Carboniferous-Permian) and Odonata (Permian-Recent). The last two orders are closely related, and it is probable that the Odonata evolved from primitive Protodonata members of both orders were predaceous, and it is in the Protodonata that the largest known insects (the Meganeuridae) occur, some of which had a wingspan of up to 75 cm. Though not known as fossils, it is presumed that juvenile Protodonata were aquatic. The remaining fossil order of Paleoptera, the Diaphanoptera (Upper Caroniferous-Permian), is of interest because its members were able to fold their wings over the body rest, though the mechanism used was not apparently the same as in Neoptera. Features of their venation and mouthparts indicate that Diaphanoptera were related to the Paleodictyoptera-Megasecoptera, assemblage. In contrast with the Paleoptera, which were inhabitants of open spaces, the Neoptera evolved toward a life among overgrown vegetation where the ability to fold the wings over the back when not in use would be greatly advantageous.

The early fossil record for Neoptera is poor, but from the great diversity of fossil forms discovered in Permian strata it appears that the major evolultionary lines had become established by the Upper Carboniferous period. Two major schools of thought exist with regard to the origin and relationships of those evolutionary lines. The more widely accepted view, proposed by Martynov (1938), is that, shortly after the separation of ancestral Neoptera from Paleoptera, three lines of Neoptera became distinct from each other. Based on his studies of fossil wing venation Martynov arranged the Neoptera in three groups, Polyneoptera (orthopteroid orders), Paraneoptera (hemipteroids orders,) and Oligoneoptera (endopterygotes). In a modification of this view Sharov (1966) proposed that the Neoptera and Paleoptera had a common ancestor (that is, the former did not arise from the latter) and, more importantly, that the Neoptera may be a polyphyletic group. In his scheme each of the three groups arose independently a consequence of which must be the assumption that wing folding arose on three separate occasions. Ross (1955), from studies of body structure, and Hamilton (1972), who examined the wing venation of a wide range of extant species as well as that of fossil forms, concluded that there are two primary evolutionary lines within the Neoptera, the Pliconeoptera and the Planoneoptera.

According to Hamilton, the ancestral neopteran group was the Protobloattoidea from which arose the Protorthoptera. The latter is the largest and most diverse of the fossil orders and may even be polyphyletic. From this group evolved the plicopteran line, the

Caloneurodea and the Hadentomoida. The latter order contained the ancestors of the Planoneoptera. As the figure and the table indicate the Pliconeoptera of Hamilton corresponds approximately to the Polyneoptera of Martynov, that is includes the orthopteroid orders except for the Plecoptera, Zoraptera, and a few fossil orders considered Planoneoptera by Hamilton, and the Planoneoptera includes two main groups that are approximately equivalent to the Paraneoptera and Oligoneoptera in Martynov's scheme. In other words, both schools agree that there are three major groups within the Neoptera but differ with regard to the relationships among these groups. The most primitive group of Neoptera is the orthopteroid complex, whose members are characterized by chewing mouthparts, presence of cerci, a large number of Malpighian tubules, separate ganglia in the nerve cord, complex wing venation (typically including a large number of crossveins) that differs between fore wings and hind wings, and a large anal lobes on the hind wing that folds like a fan along numerous anal veins.

Most, if not all, of the neopteran fossils discovered in Carboniferous starta are orthopteroid. They include members of the orders Protoblattoidea (Upper Carboniferous), Protorthoptera (Upper Carboniferous-Permian), Miomoptera (Upper Carboniferous-Permian), Caloneurodea (Upper Carboniferous-Permian), Dictyoptera (Upper Carboniferous-Recent), and Othoptera (Upper Carboniferous-Recent). As their name indicates, the Protoblattoidea were thought originally to have been ancestral to the Dictyoptera or at least to have had a common ancestry with this group. According to Hamilton (1972c). however, venational similarities between the two order are superficial and do not indicate a close phylogenetic relationship. The Protoblattoidea are apparently the most primitive neopteran group as indicated by the archedictyon arrangement of crossveins and the more or less complete fluting of the wings.

The Protorthoptera, which is Hamilton's scheme descended from the Protoblattoidea, probably resembled the latter in general features and habits but may be distinguished by their slightly reduced (polyneurous) venation. As noted above, this order is probably a "mixed bag" of insects with a polyphyletic origin. Probably it was from primitive members of this order, which soon became extinct, that the ancestors of most recent orthopteroid orders arose. Some protorthopteran fossils are considered sufficiently distinct from other members of the order Protorthoptera that Hamilton (1972c) places them in a separate order, Hadentomoida, which contains the ancestors of the Planoneoptera.

The Dictyoptera underwent a massive radiation (the Upper Carboniferous is often to as the "Age of Cockroaches" so common are their fossil remains from this period) and the order remains quite extensive today. Within this order two trends can be seen. The cockroaches became omnivorous or saprophagous, nocturnal, often secondarily wingless insects, whereas the mantids remained predaceous and diurnal in habit. Although Isoptera are known as fossils only from the Cretaceous period, comparison of their structure and certain features of their biology with those of primitive cockroaches (some of which are subsocial) indicates that they are derived from blattoidlike ancestors. Indeed, the method of wing folding in the primitive termite family Mastoter-mitidae resembles that of fossil cockroaches rather than that of recent ones.

The order Orthoptera was already widespread by the Upper Carboniferous, its members being readily recognizable by their modified hind legs. Early in the evolution of the Orthoptera a split occurred, one line leading to the Ensifera (long-homed grasshoppers and crickets), the other to the Caelifera (short-horned grasshoppers and locusts). The affinities of the two remaining orders of fossils found in the Upper Carboniferous, the Miomoptera and Caloneurodea, are problematical. The Miomoptera were small to very small insects with some orthopteroid features (for example, chewing mouthparts and cerci) but with a wing venation somewhat like that of the hemipteroid order Psocoptera. Members of the Caloneurodea had chewing mouthparts, short cerci, and orthopteroid wing venation of which basis Carpenter (1977) assigns the order to the orthopteroid group. The affinities of several orders of orthopteroids are debatable.

The Protelytroptera (Permian) were small insects whose fore wings were hard protective sheaths rather similar to the elytra of beetles, under which the hind wings folded at rest. It is possible that they were ancestral to the Dermaptera, which are first known as fossils from the Jurassic period. The Dermaptera have been placed by some authors close to the Orthoptera, Plecoptera, Embioptera, and even the nonorthopteroid Coleoptera. Other authors have suggested that they split off early from either Protoblattoidea or Protorthoptera and are only very distantly related to the other modern orthopteroid orders. As a result of his comparative morphological study. Giles concludes that the nearest living relatives of earwigs are the Grylloblattodea, and the two groups probably had a common ancestor. The Grylloblattodea show an interesting mixture of orthopteran, dictyopteran, phasmid, and dermapteran features, which has led several authorities to consider them as living remnants of a primitive stock from which both Orthoptera and Dictyoptera evolved. However, the presence of several specialized features combined with the fact that they are not known as fossils, probably precludes their being directly ancestral to any of the other orthopteroid orders. Kamp's (1973) numerical analysis has shown that considerable similarity exists between the Dermaptera and Grylloblattodea, which supports the conclusion reached by Giles (1963) of common ancestry for the two groups.

The fossil record of the Phasmida is poor, specimens being known only from the Triassic onward. Kamp's (1973) study indicates that the order Phasmida has a close phenetic affinity with the Dermaptera and Grylloblattodea, and this author suggests that the three orders may form a natural group. Although apparently a very ancient group with a fossil record extending back to the Permian period, the relationships of the Embioptera are obscure. They do not appear to be close to any other orthopteroid group and are generally assumed to be an early offshoot of prothorthopteran stock. Plecoptera (Permian-Recent) are well represented from the late Permian and probably had, as their ancestors, members of the Protoperlaria (Permian), which themselves are believed by most authorities to have evolved from Protorthoptera. Hamilton (1972b) notes, however, that unlike that of typical orthopteroids, the anal lobe of Plecoptera has only four or five veins, and, as the hind wing folds, the lobe rolls under the rest of the wing rather than folding fanwise. Thus, Hamilton places the plecopteroid orders in thePlanoneoptera as an early of`hoot of the Hadentomoida.

The phylogenetic position of the order Zoraptera is uncertain. Members of the order are not known as fossils, but in some respects they resemble cockroaches, which has led to placement of the order in the orthopteroid group. On the other hand, other morphological features, including wing venation, led Ross (1955) and Hamilton (1972c) to suggest that the affinities of the order lie with the hemipteroids, perhaps as an offshoot of the Hadentomoida.

The hemipteroid orders, which with one possible exception have living representatives, share a number of features. They possess suctorial mouthparts, a fusion of ganglia in the ventral nerve cord, and few Malpighian tubules. They lack cerci, and the anal lobe of the hind wing is reduced, never having more than five veins. When the hind wing is drawn over the abdomen, if folds once along the anal or jugal fold, not between the anal veins as in the orthopteroids. The wing venation of hemipteroids is much reduced due to fusion of primary veins and almost complete loss of crossveins and, when both fore wings and hind wings are present, is basically similar in each. The Glosselytrodea (Lower Permian-Triassic) are generally considered to have been endopterygotes, probably having affinities with the Neuroptera. Hamilton (1972c) concludes, however, that members of the order had no venational features that would link them with neuropteroids but do have feature in common with Hemiptera. As a result, Hamilton considers them as ancestral to the remaining hemipteroids. Both Hemiptera and Psocoptera are known from the Lower Permian. These early Hemiptera belonged to the suborder Homoptera, which radiated considerably in the Permian, Triassic, and Jurassic periods. Undoubted fossils of the suborder Heteroptera are known from the jurassic period, but opinion differs as to the point at which they separated from the Homoptera.

Although fossil Thysanoptera are first seen in the Upper Permian period, the record is generally poor. Ross (1955) suggests that the order stand closest to the Hemiptera. Fossil Phthiraptera have not been found, but they probably evolved from psocopteran ancestors a both Psocoptera and the more primitive, chewing lice possess as unique, rather complex hypopharynx. Members of the suborder Anoplura (sucking lice) apparently evolved in the Cretaceous from ancestral chewing forms, and are found ony on plecental mammals. The main feature which unites members of the endopterygote orders is the presence of a pupal stage between the larval and adult stages in the life history. The wing venation of endopterygotes, though planoneopteran, is quite diverse and shows no obvious evolutionary trends among the orders.

Despite the occurrence of the pupal stage there has been considerable difficultly in deciding whether the group had a monophyletic or polyphyletic origin. The difficulty arise because, whereas the orders Megaloptera. Rhaphidioptera, Neuroptera, Mecoptera, Lepidoptera, Trichoptera, Diptera and Siphonptera show affinities with each other and form the "panorpoid complex" or Tillyard (1935), the remaining orders (Coleoptera, Hymenoptera, and Strepsiptera) are quite distinct, each bearing little similarity to any other endopterygote group. The consensus of opinion is that the endopterygotes form a monophyletic group, though the major subgroups were formed at a very early date. Many endopterygote orders were well established in the Lower Permian period. At that time, the Mecoptera was a diverse group that included many more families than presently exist.

The neuropteroid orders (Megaloptera, Neuroptera, and Raphidioptera) were clearly distinguishable. Trichoptera were abundant, and primitive Coleptera (suborder Archostemata) existed. Both the Mecoptera and neuropteroid group have on different occasions been considered an ancestral to the remaining endopterygotes on the grounds that some living representatives of both groups possess a number of very primitive features. Martynova(1961) suggested that it was from the neuropteroid group that the remaining endopterygotes arose, and this suggestion is supported by Hamilton's (1972a-c) studies of wing venation. The neuropteroid orders were originally included by Tillyard in the panorpoid complex in view of the several primitive features that they share with the Mecoptera. The modern view is that the neuropteroid group is a heterogeneous assembly of insects belonging to there distinct orders and, further should be excluded from the panorpoid complex. Probably the Mecoptera and neuropteroids evolved from a common ancestor in the Upper Carboniferous period. The extant Mecoptera are the survivors of a formerly quite extensive group from which two distinct lines arose in the Permian period, one leading to the Diptera and the other to the Trichoptera and Lepidoptera.

The close link between Diptera and Mecoptera has been confirmed by the discovery of Permian fossils having a mixture of Dipteran and mecopteran features. These fossils, although having a wing venation rather similar to the modern Tipulidae (crane flies), still possess four wings. The metanotum is, however, reduced, which suggests that the trend toward a single pair of wings seen in the modern Diptera had already begun. The phylogenetic position of the Siphonaptera is unclear, Fossil fleas are known only from the Lower Cretaceous onward, and comparative studies of living fleas are of little use, since these insects are so highly modified for their ectoparasitic mode of life. Fleas resemble primitive Diptera in certain larval fearures and in the nature of their metamorphosis. For this reason some authors have suggested a dipteran ancestry. Hinton (1958) and Holland (1964), however, argue that the Siphonaptera probably evolved from ancestors which were similar to the modern Boreidae, a rather distinct family of Mecoptera. The argument is based on a number of similarities in larval structure and the fact that within the panorpoid group, only the adults of Siphonaptera and Boreidae have panoistic ovarioles. The Permian fossil Belmontia has wings with a blend of trichopteran and mecopteran features, indicating the affinity of these two orders. Although the Trichoptera is a long-established order and most modern families can be traced back to the Triassic and Jurassic periods, the use of "case" in which to shelter seems to have been a relatively recent development. These structures are known a fossils only from the end of the Cretaceous period. Fossil Lepidoptera are known with certainly from the Cretaceous, although it is generally believed that the group split off the trichopteran line during the early Triassic period.

There is no doubt that the Lepidoptera and Trichoptera are closely related, and this is attested by their many common terrestrial ancestor was in the adult stage trichopteran and in the larval stage lepidopteran in character. In the subsequent evolution of the Trichoptera the larva became specialized for an aquatic existence, but the adult remained primitive. Along the line leading to Lepidoptera the larva retained its primitive features, but the adult became specialized, especially in the development of the suctorial proboscis. The primitive lepidopteran family Micropterygidae is a relic group close to the point of separation of the Trichoptera and Lepidoptera. Members of this group have, as adults, functional biting mouthparts. In addition some species have a wing venation very similar to that of Trichoptera. The paucity of fossils makes it difficult to establish the point at which the trichopteran and lepidopteran lines separated. However, the variety of structure seem among the Eocene fossils and the distribution of certain modern groups indicate that the order is a very ancient one which in its infancy had a rather restricted distribution. The great adaptive radiation of the group probably came at the end of the Cretaceous period and was correlated with the evolution of the flowering plants.

Remains of genuine Coleoptera are known from the Upper Permian period. Somewhat earlier elytralike remains which were assigned to the order Protocoleoptera are now known to be the fore wings of Protelytroptera, an orthopteroid group. Although in the earliest beetles the elytra were not yet thickened and strengthened for their modern function as protective sheaths, they were still sufficiently specialized as to be of little use in phylogenetic studies. The origin of Coleoptera, is therefore, uncertain. Handlirsch and Zeuner concluded that they evloved from protoblattoid ancestors, which implies that the endopterygotes have at least a diphyletic origin. Crowson (1960) and others claim on the basis of larval form and mode of development that the Coleoptera are derived from neuropteroid stock. According to Crowson this claim is substantiated by the Lower Permian fossil Tshekardocoleus, which is intermediate in form between primitive Coleoptera and Megaloptera.

Fossil Hymenoptera are known from the Triassic period, but these were already quite specialized individuals, clearly recognizable as belonging to the suborder Symphyta. The more advanced Hymenoptera of the suborder Apocrita, which contains the parasitic and stinging forms are not known as fossils until the jurassic and Cretaceous periods. The great adaptive radiation of the Hymenoptera was, like that of the Lepidoptera, clearly associated with the evolution of the flowering plants. Although early paleonotologists suggested that the Hymenoptera evolved from Protorthopteran stock, a more likely origin is from neuropteroid ancestors, judging by the similarity between the larvae of Symphyta and those of panorpoid insects, and the ease with which the wing venation of Hymenoptera can be derive from the pattern found in Megaloptera.

The position of the Strepsiptera, highly modified endoparasitic insects, is uncertain. The earliest fossils are already well-developed forms from the Oligocence period and therefore speculation on their origin is based on comparative morphology, Jeannel (1960) argues that the Strepsiptera arose from an ancient stock derived from very primitive Hymenoptera. However, most authorities, including Crownson (1960), place the Strepsiptera near (or eve within) the Coleoptera by virtue of their several common features, of which the most obvious is the use of the hind wings in flight. In both groups extensive sclerotization of the sternum (rather than the tergum) occurs. The first instar larva of Strepsiptera resembles closely the triungulin larva of the beetle families. Meloidae and Rhipiphoridae. In addition the endoparasitic forms of the Rhipiphoridae have habits that are generally similar to those of Strepsiptera.

Origin and Function of the Pupa

As noted in the previous section the Oligoneoptera (encopterygote orders) are characterized by the presence of a pupal stage between the juvenile and adult phases in the life history. The development of this stage, which serves a variety of functions, is a major reason for the success (that is, diversity) of endopterygotes. Several theories have been proposed for the origin of the pupal stage.

Berlese's Theory

During its development the insect embryo passes through three distinct stages. In the first stage (protopod) no appendages are visible: this is followed by the polypod stage (in which appendages are present on most segments; and finally the oligopod stage (when the appendages on the abdomen have been resorbed). The Berlese theory suggests that the eggs of exopterygotes, by virtue of their greater yolk reserves, hatch in a postoligopod stage of development, whereas the eggs of endopterygotes, which have less yolk, hatch in the polypod or oligopod stage. Thus in this theory the abdominal prolegs must be considered homodynamous with the thoracic legs. The larva of endopterygotes corresponds to a free-living embryonic stage and the pupa represents the compression of the (postoligopod) nymphal stages into a single instar.

Berlese's theory has been criticized by Hinton (1963b) on the grounds that there is no evidence that the eggs of exopterygotes have a better supply of yolk than those of endopterygotes. Hinton also claims that the abdominal prolegs are not homodynamous with the thoracic legs but are secondary larval structures. Heslop-Harrison (1958), in proposing a modified version of Berlese's ideas, argues that Hinton's claim is not justified, especially in view of the polypodous nature of the ancestors of insects.

Novak's Theory

Novak's (1966) theory is based on Berlese's proposal that the eggs of endopterygotes hatch at an earlier stage of embryonic development than do those of exopterygotes. According to Novak in exopterygotes, which supposedly hatch in the postoligopods stage, differentiation of adult tissues begins before an effective concentration of juvenile hormone is reached in larval life so that, specifically, the wing rudiments can develop on the body surface. (Juvenile hormone inhibits adult tissue differentiation. In endopterygotes, however which hatch earlier, production of juvenile hormone begins before adult tissues have started to differentiates. Thus, the adult tissue cannot begins to differentiate until the hormone titer falls to a sufficiently low level in the final instar larva. Thus, two molts are necessary; one to evaginate the undifferentiated wing discs so that the wings can develop in a suitably large space, and another to permit expansion of the fully developed wings.

Poyarkoffs Theory

This theory has been supported by several authorities, including Hinton (1948) Snodgrass (1954) and DuPort (1958). The advantages of this theory over Berlese's that it provides a causal explanation for the origin ad function of the pupal stage. According to this theory, the eggs of both endopterygotes and exopterygotes hatch at a similar stage of development; the adult stage in the exopterygote ancestors of the endopterygotes became divided into two instars, the pupa and the imago; the subimago of Ephemero - ptera and the "pupal" stage of some exopterygotes are equivalent to the endopterygote pupal stage; the pupal stage evolved in response to the need for a mold in which the adult systems, especially flight musculature, could be constructed. The second (pupal-imaginal) molt was then necessary in order that the new muscles could become attached to the exoskeleton. In Poyarkoff's theory there is no difference between the endopterygote larva and the exopterygote nymph. In supporting Poyarkoff's theory, Hinton (1948) pointed out that the pupa (especially that of primitive endopterygotes such as Neuroptera) resembles the adult form rather than that of the larva, supernumerary molts could be artificially induced under certain conditions in exopterygotes, and in Ephemmeroptera a "natural" adult molt (form subimago to imago) occurs. Unfortunately no evidence was available to support the idea that the function of the pupal stage was to serve as a mold for development of the adult wing musculature. This theory and Hinton's (1948) support of it have been criticized by Heslop-Harrison (1958) and Hinton (1963b). Heslop-Harrison claimed that the explanation given for the origin of the pupal stage is teleological; in other words, Poyarkoff's explanation is that the pupal stage arose in fulfillment of a "need".

It is further suggested by Heslop-Harrison that Hinton's (1948)supporting evidence was carefully selected and does not represent the true situation. For example, Hinton's claim that the pupal stage resembles generally the adult was refuted by Heslop-Harrison, who pointed out several similarities between the pupal and larval instars. Hinton (1963b) stated that there is direct evidence against Poyarkoff'd idea concerning the function of the pupal stage.

First, it has been shown that tonofibrillae (microtubules within the epidermal cells which attach muscles to the exoskeleton) can be formed long after the pupal-adult molt has occurred. Second, even in highly advanced endopterygotes the fiber rudiments of the wing muscles are present at the time of hatching. These develop in the larva in precisely the same way as the flight muscles of many primitive exopterygotes. In other words, no molt is required.

Heslop-Harison's Theory

Implicit in the theories of Berlese, Poyarkoff, and Hinton (1963b) is the evolution of endopterygotes from exopterygote ancestors. Heslop-Harrison believed, however, that the earliest forms of both groups were present at the same time and evolved from a common ancestor. This ancestor had a life history similar to that of modern Isoptera and Thysanoptera, namely, Egg→Larval Instars (showing no sings of wings)→Nymphal Instars (having external wing buds)→Adult. Heslop-Harrison proposed that in the evolution of exopterygotes the larval instars were suppressed, and the modern free-living juvenile stages correspond the nymphal instars of ancestors. In the evolution of endopterygotes the nymphal stages were compressed into the quiescent prepupal and pupal stages of modern forms. (The prepupal stage is a period of quiescence in the last larval instars prior to the molt to the pupa. In other words the prepupal stage is not a distinct instar.) Thus, Berlese's original concept that the pupa comprised the ontogenetic counterparts of nymphal instar is divided into two phases. In the most primitive condition the first of these phases is an active one where the insect feeds and or prepares its "pupal" chamber.

In the most advanced condition both phases are inactive, and there are, for all intents and purposes, distinct prepupal and pupal stages, as in true endopterygotes. The main general criticism of Heslop-Harrison's theory is that it lacks supporting evidence. More specific criticisms are that (1) the fossil record indicates that the earliest insects were only exopterygote, and most authorities believe that the endopterygotes evolved from exopterygote stock (perhaps the Protorthoptera); (2) the Isoptera and Thysanoptera on which Heslop-Harrison's "primitive life history" was based are two highly specialized exopterygote orders; and (3) the implied homology of the endopterygote pupa and the last juvenile instars of the exopteygote Homoptera studied by Heslop-Harrison is not justified.

Hinton's Theory

Perhaps the attraction of Hinton's(1963b) theory is its simplicity. It avoids the "supperssion of larval," "compression of nymphal" and "expansion of imaginal" stages, found in the earlier theories and provides a simple function explanation for the evolution of a pupal stage. In Hinton's theory the pupa is homologous with the final nymphal instar of exopterygotes, and the terms "larva" and "nymph" are synonymous. Hinton proposes that, during the evolution of endopterygotes, the last juvenile stage (with external wings) was retained to complete the link between the earlier juvenile stages (larvae with internal wings) and the adult, hence the general resemblance between the pupa and adult in modern endopterygotes. Initially, the pupa would also resemble the earlier instars (just as the final instar nymph of modern exopterygotes resembles both the adult and the earlier nymphal stages). Once this intermediate stage had been established it is easy to visualize how the earlier juvenile stages could have become more and more specialized (for feeding and accumulating reserves) and quite different morphological from both the pupa and the adult (the reproductive and dispersal stage). At the same time the pupa itself became more specialized. It ceased feeding actively, became less mobile, and was concerned solely with the metamorphosis from the juvenile to the adult form. Concerning the functional significance of the pupal stage, Hinton suggested that, as the endopterygote condition evolved, there was insufficient space in the thorax to accommodate both the "normal" contents (muscles and other organ systems) and the wing anlage. Thus, the function of the larvalpupal molt was to evaginate the wings. This would permit not only considerable wing growth (as greatly folded structures within the pupal external wing cases) but also the enormous growth of the imaginal wing muscles within the thorax.

The latter is facilitated, of course, by histolysis of the larval muscles (a process which is often not completed for many hours after the pupal-adult molt). The function of the pupal-adult molt is simply to effect release of the wings from the pupal case. The original function of the pupal stage was, then, to create space for wing and wing muscle development. But, once a stage had been developed in the life history in which structural rearrangement could take place, the way was open for increasing divergence of juvenile and adult habits and, subsequently, a decrease in the competition for food, space, etc. between the two stages.

The Success of Insects

The degree of success achieved by a group of organisms can be measured either the total number of organisms within the group or, more commonly, as the number of different species of organisms which comprise the group. On either account the insects must be considered highly successful. Success is dependent on two interacting factors: (1) the potential of the group for adapting to new environmental conditions and (2) the degree to which the environmental conditions change. Since success measured as the number of different species is a direct result of evolution, the environmental changes that must be considered are the long-term climatic changes that have occurred in different parts of the world over of several hundred million years.

The Adaptability of Insects

The basic feature of insects to which their success can be attributed must surely but that they are arthropods. As such they are endowed with a body plan that is superior to that of any other invertebrae group. Of the various arthropodan features the integument is the most important, since it serves a variety of functions. Its lightness and strength make it an excellent skeleton for attachment of muscles as well as a "shell" within which the tissues are protected. Its physical structure (usually including an outermost wax layer) makes it especially important in water relations of arthropods. Because they are generally small organisms, arthropods in almost any environment face the problem of maintaining a suitable salt and water balance within their bodies. The magnitude of the problem (and therefore, the energy expanded in solving it) is greatly reduced by the impermeable cuticle. Arthropods are segmented animals and therefore have been able to exploit the advantages of tagmosis to the full extent. Directly related to this is the adaptability of the basic jointed limb, a feature exploited fully by different groups of arthropods. Since all arthropods possess these advantageous features, the obvious question to ask is "Why have only insects been so successful?" or, put differently, "What features do insects possess that other arthropods do not?" Answering this question will provide only a partial answer, since, as was stressed above the environmental changes that take place are also very important in determining degree of success. Take, for example, the Crustacea Compared to other invertebrate groups they must be regarded as successful (at least 26,000 species have been described), yet in comparison to the Insecta they come a distant second. Although this is due partly to their different features, it must be due also to the different habitats in which they evolved. The Crustacea are a predominantly marine group. In other words, they evolved under relatively stable environmental conditions. Furthermore, it is likely that when they were evolving the number of niches that were available to them would be quite limited because most were occupied by already established groups. The insects, on the other hand, evolved in a terrestrial environment subject to great changes in physical conditions.

They were one of the earliest groups to "venture on land" and therefore had a vast number of niches available to them in this new adaptive zone. Most insects, modern and fossil, are small animals. A few early forms achieved a large size but became extinct presumably because of climatic changes and their inability to compete successfully with other groups. Small size confers several advantages on an organism. It facilitates dispersal, it enables the organism to hide from potential predators, and it allows the organism to make use of food materials that are available in only very small amounts. The great disadvantage of small size in terrestrial organisms is the potentially high rate of water loss from the body. In insects this has been successfully overcome through the development of an impermeable exoskeleton.

The ability of fly was perhaps the single most important evolutionary development in insects. With this asset the possibilities for escape from predators and for dispersal were greatly enhanced. It would lead to colonization of new habitats, geographic isolation of populations, as a result, formation of new species. Wide dispersal is particularly important for those species whose food and breeding sites are scattered and in limited supply. The reproductive capacity and life history are two related factors that have contributed to success of insects. Production of large numbers of eggs, combined with a short life history, means a greater amount of genetic variation can occur and be tested out rapidly within a population. The net result is (1) rapid adaptation to changes in environmental conditions, and (2) rapid attainment of genetic incompatibility between isolated populations and formation of new species. The evolution of a pupal stage between the larval and adult stages has led to a more specialized (and, in a sense, a more "efficient") life history.

The main function of the larva is accumulation of metabolic reserves, whereas the adult is primarily concerned with reproduction and dispersal. This means that insects can utilize food sources that area available for only short periods of time. Although the pupa has as its main function the transformation of the larval to the adult form, it has in many species become a stage in which insects can resist unfavorable conditions. The development, and the restriction of feeding activity to one phase of the life history, have facilitated the expansion of the insect fauna into some of the world's most unfriendly habitats. Several feature of insects have contributed therefore to their success (diversification). It is important to realize that these features have acted in combination to effect success, and, furthermore, little of this success would have been possible except for the unstable environmental conditions in which the insects evolved.

The Importance of Environmental Changes

The importance of environmental changes in the process of evolution, acting through natural selection, is well known. These changes can be seen acting at the population or species level on a short-term basis, and many examples are known in insects, perhaps the best two being the development of resistance to pesticides and the formation of melanic forms of certain moths in areas of industrial air pollution. Of greater interest in the present context are the long-term climatic changes that have taken place over millions of years, for it is these that have controlled the evolution of insects both directly and indirectly through their influence on the evolution of other organisms, especially plants. Although life began at least 2.5 billion years ago, it was not until the Silurian period (about 425-500 million years ago) that the first terrestrial organisms appeared, an event probably correlated with the formation of an ozone layer in the atmosphere, which reduced the amount of ultraviolet radiation reaching the earth's surface.

The earliest terrestrial organisms were simple low-growing land plants that reproduced by means of spores. They were soon followed by mandibulate arthropods (scorpions, myria- pods, and wingless insects) which presumably fed on the plants, their decaying remains, their spores, or on other small animals. In these early plants, spores were produced on short side branches of upright stems. An important evolutionary development was the concentration of the sporangia (sporeproducing structures) into a terminal spike. Whether these spikes were particularly attractive as food for insects and whether, therefore they may have been important in the evolution of flight is a matter for speculation. During the Devonian and Lower Carboniferous periods, a wide radiation of plants occurred. Especially significant was the development of swamp forests that contained, for example, tree lycopods, calamites, and primitive gymnosperms.

The evolution of treelike form though in part due to the struggle for light, may also have enabled the plants to protect (temporarily) their reproductive structures against spore-feeding insects and other animals. In contrast to the humid or even wet conditions on the forest floor, the air several meters above the ground was probably relatively dry. Thus, the evolution of trees with terminal sporangia several meters above the ground may have been an important stimulant to the evolution of a waterproof cuticle, spiracular closing mechanisms, and eventually, flight. The trees, together with the ground flora, would provide a wide range of food material. As noted earlier, winged insects appeared in the Lower Carboniferous. Many of these were mandibulate (for details see previous section) and fed on soft parts of plants or litter on the forest floor. Others, for example, some Paleodictyoptera, had mouthparts in the form of a proboscis which has led to the suggestion that these insects were adapted to feeding on liquids either freestanding or as sap in plant tissues. Smart and Hughes (1973) believe, however, that the proboscis might have been used as a probe for extracting pollen and spores from the plants' reproductive structures. They argue that not until the Upper Carboniferous did plants evolve which had phloem close enough to the stem surface that it was accessible to Hemiptera. Yet other insects such as the Protodonata, Megasecoptera and later. Odonata, were predators. Some of these were very large, and it has been suggested that the evolution of large size was a result of competition between these inset and the earliest terrestrial vertebrates, the Amphibia. Certainly large size would be favored by the year round, uniform growing condition.

In addition to the forest ecosystem, there were presumably other ecosystems, for example, the edges of swamps and higher ground, that had their complement of insects. However, such ecosystems did not apparently favor fossilization and their insect fauna is practically unknown. Toward the end of the Carboniferous period other climatic changes of importance in insect evolution took place. At this time extensive mountain ranges were formed and over many parts of the earth the climate became cooler and drier. These changes created not only many new habitates but also barrier that prevented gene flow between population. By the end of the Premian period most of the older orders were extinct and had been replaced by representatives of the modern orders. In the Triassic and early Jurassic periods the gradual radiation of insects continued but was largely overshadowed by evolution of the reptiles. The latter occurred in such large numbers and in such a variety of habitats that the Triassic period is generally known as the "Age of Reptiles." Many of them were insectivorous, and this acted as a selection pressure favoring small size, which is a general feature of fossil insects from this period. Early in the Triassic period the first bisexual flowers appeared.

The occurrence together of male and female structures immediately leads to the possibility of a role as pollinators for insects that feed on the reproductive parts of plants. Because of the risk of having their reproductive structures eaten, these early plants probably produced a large number of small ovules and much pollen. The insect fauna of the Triassic period still included the "orthodox" plant feeders such as Orthoptera and some Coleoptera, the plant-sucking Hemiptera, and predaceous species (Odonata and Neuroptera). However, there were also a large number of primitive endopterygotes, mostly belonging to the panorpoid complex, Hymenoptera, and Coleoptera, which as adults were mandibulate and therefore were potential "mess-and-soil" pollinators. By the middle of the Jurassic Period the decline of the reptiles had begun and the earth's climate had become generally warmer. As a result the insect fauna increased both in mean body size and in variety. A good deal of mountain formation occurred at the end of the Jurassic, creating new climatic conditions in various parts of the world including, for the first time, winters. The Cretaceous period was an extremely important phase in the evolution of insects, for it was during this time that vast adaptive radiations of several endopterygote orders took place. In some instances it must be assumed that these radiations directly paralleled the evolution of angiosperms, although, it must be emphasized, the fossil record of angiosperms flowers is sparse. As a result of this coevolution, some extremely close interrelationships have evolved between plants and their insect pollinators. For other orders, the radiation was only indirectly due to plants; for example, a large variety of parasitic Hymenoptera appeared, correlated with the large increase in numbers of insect hosts.

The Cretaceous period was also rather active, geologically speaking, for a good deal of mountain making, lowland flooding (by the sea), and formation and breakage of land bridges took place. All these processes would assist in the isolation and diversification of the insect fauna. The decline of the reptiles became accelerated during the late Cretaceous period, and they were succeeded by the mammals and birds. These groups became very widespread and diverse during the Tertiary period. Paralleling this diversification was the evolution of their insect parasities. Throughout the Tertiary period the climate seems to have alternated between warm and cold. In the Paleocene epoch the climate was cooler than that of the Cretaceous. Thus cold adapted groups became widely distributed.

A warming trend followed in the Eocene so that cold- adapted organisms became restricted to high altitudes, while the warm-adapted types spread. It appears that by the end of the Eocene period (approximately 36 million years ago) most modern tribes or genera of insects had evolved. In the Oligocene the climate became cooler and remained so during the Miocene and Pliocene epochs. However, in these to epochs new mountain ranges were formed, and some already existing ones were pushed even higher. The Tertiary period ended about one million years ago and was followed by the Pleistocene epoch of the Quaternary period. Temperatures in the Pliestocene(Ice Age) were generally much lower than in the Tertiary, and four distinct periods of glaciation occurred, at which time most of the North American and European continents had a thick covering of ice. Between these periods warming trends caused the ice to recede northward. Accompanying these ice movements were parallel movements of the fauna and flora. However, the overall significance of the Ice Age in terms of insect evolution is uncertain at the present time.

7

Insect Hormones

The insect hormones are those which are effective and also well known by entomologists influence pre and postembryonic development. Three of these so-called metamorphosis hormones are recognized: (A) the activation hormone manufactured by specific neuroscecretory cells in the insect brain, which conditions the reactivation of the body after each moult and the production of the other two metamorphosis hormones; (b) the moulting hormone produced by the prothoracic glands or corresponding tissues, which conditons the moulting process and thus, indirectly, growth and morphogenesis; (C) the juvenile hormone produced by the corpora allata, which conditions the growth of the larval parts in all metabolous insects (pterygota), the function of the ovarial follicles of adult females in most insects, and many other structures and functions of the insect body, which are unable to evolve in its absence. All three hormones are essential for the normal course of postembryonic development. The activation hormone is a typical neurohormone, the other two are true glandular hormones order, and sometimes even class unspecific. They are dependent on one another both in their production and function.

Terminology

Metamorphosis

Metamorphosis implies that the single part of hte postembryonic development which takes place in the absence of a morphogenetically active concentration of *juvenile hormone*, i.e. the last larval instar of Hemimetabola and both the last larval and the pupal instars of Holometabola. This is corroborated with most other physiological works but in disagreement with some morphologists. But it must not be denied that a part of the adult differentiation is recognized at the beginning of each larval instar, there is no more reason to call this metamorphosis than there is in the case of the similar successive postembryonic differentiation in ametabolous insects (Apterygota).

The Postembryonic Development

The postembryonic development of insects can thus be divided into two parts or stages: the larval stages (or larval development) beginning with the hatching of the larva and ending with the penultimate larval instar (the ecdysis from the penultimate to the last larval instar); and the later metamorphosis stage (or metamorphosis), ending with the development of the adult insect, i.e. the last larval instar of Exopterygota and the same including the pupal instar in Endopterygota.

Instar

Both the larval and the metamorphosis stages are sub-divided into instars. An *instar* is that part of the postembryonic development which takes place between two successive ecdyses. The term is used in both the morphogenetical and the chronological sense of the word. Yet another definition was suggested by Hinton (1958), and consequently accepted by many other entomologists, according to which the instar begins at the moment of deposition of the new cuticle and ends with the moult, which is the next separation of the cuticle from the epidermis. based on the morphological description this has the advantage that the term 'instar' refers to one and the same cuticle whereas in the earlier definition, each instar starts inside the old cuticle which is replaced by a new one during the instar. It has also the advantage of drawing hte attention of morphologists to those stages in the moulting process which take place beneath the opaque cuticle. Notwithstanding the fact that postembryonic development is further divided into a succession of overlpapping cyclic processes, so that the determination of the limits of individual instars is to some extent arbitrary, the first-mentioned definition of the instar seems most appropriate for the purposes of this book for the following reasons:

1. The strcture of the body, even though determined by the newly deposited cuticle, particularly that of heavily sclerotized parts such as the head capsule, is also changed at each ecdysis but not at the moults.

2. The detachment of hte cuticle at apolysis can only be observed in histological sections (rather difficult to make because of the hard sclerotized cuticle). The moment of ecdysis is therefore as much preferable as the dividing point from a practical point of view.
3. Ecdysis is linked with a series of other processes, such as the resorption of the moulting fluid, the occupation of the new shape of the body at the distention of the newly formed cuticle (even though this was laid down some time previously), the deposition of the wax and cement layers of the epicuticle, the tanning and hardening of the new cuticle, etc. It is therefore the most significant and visible section of each moulting cycle.
4. Inspite of the fact that successive moults are connected with an uninterruped succession of various physiological processes, most of these are, nevetheless, clearly arranged in cycles within each interecdysis, but not respiration and hormone production, resulting in swelling of the during moults. By selecting the detachment of the cuticles as the start of an instar, all of these clearly logical processes are divided between two different instars.

Most of these reasons remain acceptable even in such cases as the last larval instar of higher Diptera. There, although the chief characteristic of ecdysis, the shedding of the exuvium, is eliminated, as is the tanning of the newly formed pupal cuticle, other distinguishing features, such as, for example, the removal of the moulting fluid and the formation of the epicuticular layers, remain; and these are enough to enable the instars to be distinguished.

Pharate instar

If we do not accept Hinton's definition of the term 'instar', the term 'pharate instar' has a restricted meaning. It seems to be justified only in cases in which a new instar has been formed completely and the moulting fluid has been absorbed, as in the puparium of higher Diptera, for example. The determining factor is thus the extistence of air between the old and the new cuticle.

Development Stage

Generally the term stage of development is applied for a section of ontogenetical development characterized by significant biochemical or morphological processes or by both, such as the protoped, polypod, oligopod, and postoligopod stages in embryogenesis, the larval stage (comprising the first to penultimate larval instar), the metamorphosis stage (comprising the last larval and the pupal instars), the maturation stage, the stage of sexual maturity, etc.

Moulting Process

There are several biochemical, physiological and morphological changes preceding each ecdysis constitutes the moulting process or moulting. This begins with the swelling of the epidermis followed by detachment of the old cuticle, usually referred to as the moult or apolysis. Then the moulting fluid is produced, in which the inner layers of the cuticle are gradually digested, and a new cuticle is laid down by the epidermal cell. The process ends in ecdysis which is preceded by the resorption of the moulting fluid and followed by the tanning and hardening of the exocuticle and the secretion of the wax and cement layers of the epicuticle. the laying down of the inner layers of the endocuticle by the epidermal cells persists after the ecdysis and usually does not stop until immediately before the start of the next moulting process. Thus the moulting process is terminated.

Few years back researchers began to use the terms 'larval-larval moult' in place of 'larval moulting', 'larval- pupal moult', instead of 'pupal moulting' and 'pupal-imaginal moult' instead of 'imaginal moulting. Apart from the linguistic aspect, this seems completely unnecessary, in so far as we taken into account that the above terms express the result of the relevant moulting process. In any case the meaning is usually amply evident from the cortex. However, the term 'emergence' for imaginal ecdysis is exitrely adequate.

The term 'moult' does not mean 'moulting process' since it has been accepted as the English term for 'apolysis', i.e. only one of the initial phases of the moulting process.

Larval moulting-any moulting which results in larval form.

Pupal moulting-moulting to pupal form. When a second pupa is produced as a result of experimental JH treatment, we tern it as 'second pupal moulting'.

Imaginal moulting or *emergence-moulting* which culminates in the adult form.

Growth

By growth is generally accepted as the increase in the amount of living matter of the body (organized protein molecules) unlike, e.g., the increase in weight by accepting food or ingesting water.

The following lines piont the differnet types of growth may be distinguished in insects:

Isometric growth (harmonic or proportionate growth)

The type of growth where there is an equal rate of increase in all parts of the body, i.e. there are no morphological changes. The shape of the body at the end of an isometric growth period is a homothetic figure of that at the beginning.

Anisometric growth (disharmonic or disproportionate growth)

The type of growth where there is a different rate of increas in different parts of the body, with a resulting changes of shape. two kinds of anisometric growth are distinguished in this book (cf. Novak, 1960, 1962). These are:

(i) *Allometric growth* (or anisometric growth in the restricted sense). In this kind of growth parts of the body grow although at different rates, as for instance in postembryonic development in mammals.

(ii) *Gradient growth.* In this type of growth, only few parts of the body, the gradients, grow. Growth of the remaining parts is stopped and eventually they are wholly or partly removed either by histolysis or phagocytosis. This occurs, for example, in the embryogenesis of most animals an in the metamorphosis of insects. Gradient growth is the significant feature in morphogenesis.

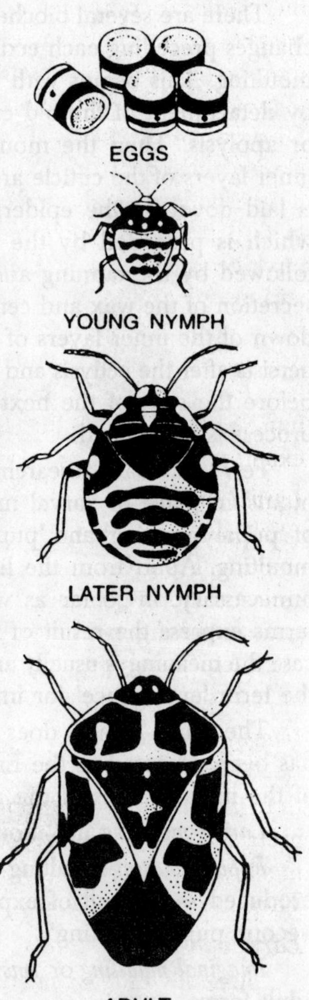

Fig. 7.1. Stink bug (Hemiptera). Stages of life history exhibiting heterometabolic metamorphosis or to say gradual metamorphosis.

Differentiation

The important aspect of development, the sequence of changes which occurs in the body in the given period of ontogenesis, is irrespective of growth. The three following types are discerned (1) chemical (biochemical) differentiation, the sequence of changes in the chemical composition of the body irrespective of changes of weight, shape, function, etc., (2) morphological differentiation or morphogenesis, the sequence of changes in the shape of the body, (3) functional differentiation, the sequence of development of various functions of the body. These three types of classifications are all closely related in normal development. In ontogenesis, chemical differentiation usually precedes morphological differentiation, which in turn precedes functional differentiation. On the other hand in phylogenesis, functional differentiation usually precedes chemical and morphological differentiation of the corresponding part of the body. (This follows from the action of natural selection and its role in evolution: it means that structures of the body can be favoured by selection and thus evolved and specialized for a given function only when this function already exists and is to some extent useful for survival to some extent.)

Larval structures

They are the parts of the body which can grow only the larval stage (see above), i.e., in the presence of a certain minimum concentration of juvenile hormone. They lose the ability for further growth during metamorphosis, and usually undergo partial or complete histolysis at that time (cf. Novak, 1951b, 1956). They include the greater part of the larval body is most pterygote insects.

Imaginal structures

Those parts of the body the growth of which is independent of the presence of juvenile hormone in the beginning of metamorphosis. Their development is accelerated durign metamorphosis when the growth of the larval structures ceases. They include the greater part of the imaginal body.

Larvo-imaginal structures

Are fully differentiated and functional in the larval period, but they remain and continue to function in adults.

Larva

The term larva is used here in its broadest sense, as in Wigglesworth (1939) and Hinton (1958), but not Jeschikov (1941) and Snodgrass (1954), to include the entire postembryonic stage of development which precedes metamorphosis in all insects. The term 'nymph', as a special type of larva, is retained for the larval stage of Hemimetabola and 'larvula' is kept for the first larval instars of Odonata differing from the later ones. 'Eonymph', mesonymph' and 'pronymph' in the sense of Berlese (1913), and Slama (1960) refer to the pre-pupal stages of sawflies. Pre-pupa is the term for the latter portion of the last larval instar of Holometabola, which is distinguished by the advanced pupal moulting process with plenty of moulting fluid and retraction of the body.

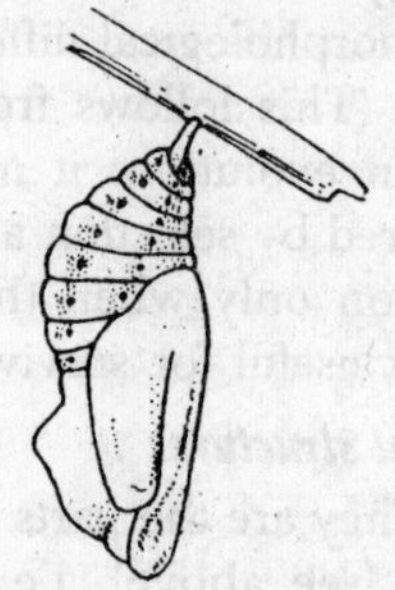

Fig. 7.2. Butterfly (Lepidoptera). Stages of life history exhibiting holometabolic (complete) metamorphosis.

Pupa

The term pupa is used for the last postembryonic instar (second metamorphosis instar) in Holometabola, which is the period between the pupal and imaginal ecdyses. Internal morphological changes, which vary considerably in extent according to the insect, occur during this instar, whilst most of the external changes are occur in the first metamorphosis instar (last larval instar).

Hemimetabola and Holometabola

These terms are used as in the taxonomic sense of Imms (1946) and Obenberger (1952), but not Weber (1952), but not Weber (1954), Snodgrass, etc. *Hemimetabola* (= Heterometabola, =

Exopterygota) have only one metamorphosis instar, *Holometabola* (= Endopterygota) have two. The prefix 'hemi' does not refer to the degree of morphogenesis which, on average, is notmuch different in the two subclasses, but to the number of metamorphosis instars.

Abbreviations

Abbreviations are generally used for frequently recurring terms, for the purpose of saving space and also for greater clarity. The advantage of this was soon relaized by chemists, who introduced a generally recognized and employed letter-symbol for every element (chemical compound). The same need is increasingly felt by specialists in insect endocrinology, but no rational abbreviation system which would be generally accepted has so far been created.

The system applied throughout this book, has the following rules Capital letters are employed for the hormones forming the main subject-matter and small letters for the often no less useful abbreviations of various endocrine organs. Differences in the origin of the same principle are expressed by adding a small letter or letters after the hormone symbol. The same primary symbol is always used for one principle (substance, effect). Consequently, instead of CAH, used by some authors for the corpus allatum hormone, and JHa for JH analogues, the common symbol JH is used for both, JHca denoting corpus allatum JH and JHa juvenile hormone analogues. JH is used alone where effects of the principle are discussed in general, irrespective of its origin. The abbreviations used by other authors are given in the list of synonyms at the beginning of each section dealing with the relevant hormones.

ca–corpora allata (corpus allatum).

cc–corpora cardiaca (corpus cardiacum).

nsc-neurosecretory cells.

ncc I, II, II–nervi corporum cardiacarum, first, second, and third pari.

agl–apolytic glands.

pgl–prothoracic glands.

rgl–ring gland.

vgl–ventral gland.

nho–neurohaemal organs.

mnho–metameric neurohaemal organs.

cns–central nervous system.

No abbreviations are used in this book for organs not associated with endocrine activity.

AH–activation hormone.

JH–juvenile hormone.

JHca–corpus allatum hormone.

JHa–juvenoids (juvenile hormone analogues).

MH–moulting hormone (ecdysone)

MHpg–prothoracic gland MH.

MHd–ecdysoids (ecdysone derivatives).

GH–gradient factor ((the distinction formerly made between the GF of insect metamorphosis and the general gradient factor gf has been abolished, because it is nwo assumed that they are identical).

BF-blastomo-factor

To facilitate easy reading the terms are also given together with the abbreviations by which theya re represented, wherever they appear for the first time in each chapter.

THE ACTIVATION HORMONE (AH)

The initial statement regarding a hormonal function of the insect brain was made by Kopec (1917, 1922) whose experiments with Lymantria dispar caterpillars proved conclusively that the brain is necessary for pupation. This was the first proof for an internal secretion of any kind in insects. Tauber (= Taabor) (1925), working independently, concluded, from several years' work on blood transfusion and ligaturing in Deilephila euphorbiae caterpillars, that insect metamorphosis is controlled by endocrines. More indication for the presence of the humoral control of moulting in other species of Lepidoptera is contained in the papers by Koller (1938) and Buddenbrock (1950).

But the confirmation of the existence of a special hormone, produced by the brain which controlled moulting, was achieved by Wigglesworth (1936) from his classic experiments with decapitation, parabiosis and transplantation in the blood-sucking bug Rhodnius prolixus. This was the genuine beginning of insect endocrinology which has now evolved into a special, very extensvely studied, biological discipline. It was the same author who, in collaboration with Hanstrom (1943), first discovered the source of the activation hormone in the neurosecretory cells of the brain. Another comtemporary entomologist Fraenkel (1935) demonstruted a similar

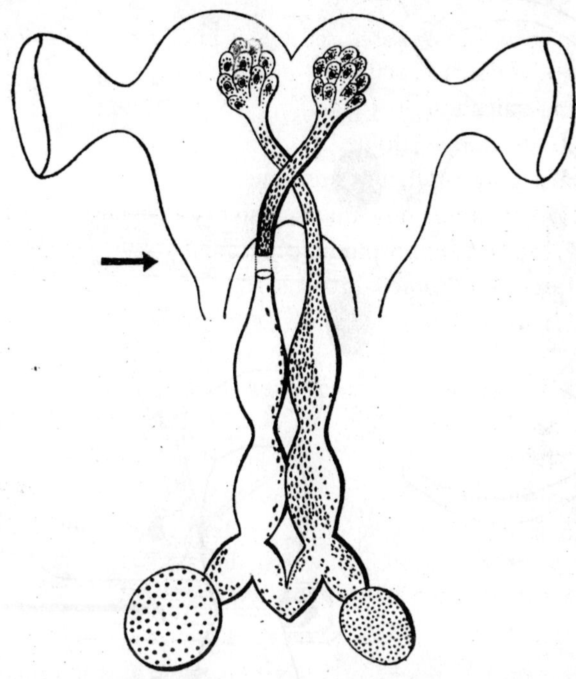

Fig. 7.3. The effect of breaking one of the two nervi corporum cardiacarum. Note the proximal accumulation of neurosecretory granules in the nerve stump, its disappearance from the corpus cardiacum and the enlargement of the corresponding corpus allatum.

hormonal control of pupation in Calliphora larvae by his well-known ligaturing experiments. Similar onclusions were reached by Bounhiol (1938) from his experiments. Similar conclusions were reached by Bounhiol (1938) from his experiments with Bombyx mori; by Bodenstein (1938), on the basis of several years' experimental work on the transplantation of appendages from young to fully grown caterpillars; by Kuhn and Piepho (1940) for Galleria mellonella, and by Hadorn (1939), Hardon and Neel and by Hadorn and B. Scharrer (1938) for Drosophila. Since then the number of papers on endocrine control of moulting has increased in number.

Fukuda (1944) was the first to prove that the AH does not affect the moulting process directly but does so by controlling the secretion of the prothoracic glands. These findings were confirmed and closely examined by Williams (1952) and his colleagues.The suggestion of the possible function of the corpora cardiaca in neurosecretion was initially made by De Lerma (1933, 1934) and

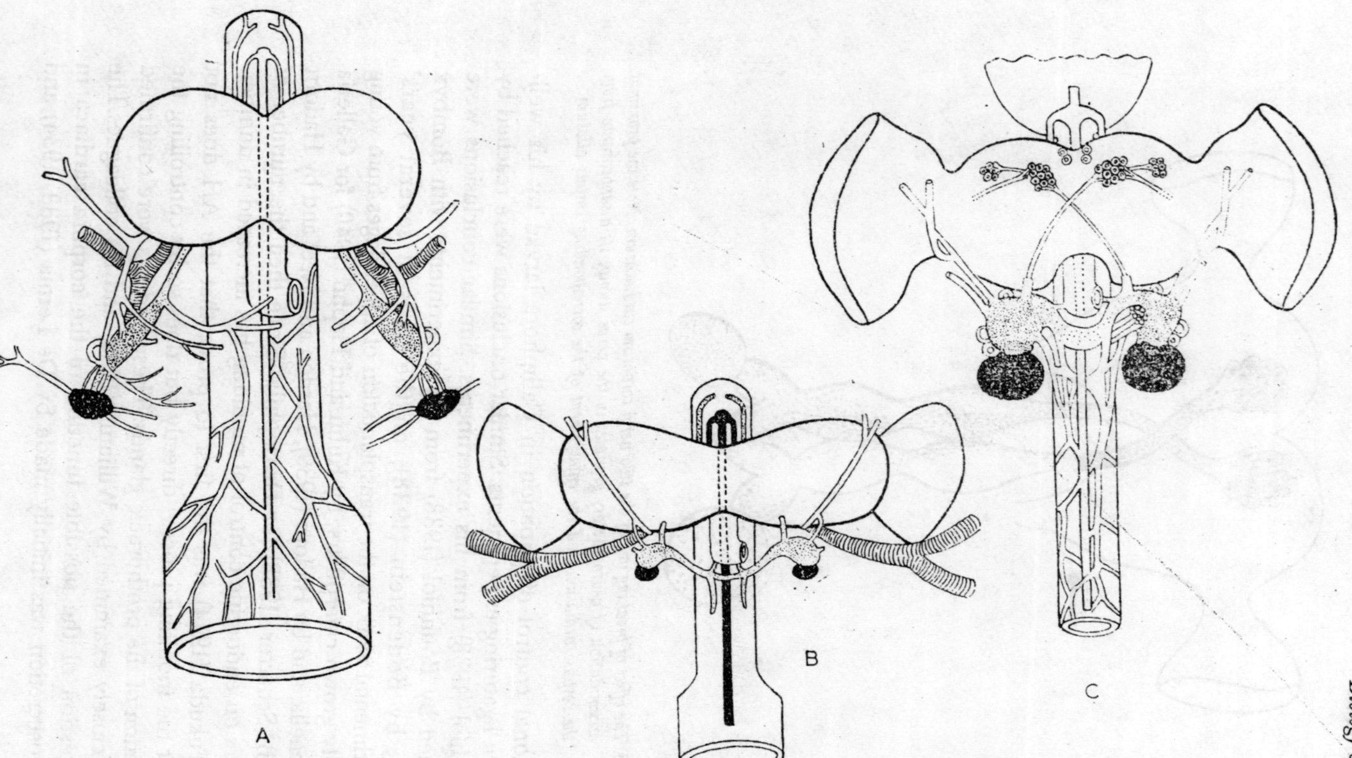

Fig. 7.4. Scheme of the brain and retrocerebral endocrine system in three developmental stages of Lemon Butterfly (Papillio demoleus). A–Last instar larva; B–Early pupa; C–Adult corpora cardiaca (dotted), corpora allata (black).

its connection with the neurosecretory cells of the brain was elucidated by Hanstrom (1942), by Pflugfelder (1937, 1938c, d) and, particularly, by B. Scharrer (1937, 1941) and B. and E. Scharrer (1937, 1944) who confirmed it by a detailed histological and experimental study, that the neurosecretory cells of the brain with their axons, together with the corpora cardiaca, form a single neurosecretory system corresponding with that of the neurosecretory cells of the brain with their axons, together with the corpora cardiaca, form a single neurosecretory system corresponding with that of the neurosecretory cells of the hyopthalamus with their axons and the neurohypophysis, in vertebrates.

Since that time the neurosecretory function of the insect brain and corpora cardiaca, and its connection with moulting and metamorphosis, has been found whenever it has been seriously looked for, that is, in practically all the chief orders of metabolous insects (pterygota). There is also histological evidence for its existence in Apterygota. Number of resarch papers by Gersch and his school on the chromatophorotropic and myotropic activity of insect brain extracts has raised the question as to whether some of the newly discovered (e.g. neurohormone C and neurohormone D of Gersch and Unger, 1957a; cf. 1960) could be identified with AH. This problem has been discussed compreensively by Raabe (1959). The possibility of the neurohormone D and AH being identical was suggested by Gersch (1962) on the basis of experimental evidence.

Recent investigations by the same author and his colleagues showed that the activation hormone is different from neurohormone D, and that it is composed of two components, with-different effects. These are the activation factor I (AH_1), which stimulates the synthesis of every kind of RNA, and the activation factor II (AH_{11}), which raises membrane potentials and thus has a positive effect on permeability (Gersch et al., 1969; Gersch and Sturzebecher, 1968, 1972; Gersch et al., 1973; Baumanna and Gersch, 1973). Both effects seem to be quite general in character, so that they might account for most of the specific AH effects, e.g. stimulation of the secretory activity of various glands metabolism, water balance, diapause development. Nevetheless, it is yet probable that AH_1 and AH_{11} could be two separate neurohormones produced by different nsc.

A number of other functions have been attributed to many known nsc, both of the brain and singel ganglia of the ventral nerve cord; these will be discussed later. B. and E. Scharer (1944) came to the conclusion that a close analogy exists between the neurosecretory system of three large groups of animals, Crustacea, Insects and Ver tebrates, are in full agreement with such a possibility. They confirm the earlier views of the above mentioned researchers, Hanstrom (1939), and others. The relationship between AH production and secretory activity of hte corpora allata, which regulates the activity of the ovarial follicle cells (studied in detail by Johansson, 1958), is in agreement both with the general nature of the AH effects and with the features of neurosecretory system common to the three groups of animmals mentioned.

The Neurosecretory Cells of the Brain (NSC)

Morphology

The neurosecretory cells of the brain, the source of the AH, are usually arranged in two groups, placed symmetrically on the upper surface of each hemisphere near the median furrow in the parts intercerebralis protocerebri. There are 4 to 15 or more main nsc in each group (4 in Bombyx mori, 6 in Calliphora larva, 7 in Pyrrhocoris apterus, 7 to 8 in various Hymenoptera, 15 in cockroaches and up to several hundreds in Orthoptera). They are often quite visible in a living brain dissected in Ringer because of thier milk-white or slightly bluish colour. The transparent nucleus often appears as a dark spot in the middle of each cell. The cells are more obvious when examined under dark-ground illumination, appearing shining white against the dark background. The axons of these two groups of cells form a pair of nerves-the nervi corporum cardiacarum 1 (interni). These two nerves cross inside the brain in the mid-line and each enters the corpus cardiacum on the opposite side the body. External to each of the two above-mentioned groups of NSC is, in most insects, a smaller group, usually of 2 to 4 cells, the axons of which form the nervi corporum cardicarum II (externi), each nerve entering the corpus cardiacurn on the siimilar portion of the body.

The neurosecretory material has been detected in histological preparations and under dark-ground illuminationg. Severing the nervi corporum cardiacarum led to an accumulation of nurrosecretory material in the distal end of the portion connected to the brain and the disappearance of granules in the end on the other side of the cut. Resembling discovery was made by E. Thomesen (1954), using dark-ground illumination, when the cardiac-

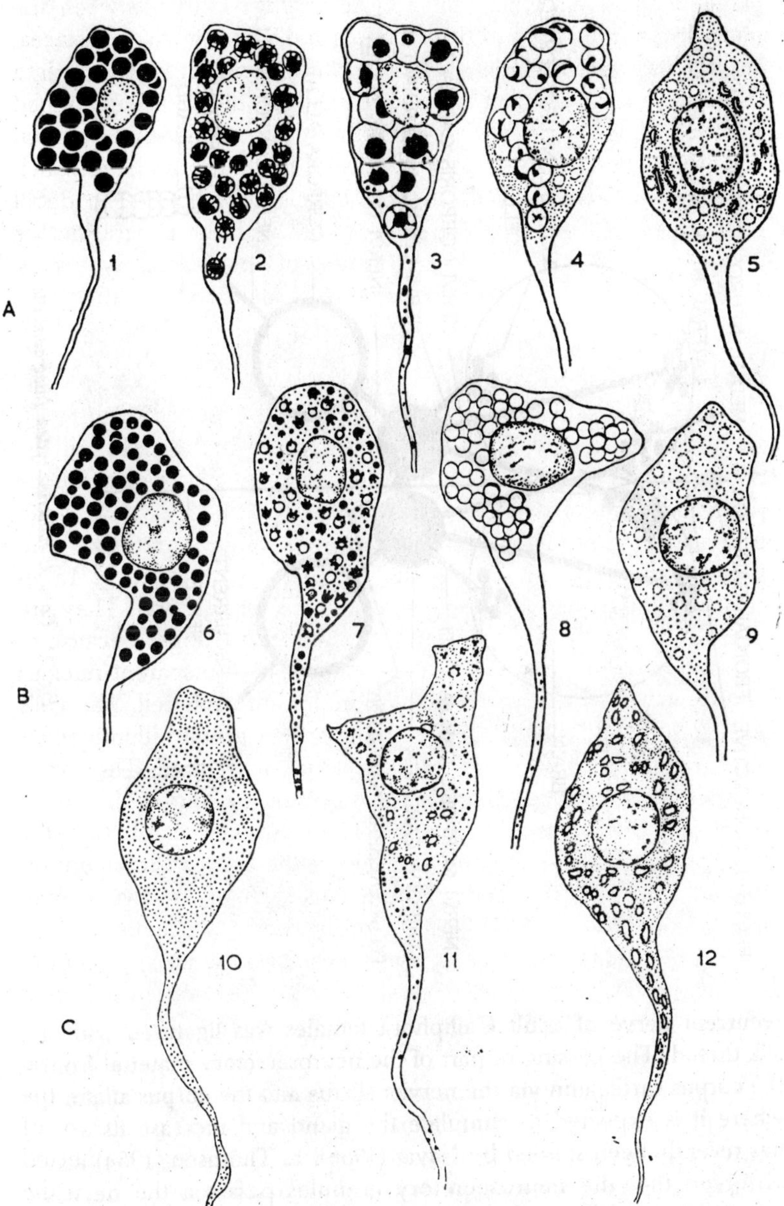

Fig. 7.5. Scheme of developmental changes in three types of neurosecretory cells of the pars intercerebralis in Lampyris noctiluca. A–nsc with large granules; B–nsc with medium sized granules; C–nsc with fine granules.

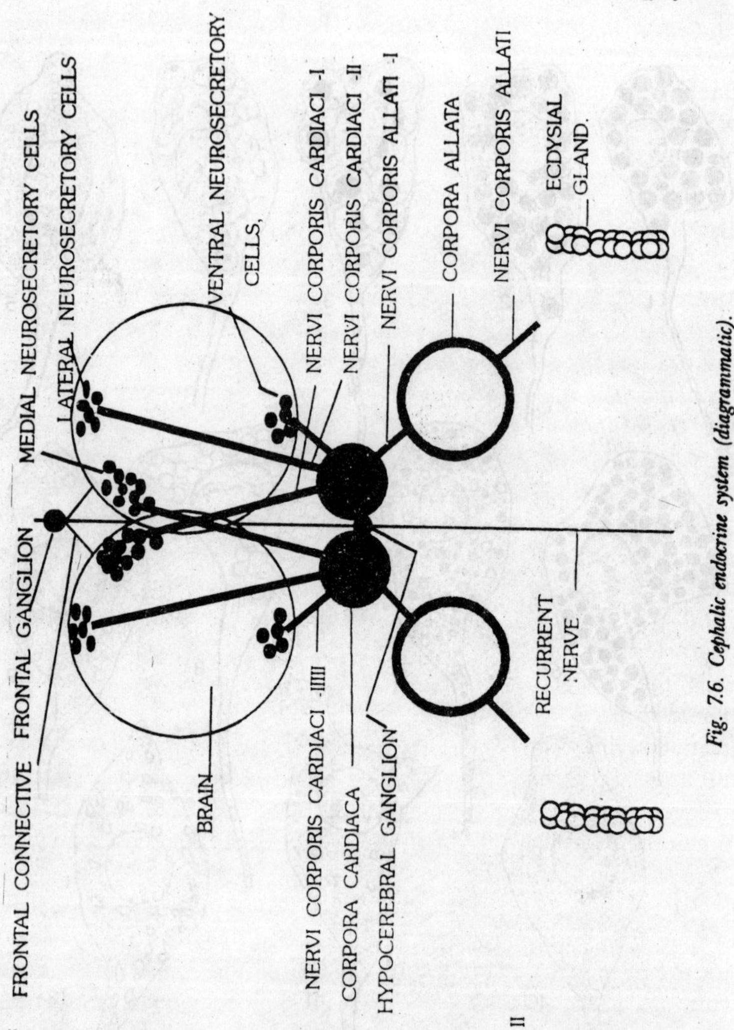

Fig. 7.6. Cephalic endocrine system (diagrammatic).

recurrent nerve of adult Calliphora females was ligatured with a silk thread. The passing of part of the neurosecretory material from the corpus cardicaum via the nervus allatus into the corpus allatum, where it is expected to stimulate the gland and increase its size, has recently been studied by Nayar (1958). E. Thomsen (1954) also observed that the neurosecretory granules pass via the nervus recurrence into the walls of the dorsal aorta, which is innervated by this nerve. From here the neurosecretory granules appear to more directly into the blood stream. In many circumstances distinct cyclicity of the amount of neurosecretion was detected. For example Baehr (1969) found accumulation of neurosecretory granules in one of the three different types of A- nsc in about 12 min after a blood meal and their release at the of ecdysis, immediately after sucking blood and during egg laying.

Histology

The use of special staining techniques, such as Gomori (chrome haematoxylin-phloxin;) and Paradlehyde Fuschsin, renders the nsc exremely visible in histological prepartions and the neurosecretory materail may readily be followed along their axons. Using histochemical methods, a considerable amount of acid phosphase was found in the cytoplasm of most of the nsc in the bug Iphita limbata, whilst it was nonexistent in the cytoplasm of the other nerve cells. On the other hand, the phosphates content of the nuclei of the neurosecretory cells was lower than that in the neighbouring nerve cells. Based the basis of his detailed histological studies on the brain of the Iygaeid bug Oncopeltus fasciatus Dallas, Johansson (1958a, b) distinguished four different types of cells, the staining characteristics of which suggested a neurosecretory function. A-cells, the most conspicuous of all, stain dark purple-red or dark-blue with paraldehyde fuchsin and blue-black with Gomori chrome haematoxylin. They enclosed a large number of granules which full most of the perikaryon. The cells are pear-shaped or sometimes irregular in shape, and occur immediately beneath the neurilemma. When the livecells were examined under dark-ground illumination, their shining bluish-white colour makes them quite visible. Their axons form the nervi corporum cardiacarum I. B-cells stain green or blue-green with paraldehyde fuchsin and red with Gomori. They are somewhat smaller than the A cell and more irregular in shape. They are without granules and occur deeper in the mass of ganglion cells. C-cells stain purple-red with paraldehyde fuschsin and reddish with chrome haematoxylinphloxin. They are found in the same layer as the B-cells, are irregular in shape and contain a flaky cytoplasm. D-cells have a fine granular cytoplasm which stains plae purple-red with paraldehyde fuchsin and pale blue-black with chrome haematoxylinphloxin. They are to some exten larger than the A-cells. Removal of the A-cells resulted in reduced fertility, and delayed, but did not prevent, egg laying. After extirpation the corpus allatum bwcame smaller.

Baehr, distinguished seven different groups, used different letter for the individual types, and so did Girardie and Girardie (1966), and Joly (1970) who followed Girardie. Changes in the histological appearance of the neurosecretory system of the Cholorado Beetle Leptinotarasa decemlineata were similary detected by Schooneveld (1966-1972).

Ultrastructure

Because the ultrastructure fo the pars intercerebralis was first described by B. Scharrer (1962) in the cockroach Leucophaea moderae, it has been well-considered by various other researcher e.g. Willey and Chapman (1962) in Blaberus craniifer, Normann (1965) and Bloch et al. (1966) in Calliphora erythrocephala, and Girardie (1967) in Locusta migratoria, etc. Diverse types of cells and neurosecretion were described.

Embryogenesis

The lone information on the origin and development of the neurosecretory cells dring embryogenesis are available in the papers by Jones (1953, 1956a, b). He discovered that in the embryos of Locustana pardalina, after diapause when the mitotic activity has occured again, several large cells can be viewed in those places in the brain where typical neurosecretory cells will later occur. These cells have a particular connection with the function of the ventral glands of the head and with moulting. They are probably the promary active source of the hormone in the insect embryo.

Phylogenesis

There is plenty of research material on the source and evolution of the neurosecretory cells during phylogenesis. It has been the neurosecretory cells that are found in all insects including Apterygota and also in other arthropods such as Crustacea Xiphosura Chilopoda Symphyla in Annelida and, phylogenetically 'lowest', in Turbellaria Polyclada. Hanstrom (1940, 1953) presuppsed that neurosecretory cells originated with the so-called lateral frontal organs in the lower Crustacea, which lie in the epidermis quite separate from the brain. In the higher Crustacea they are very close to the brain and form the well-known pair of frontal organs. The neurosecretory cells of these are lined by their axons to the corresponding sinus gland at the side of the brain, where the secreted material is accumulated.

In Apterygota, they form two group of cells on the surface of the protocerebrum which are enclosed by a connective tissue membrane. From all these groups a nerve arises which passes through the brain in a typical chiasma and reaches the corpus cardiacum of the opposite side. The neurosecretory cells nad the granules in their axons give the same Gomori-positive reaction as the neurosecretory cells of Pterygota. The typical neurosecretory cells of Pterygota develop from these two groups by a process of inclusion into the brain mass during which the connective tissue membrane vanishes.

It is clear from the published research work that the activation hormone is phylogentically the oldest of the three metamorphosis hormones. As regards the origin of the neurosecretory cells of the pars inter- cerebralis, Clark (1955) derives their function from the original epidermis cells from which they have evolved as nervous tissues and accepts their secretory function to be primary and thier nervous function secondary. He indicates oints out that they are neurosecretory cells found in the most primative, and phylogenetically oldest parts of the brain.

The Corpora Cardiaca (cc)

Morphology

It was Lyonet (1762) who was the first mention of cc in the literature is without doubt that Lyonet (1762) in this work on the anatomy of Cossus cossus. They include of a pair of bodies situated immediately behind the brain, between the anterior end of the dorsal

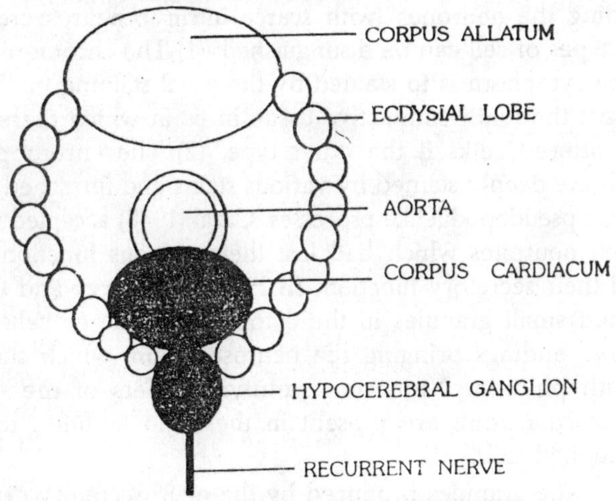

Fig. 7.7. Weismann's ring or ring gland in cyclorrhaphan larva (Diptera).

vein and the oesophagus, in front of the corpora allata with which they are linked by the nervi allati. They are sometimes fused medially and in some cases they are also fused with the corpora allata. In the ring-gland of Diptera Cyclorhapha, they form the lower median part of the ring, which is characterized by the small size of the cells, and fuse in the middle with the hypocerebral ganglion and at the sides with the R-cells. They are as mentioned earlier, innervated by two pairs of nerves from the brain, the nervi corporum cardiacarum I and II (interni and esterni) which are constituted respectively of the axons of the median and lateral groups of neurosecretory cells of the pars intercerebralis protocerebri.

The cc connected medially with the hypocerebral ganglion by a nerve bridge and with the suboesophageal ganglion by the nervi allati. They send one or paired visceral nerves along the digestive tube to the visceral ganglia. The structure of the nervus allatus in Gryllodes sigillatus was described by Awasthy (1968) and in Gryllus domesticus by Belyaeva (1966), and by Theodorescu and Novak (1973). The cc is composed of two parts, a nervous part (originally, cc were supposed to be normal ganglia, the pharyngeal ganglia'), and a glandular part. They differ, at first glance, from ca and from all nervous ganglia by their bright, milky-white, opalescent colouration remembering that of the A-nsc.

Histology

Including the neurones (with scarce intrinsic neurosecretion) two other types of cell can be distinguished: (1) The chromophobic cells whose cytoplasm is to stained by the usual staining methods. They consist the main connective tissue in cc in which there are, usually, scattered ceiks if the other type. (2) The chromophilic cells which are deeply stained by various stains and furnished with characteristic pseudopodia-like processes. Cazal (1948) accepted them as modified neurones which had lost their nervous function and intensified their secretory function. In Plceoptera, Arvy and Gabe (1954) found small granules in the cytoplasm of the cc cells and in the nerve endings bringing the neurosecretion, which stained deeply with phloxin. Numerous ramifying fibers of the nervi corporum cardiacarum are present in the cc in addition to the alone-mentioned cells.

Some of the granules produced by the neurosecretory cells of the brain are stored here, but the remainder reach the surface where they are washed off by the haemolymph. A small amount of neurosecretory matrial has been shown to pass through the cc into the ca, via the nervi allati in several species. Besides to the neurosecretory granules, Nayar (1956, 1957b) found a secretion, which he did not accept to beleive to belong to neurosecetory origin, in the chromophilic cells of Iphita limbata. De Lerma (1956) also advocates the existence of a special secretion of the cc in Hydrous piceus. It occurs in the form of homogeneous acidophillic inclusions which are somewhat abundant in the chromophilic cells of the cc. No intracellular Gomori-positive granules' were found within these two types of corpora cardiaca. Arvy ad Gabe (1953a, b, 1954) also insisted on the occurrence of a true intracellular neurosecretion in the neurosecretion in the cc on the basis of their great knowledge of many group of insects. Johansson (1958) was able to differentiate three different types of celll in the cc of Oncopeltus fasciatus. Couple of these are large chromophilic cells which occurs in different parts of the cc and differ in their staining characteristics with respect of paraldehyde fuchsin. Those of the frontal and lateral portions of the gland stain green where as those of the hind and median portions take on a pale red colour. The third type are interstitial chromophobic cells. These are dispersed through-out cc but are most abundant medially. It is remarkable that Johansson did not find any accumulation of neurosecretory material in the cc of Oncopeltus, where the granules are stored only in the walls of the aorta dorsalis as in Iphita limbata changing from the condition in Rhodnius prolixus where they are stored in the cc.

In Hydrocyrius columbiae granules are also stored in the walls of the aorta as they are in Pyrrhocoris apterus. Phylogenetically, it is apparently a secondary utilization of the tissue which capacites the neurosecretory material to pass into the haemolymph at a quicker rate. Mayer and Pflugfelder (1958) discovered only two kinds of cells in cc of Carausius morosus: the osmiophilic cells containing granules of neurosecretory material and generally forming well-delimited groups in the corresponding part of the gland; and the osmiophobic cells containing a system of vacuoles, special lamellated granules and peculiar ablong mitochondria. In the nerve endings inside the gland, granules 500 to 2000Å in size were found, which conforms to those known from the neurohypophysis of vertebrates and are of neurosecretory origin. M. Cazal (1971) studied the structure of the cc in Locusta migratoria in detail, while Divakar and Novak made an exact study of thier morphology and histology in Schistocerca gregaria. Their histology during the penultimate and last larval instars, and the pupal instar in Galleria mellonella were indicated by Karachali and Novak (1973).

Ultrastructure

A much of the research to papers has been published on the ultrastructure of the cc in various insects species, the first being those by Pflugfelder (1958) in Dixippus morosus and by B. Scharrer (1963, 1968) in Leucophaea maderae. Other authors who examined the ultrastructure of the cc were Johnson (1966) in Calliphora stygia, King, Surrinder, et al. (1966) in larval Drosophila melanogaster, King, Aggarwal et al. (1966) in adult females of the wild type and a mutant of the smae species, Normann (1965) in Calliphora erythrocephala, M. Cazal (1971), in Locusta migratoria and Cassier and Fain-Maurel (1971), in the same species, in is probably the most comprehensive and recent study on this subject. There are two methods in which neurohormones can be released into the circulating fluid, Cassier and Fain-Maurel regard the one which attributes the passage of intact granules by exopinocytosis as the more likely in Locusta. They observed no fragmentation of the granules nad found four types of neurosecretory granules.

Function

Johansson (1958) assumed accepted the cc to has two functions: the of the products of diverse groups and types of neurosecretory cells, and in addition, the production of a separate hormone of their own. A similar suggestion was made by Cameron (1953) previously he demonstrated that the cc produce a special active substance (orthodiphenol) which is quite different from neurosecretion and affects the heart-beat and peristaltic movements. Such a result was obtained by Altmann (1956), who examined the myotropic effects of corpora cardiaca extracts of the bug Iphita limbata on the gut peristalsis of another bug Aspongopus janus Fabr, and their chromatophorotropic effects on the red pigement cells of the decapod Coridina laevis Heller. The neurosecretory cells of the brain of the same species had no such effect. ohansson's (1958) experiments on the influence of the cc on reproduction failed to give postive results. Recent discovery clarifies as to the intrinsic endocrine activity of the cc both in their glandular and their neurosecretory (intrinsic nsc) components, also as their neurohaemal function for the brain neurosecretory cells. These three functions are perhaps all mediated by more than one active substance.

In addition to AH_I and AH_{II}, cc were discovered; by many authors, to contain the following substance; polypeptides with myotropic and chromatophorotropic activity (e.g. the neurohormones C and D of Gersch and his school), hyperglycaemic and hypertrehalosaemic action produced by extracts of both neurohaemal and glandular components (so that at least two different substances are involved), substance affecting nervous function, substances having both a diuretic and an antidiuretic effect on water metabolism (most of them undoubtedly of peptide nature, see M. Cazal, 1971) and a number of pharmacodynamic substances such as serotonin, orthodiphenol, etc More information is required to identify all these substances and to consider their activity spectra.

Embryogenesis

The functioning of cc during the embryonic period was studied by Weismann (1926) and Plugfelder (1937) in Carausius morosus; by Mellanby (1936) in Rhodnium prolixu, by Roonwal (1937) in Locustana paradalina; by Poulson (1937) in Drosophila, and by several other research. Their findings, cc, together with the hypocerebral ganglion, originate as unpaired dorsal evaginations of the oesophagus. Only later are they are separated and their cells differentiated into nervous and secretory ones. Initially, they are espected to have been normal visceral ganglia which secondarily attained a neurosecretory function and became innervated by the nervi corporum cardiacarum.

Phylogenesis

The cc are well formed in the Apterygota with the possible probable exception of Collembola (cf. Cazal, 1948). They are more developed in this group than the ca and have been described in Hapygidae and Lepismatidae. They are rather reduced in Campodeidae and are found in other Diplura and in Thysanura. It is probale that cc have developed during the phylogenesis of Apterygota, but earlier than ca. If Cazal's (1948) obsevation that the corpora cardiaca are absent in Collembola is confirmed, this will agree with the conclusion of B. Scharrer (1952) who comfirmed it from experimental evidence, that the cc merely store temporarily the secretion from the neurosecretory cells of the brain and are not necesary for the action of the *activation hormone.* Thus, for instance, transplantation of a brain with active neurosecretory cells may break diapause in cecropia pupae without simultaneous transplantation of the cc. There is as yet no specific evidence for the presence of a special cc hormone as was stated by L'Helias (1956b).

The Direct Effects of Activation Hormone (AH)

AH is more complex in contrast with other hormones. It is difficult to determine whether particular action of AH is related to that of hte other brain neurohormones. This is often unasweable problem, so that the present classification is arbitrary and is based on the author's opinion, but supported by the date on AH_1 and AH_{11} characteristics.

The activation of the prothoracic glands

The first rescarch papers by Kopec (1917) and Wigglesworth (1934, 1940) pointed out the necessity of the brain for moulting, pupation and metamorphosis. The discovery by Fukuda (1940) of the function of pgl in *Bombyx mori* demonstrated that the effect on moulting is an indirect one and depends upon the activation of these glands. This was confirmed by Williams (1952), who found that active pgl alone are able to induce moulting in isolated pupal abdomens of cecropia while an active brain can only produce the same effect in the presence of (inactive) pgl. Wigglesworth (1952b) came to the same conclusion with *Rhodnius prolixus* and by several other authors with various insects. Apparently some of the recent finding seem to contradict this. For example, Johansson (1958), discovered that complete extirpation of all ten of the type A neurosecretory cells of the pars intercerebralis in *Oncopeltus fasciatus*, carreid out not more than four hours after the last larval moult, did not prevent imaginal moulting, although this process started two or three days later than in the controls.

Undoubtedly, this is because of the secretion of the A-cells, which is stored in the form of granules in the walls of the aorta, as mentioned above, and, also may be, in the A-nsc of the ventral nerve cord. According to Raabe (1959b), the way in which pgl are stimulated by AH has not yet been explained. A possible explanation was suggested by Church (1955), who considers it might be directly concerned in the syntesis of ecdysone. The findings of E. Thomsen and I. Moller on the influence of AH on intestinal protease activity in *Calliphora erythrocephala*, as well as other effects, would however suggest an action or more ordinary nature, as would also its relationship with the *antidiuretic hormone* of the vertebrate hypothalamus.

The activation of the corpora allata

AH's effect on the function of ca is now adequately demonstrated. AH is necessary for the reactivation of ca in a manner similar to that of pgl. This was suggested in some of the earlier accounts and is quite evident from the work of Johansson (1958a, b) de Wilde (1958a-c) and his collaborators, Nayar (1957). Johansson (1958a) observed the effect of extirpation of the A-cells of the brain on the volume of ca and found that complete exitrpation of these cells results in extraodinarily small ca. The same process was observed by E. Thomsen (1952) in Calliphora only a few A-cells are left, the ca volumes reamin normal. The proper understanding of how the ca are activated by AH has proved difficult for two reasons: (1) The action of the ca seems to depend upon nervous stimuli; (2) the activation hormone reaches the ca not only through the haemolymph, but also directly from the brain, in the form of granules, via the nervi corporum cardiacarum cc and nervi allati. Niether of these factors appears to play a primary role in the regulation of ca activity, except in the case of diapause (see below), but such has been attributed to both by different authors. The activation of ca produce the primary effect of the activation hormoneon the ovaries.

The effect on ovarian development

Including its indirect effect on ovarian development through the activation of the ca, the existence of a direct effect of AH on the development of eggs has been hinted. A detailed examination of the endocrine mechanisms controlling ovarian development in adult *Calliphora erythrocephala* showed that removal of the neurosecretory cells of the brain can be extirely compensated for by the implantation of active nsc from another specimen but not by active ca. The effects of AH deficiency are different from those of allatectomy. Implantation of cc from an active specimen has a similar, though lesser, effect than the reimplantation of neurosecetory cells. The cc alone in this species. Thomsen concludes that the function of AH in the insect organism is quite complicated and that the nsc of the brain form an 'all-over controlling centre of the endocrine system'. The removal of median nsc from the ovaries Indian Housefly *Musca nebulosa*, the ovaries failed to develop.

The effects were not similar with those of allatectomy; instead of the hypertrophy of the fat body which follows allatectomy, the fat body was reduced in size after nsc removal compared with the accessory glands, and the accessory glands were devoid of secretion. This is probably absence f the general stimulant effect of Ah. No yolk was deposited after median nsc removal in the flesh-fly Sarcophaga bullata, although yolk deposition started after their reimplantation, but not after the implantation of ca or of brain tissue minus the nsc. Removal of brain in female Bombyx mori pupae often caused the laying of diapause, instead of non-diapause eggs, which were distinguished by low carbohdrate and lipid concentrations. The impact of the nsc of the pars intercerebralis on the development of the ovaries has been.

Highnam (1962) has demonstrated that in *Schistocerca gregaria* they exert a positive control over occyte development and that copulation as well as electrical stimulation, drasrtic wounding, or enforced activity may all bring about a release of material from the nsc of the pars intercerebralis and accelerate the development of the terminal occytes in 14-day-old females reared without males. Hill (1962) found that in the same species a there is positive control of haemolymph protein concentration by neurosecretion during ovarian development. The presence of mature males to accelerates the release of material from the female neurosecertory system, resulting in rapid development of terminal occytes and an increase in the number of eggs. A positive effect of the pars intercerebralis on the development of the ovaries, as well as on postembryonic development, on the function of the ventral glands and ca, and on metabolism, water balance, and chromatic adaptation was described by Girardie (1964) in *Locusta migratoria.* Mordue (1965a-c) too, obtained similar result with adult females of *Tenebrio molitor.*

The effect on the accessory glands

The impact of cerebral neurosecretion on the male accessory glands, initially discovered by Wigglesworth (1936), and since then confirmed by various researcher working on other species of insects is, as in the case of the ovaries, primarily an indrect one through activation of the ca. No direct effects have obeen till now observed, though they may be assumed.

The control of diapause

As indicated by E. Scharrer (1952), the author of the concept of neurosecretion, the specific role of neurohormones in the animal body depends on a long-term co-ordination between the nervous system and the ininternal secretion. A classic example of this interdependence is the direct influence of AH on both larval (pupal) and imaginal diapause. In each case it is a nervous stimulus induced by external conditions which induces a temporary interruption of AH production whcih in turn results in the inhibition of the functions concerned. As demonstrated by Williams (1948) and several other entomogists a stage corresponding to natural diapause can be caused by the removal of the brain, the source of AH.

Similarly, a natural diapause can be broken at any time by implanting an active brain or other source of AH (corpora cardiaca or corpora allata; see below). It seems that this dependence is the result of a phylogenetic utilization and improvement by diapausing insects of a tendency present in all insects. This results in the control of AH production in response to a variety of unfavourable environmental conditions and it is highly adjustable. Thus the process of diapause generally occurs in the winter period in temperate regions, whereas it is common in the hot dry season in the tropics and subtropics. The same is true of the dependence of AH production on the nervous impluse produced by the distension of the abdomen in blood-sucking insects as described by Wigglesworth (1936, 1948) and by Detinova (1954). The condition in blowfly larvae first observed by Cousin (1933) may be assumed as an transitional stage in the developing process diapause.

The effect on morphogenesis

In the experiments with the cricket Pteronemobius haideni, Sellier (1956) found that the implantation of an active brain into the larva (VIIIth instar) at the start of the diapause not only in suppression of diapause, but also in a lengthening of the wings when they become adults. But there were no changes when the implantation was effected at a more advanced stage in diapause. The implantation of brains from other species of the genus *Gryllus* gave similar results. However, this seems to be much more an indirect effect depending on the time of activatoin of the ca. The entire problem of both direct and indirect effects has been discussed by Gilbert (1964) and Buckmann (1969), but many questions are yet unanswered.

The effect on the fat body

An observation made by E.Thomsen (1952) proves that the fat bodies of *Calliphora* females, whose AH source had been removed, had a higher glycogen content though the quantity of fat was reduced. The glycogen content corresponded with that following allatectomy but the reduction in the quantity of fat was grater.

The effect on intestinal proteinase activity

The two entomologist known as, E. Thomsen and I. Moller (1959a, b) studied the effect of extirpation of the neurosecretory cells of the pars intercerebralis on intestinal proteinase activity in Calliphora females. The colorimetric method of Day and Powning was used to determine this activity in gut homogenates. The proteinase active in the operated specimens was 5 to 8 times lower than in the control. The effect of the removal of AH was identical with that caused by lack of proteins in the food. As proteinases are themselves proteins, their production may be considered as protein synthesis in the gut epithelium cells, and the researchers conclude that AH affects protein synthesis in gneral. This, But this, is not the only possible explanation–a more general activation effect corresponding to that observed in other tissues, such as, for example, some kind of membrane activation, would be equally satisfactory. Corresponding results were reached independently by Strangways-Dixon (1959, 1961) using the same species. He showed that by feeding flies selectively on sugar and protein that when deprived of their neurosecretory brain cells the flies digested no proteins even when forcibly fed on them for six days; their ca and their eggs showed no increase in size.

The effect on the hind midugt esterases

E. Thomsen (1966) studied the esterase distibution pattern in the hind midgut cells of *Calliphora erythrocephala* females by applying 5-bromoindoxylacetate (as substrate) to fresh tissue. The enzyme appeared in the form of granules, filaments and 'caps' similar to those described by Wigglesworth (1958) in Rhodnius prolixus. Generally in meatfed flies, the pattern of enzymes in contact with fat droplets followed a cycle correlated with ovarian development. In flies fed on a protein-free diet, the esterase level was low. The same fining was made in flies deprived fo their medial nsc, even when they were fed on meat. It was comfirmed that the medical nsc control, or influence, esterase production in the cells.

The effect on on water balance

Altmann (1956a, b) experimented on studied the effect of extracts from different endocrine glands on the intake, retention, and excretion of water by the honeybee. Injections of ca extracts increased water consumption and decreased the viscosity of the haemolymph; cc extracts increased haemolymph viscosity and decreased water consumption. Excretion was increased by ca extracts and decreased by extracts of cc. Injections of adrenaline produced similar effects to corpora cardiaca extracts. The effects of the corpus allatum extracts could be due to juvenile hormone or to neurosecretory material reaching the gland via the nervus allatus; the latter is more probable.

In *Iphita limbata*, when distilled water was provided for drinking for some time an increased amount of neurosecretory material was found in the walls of the aorta, and nsc were practically free of granules. While in the insects which were forcibly fed on salt water for the same length of time, the release of neurosecretory material through the walls of the aorta almost stopped and the neurosecretory cells were filled with granules. There were similar results were obtained with specimens where the wax layer of the epicuticle was removed by washing with benzene or chloroform. In insects kept in an atmosphere dried with calcium chloride, neurosecretory material accumulated in the nsc; in the those kept in a more humid atmosphere, the material was released and enlarged quantity of granules was found in the walls of the aorta. Same result were acquired by Gutmann and Novak in Pyrrhocoris apterus. It may be concluded that nsc enable the animal to maintain the water balance within certain limits. The comparison with the antidiuretic hormone of the vertebrate hypthalamus is striking.

Raabe (1959a) reached to the same conclusion with *Carausius morosus* where the cc had the same effect as the neurosecretory cells of the brain. This agrees with the findings of Stutinski (1952a, b). He injected rats with extracts of the pars intercerebralis protocerebri and cc of the cockroach *Balbera fusca* and found that the urine was reduced in quantity in the same way as after the administration of pitressin. The opposite result was obained by Nunez (1956) using the beetle *Anisotarsus cupripennis.* There, the extracts of brain nsc and cc appeared to produce diuretic effects. These experiments, are yet to be analysed under comparable conditions.

Control of phase differentiation

Conforming with Girardie (1966) and Girardie and Girardie (1966), Joly (1970) assumed that phase differentiation in Locusta migratoria was controlled by special nsc in the pars intercerebralis. Three types of cells were recognized in the pars intercerebralis: A-cells, which are small and contain a large amount of Gomori-positive neurosecretion, B-cells, which are similar, but have highly phloxinophilic cytoplasm and no neurosecretory granules (A- and B-cells are regarded as being possibly two different stages of the same type of cell) and C-cells, which are much larger and contain only a few neurosecretory granules. But, no comparison with various types of cells in other insects, similarly experimented by other authors, is available. Destruction of the C-cells by electrocagulation had the same effect as ca extirpation. If performed at the beginning of the IVth (penultimate) larval instar, the result would be a well-developed adult. Destruction of the A- and B- cells had the reverse effect, i.e. it acted like ca implantation. If carried out at the beginning of the last larval instar, a green adultoid developed. The implantation of extra A- and B- cells acted like allatectomy if carried out at the beginning of the penultimate instar. The researchers accepted that the C-cells activate the ca under normal conditions, while the A- and B-cells inhibit them. There is, however, a simpler explanation, consistent with findings in other insects, which has not been sufficiently ruled out, i.e. that it is injury to the C-cells which inhibits the ca in the first case. In the second case, removal of the A- and B-cells would prolong the interecdysial period of the last larval instar, so that the individual's own JH would, unlike in normal development, remain active; while the implantation of extra cells in the IVth instar would shorten the intermoult period, with the result that the ca would not attain an active concentration in the penultimate instar. Joly's inability to find any difference between ca ultrastructure in normal specimens and those with destroyed A- and B-cells supports the second theory.

Other effects

Many Confirmed effects of AH, as well as the possibility that they may be due to a common and very general action, leaves no doubt that further research will reveal a number of other effects. It also remains to be determined whether some of the observed effects of brain and cc extracts, such as the various chromatophorotropic and myotropic actions, are due to AH or to other neurosecretory products. There is little experimental evidence for any definite information on this connection.

The Indirect Effects of AH

Probably a hormone may act both directly and indirectly on the same tissue, e.g. it may stimulate RNA production by the tissue directly and at the same time induce the production of another hormone stimulating tissue growth or function. Another aspect to be considered is that the direct or indirect effects may be of different degrees. For instance, a hormone may act on the target tissue directly or may induce another hormone to do so, or the other hormone may cause a metabolic change in which only the tissue is affected. The control of the moulting process must be considered as an indirect effect of the activation hormone.

Moulting is restricted by removal of the AH source and re-evoked by its reimplantation. For this reason, the brain hormone was originally described as the moulting hormone by Wigglesworth (1934). It was called a 'growth and differentiation hormone' (partim) by B. Scharrer (1948) and Williams (1947, 1952a). The ca hormone was also initially interpreted by researchers as a moulting hormone. Confimating of an indirect as opposed to a direct effect is provided by the fact that the effect can be produced by another hormone but not by the given hormone alone. While a mere finding that removing another organ inhibits the effect is in no way proof of an indirect effect. In such a case there is always the possibility that the effect of the hormone was prevented by breaking another link in the chain between the hormone and the part of the body responding. With the metamorphosis hormones it is particularly important to differentiate between the direct and indirect actions of a given hormone. The interaction of thier effects is very complicated, as in metamorphosis and moulting. View, a revision of many of the false on the metamorphosis hormones and their mode of action is required.

The Control of AH Production

The functional cycles of neurosecretory cells. The existence of particular 'critical' periods in each instar as far as the hormonal activity is concerned was demonstrated experimentally shortly after the discovery of the first metamorphosis hormones. During these periods, the presence of the source of the given hormone is necessary for normal development whereas after wards it is no longer necessary. This shows that there is a certain cyclicity in the function of the AH, which has been confirmed by subsequent histological research on the neurosecretory system. The course of neurosecretion and the function of cc during postembryonic development has been studied in detail by Herlant-Meewis and Paquet (1959) in *Dixippus morosus.* They discovered a decrease in the amount of neurosecretory granules in the pars intercerebralis and cc at the time of each moult. Neurosecretory activity reached its maximum after about one-third of the next intermoult period had elapsed. Even at this time also, the largest amount of neurosecretory material was seen to pass into the corpora cardiaca. This period of maximum secretory activity agrees well with what has been observed regarding the critical period for AH release in other species.

The quantity of the neurosecretion in cc decreases makedly about half way through the intermoult period, i.e. when the critical period for the juvenile hormone has been reached. Another increase in neurosecretory material was observed in the last quarter of the intermoult period. Cyclical changes in the amount and character of the neurosecretory material were observed in Phasmids and other insects by a number of other authors while a report from Fuller (1959) discribes recurrent cyclical changes in the amount of secretion in the individual neurosecretory cells of the pars intercerebralis of Periplaneta americana without any noticeable cyclicity in the neurosecretory system as a whole. There have been suggestions about the factors which govern the changes observed in neurosecretory activity.

One of the first answers was provided by Wigglesworth (1936a, 1940a) in his classic experiments with the blood-sucking butg Rhodnius prolixus. He demostrated that production of the brain hormone and thus also the start of the moulting process, and each the interrelated developmental processes, is induced by a nervous stimulus produced by distension of the abdomen due to ingested blood. This stimulus depends on the quantity, not the quality, of the fluid intake, since the same effect is caused by a corresponding volume of pure water. Several small blood meals do not, however, produce any effect, even though thier sum exceeds the necessary minimum quantity. Severing the ventral nerve cord anywhere between the brain and the abdomen prevents the passage of this stimulus. Therefore it can be accepted as a typical-case of the transformation of a nervous stimulus into an endocrine impulse effected by the nsc of the pars intercerebralis protocerebri, as claimed by E.Scharrer (1952). A very similar neurosecretory effect was found by Detinova (1954) in Anopheles.

On the other hand, no such connection was observed byNovak (1951b) in the plant-feeding bug *Oncopeltus fasciatus* and many other insects where poorly fed specimens can undergo a normal moulting process. It is also recognized that meal-worms undergo several extra moults when left without food, and starving clothes-moth larvae can moult as many as 40 times. The above-mentioned dependence of AH production on a nervous stimulus from the distended abdomen is probably a secondary phylogenetic adaptation in blood-sucking insects. It enables them to survive for long periods in the absence of a suitable host in a state of reduced metabolism corresponding to that of a true diapause. In most other insects, AH producting is automatically renewed at the beginning of each instar with the passing of the first digested food into the haemolymph.

The phylogenetic character of this dependence on the stimulus from the abdomen was shown by Larsen and Bodenstein (1959) in the mosquitoes *Culex pipiens* and *Aedes aegpti.* Whereas *A. aegypti* and normal *C. pipiens* show the usual dependence of AH production (and through this juvenile hormone production which regulates egg development and oviposition) on the distension of the abdomen by blood. AH prodution by the autogenic form, *Culex molestus,* is free of this stimulus. An exhaustive study of the function of the neurosecretory cells in insects which feed continuously throughout their growth has been made by Clarke and Langley (1962) using Locusta migratoria L. No changes were discovered in the amount of neurosecretory material in the median nsc or in their axons (nervi corporum cardiacarum), or in cc during postembryonic development or at different times during the intermoult period. The neurosecretory material is apparently produced continuously throughout growth in this species.

As the AH is presumably not used during the ecdyses, the authors assume it accumulates in the haemolymph at these times. The increased concentration of the hormone in the haemolymph at these periods would then reactive agl. A very remarkable association between the frontal ganglion and the production of neurosecretion was also found. Its extirpation resulted in the complete inhibition of further growth and moulting. The same effect was attained by cutting the frontal connectives, whereas cutting the recurrent nerve or the ventral nerve cord in front of the first abdominal ganglion had no effect. Histological examination of neuroendocrine system five days after the operation revealed a marked accumlation of neurosecretory granules in cc, the nervi corporum cardiacarum I but scarcely any in the nsc. No neurosecretory cells were discovered in the frontal ganglion. The authors suppose that the forntal ganglion plays a part in transmitting the nervous impulses from stretch receptors in the oesophagus to the nsc of the pars intercerebralis which would thus connect the release of the hormone with the intake of food.

A momentary inhibition of AH production caused by various external impulses, the mechanism s of which are not yet entirely apprehencled, is the direct internal cause of all types of insect diapause except the early embryonic one. In addition to photoperiodism, temperature and the above-mentioned distension of the abdomen in blood-sucking insects, several other factors controlling AH production have recently been described. Therefore the inter-relationship between juvenile hormone production and the carrying of oothecae by female cockroaches, incorrrectly interpreted as a direct effect on corpus allatum secretion, is undoubtedly governed by neurosecretion. Similarly, neurosecetion, or AH production, appears to be the chief mechanism influenced by the quality the food as shown by E. Thomsen (1959) and Strangways-Dixon (1959); and the effect of feeding on protein synthesis is also controled in this manner.

Thomsen and Lea (1968) examined cyclic changes in the medial nsc of *Calliphora erythrocephala* under various conditions. The nuclei and nucleodi displayed cyclic changes in volume and in the amount of neurosecretion. Neurosecretory activity rose on the first day after emergence, with acceptance of a sugar diet; it then fell again but rose once more after the fourth day in connection with the beginning of meat eating and fell again after eff laying. Allatectomy reduced nuclear and nucleolar volume, but the implantion of active ca renewed neurosecretory activity in such specimens. The authors concluded that the activity of the neurosecretory cells was regulated by ca, but the effect is undoubtedly an indirect one.

Chemical Characteristics of the AH

The researcher also who was first to identify the chemical nature of the brain neurohormone was L'Helias (1955a, 1956a). Her pterinederivative theory suggested the existence of a close relationship between AH and JH on a pterine basis. A few years later, relationship between AH and JH on a pterine basis. A few years later, however, Gersch and Unger (1962) showed that pterines had nothing in common with either of these hormones and suggested, on the basis of earlier paper that neurohormones were possibly of a peptidenature. This accords with phylogenetic considerations on their relationship to vertebrate neurosecretory material, the polypeptide character of which has now been fully accepted.

The discoveries by B. Scharrer (1952) and others appear to speak equally against the hypothesis of L; Helias (1956) on the interaction of the brain secretion with those of cc or ca (1) ext\irpation of the cc has no qualitative effect either on moulting or metamorphosis; (2) the effects of brain extirpation may be at least partly compensated by the implantation of active cc (3) the careful histological studies on the neurosecretory activity of the brain and cc in some Hymenopytera and Diptera carried out by M. Thomsen (1953a, b) give support to the conclusion that the neurosecetory material passes through cc directly into the haemolymph. Few rearchers ignored the findings of a Gersch's school or phylogenetic relationships into account, however, and expressed entirely different views on the chemical character of AH and other neurohormones. For example, isolation of the active ingredient of the AH has been reported by Kobayashi and Kirimura (1958). They used 8500 silkworm (*Bombyx mori*) pupae preserved in methanol 24 hours after pupation and centrifuged three times after homogenization 200 ml of the methanol solution obtained were concentrrated to 30ml and extracted with 145 ml ethyl either. On evaporation, the either solution yielded about 2 mg of an oily yellowish brown material; the evaportaion temperature did not exceed 38C. A injection of this substance into decerebrated permanent pupae, in which no ecdysone has previously been found, caused these to moult to adults 16 to 20 days later.

It has been claimed by Kirimura et al. (1962), that the active principle of these extracts is cholesterol and its identity with AH was taken for grantend. This view has, not much assistance from other facts known about AH. Nevertheless, when correlated with the recent finding of Karlson and Hoffmeister (1963) that cholesterol is the precursor of ecdysone, it becomes of remarkably. Carlisle and Ellis (1963) injected IVth instar nymphs of *Locusta migratoria* migratoriodies with cholesterol dissolved in either. The purified perparation and one of two samples of commercially available crystalline cholesterol were ineffective whereas the other commercial sample, hastened moulting by about 18 hours (PL V. 01). The authors found that pure cholesterol as such has no prothoracotropic effect, which they assume to be derived from impurities present in the sample concerned. They conjectured that the active ingredient is a steroid related to cholesterol, and that the natural brain hormone is similarly a steriod of this group, or possibly a triterpenoid of related configuration.

Even Gillbert (1964) expressed his opinion in favour of the steriod character of the brain hormone in his review of the question. He also agreed with the earlier authors in assuming 'a number of neurosecretory substances produced by the brain, the prothoracotropic effect being produced by one of the them and the other effects of the brain neurosecretion by others. Another researchertook a different stance Novak agreed with Gersch (1962, cf. 1964) in his assumption of the polypeptide nature of AH and claimed that most of the known effects of the brain secretion could will be produced by one and the same substance influencing membrane activity and thus the water metabolism and secretory activity of the cells. The only exceptions are neurosecretion from the lateral nsc, shown to be engaged in inducing circadian rhythms of activity and the effects of neurohormone C. the effect of cholesterol would be that of a vitamin supplying the steroid skeleton necessary for the production of ecdysone. This is necessarily lacking in decapitated or decerebrated insects unable to accept food. In such cases its supply by injection can reinduce the production of MH when at least a small amount of AH is present. but this, is not the case in late diapause when AH is absent.

Williams (1963, 1968), first thought that AH was hyaluronic acid or a related substance and later expressed the view that it might be a mycopolysaccharde. Neither of these hypotheses was confirmed by further investigations.

Contemporary research work by Gersch and his colleagues submitted definitive evidence in support of their original opinion. First of all they showed that the prothoracic gland stimulating agent is different from neurohormone D, previously described, and that it actually affects the prothoracic gland. Using electrophoretic separation on polyacrylamide gels, they further succeeded in demonstrating that the actual AH, i.e. the prothoracic gland-stimulating hormone consist of two components- one with high molecular weight, stimulating RNA synthesis, and the other, AH_{11} with low molecular weight, which raises the membrane potential and hence membrane permeability. Both components are of a peptide nature. However, It was discovered that Neurohormone D, is a peptide with a molecular weight of about 2000, to be thermo- and acid labile and to be decomposed relatively quickly by trypsin.

The Mode of Action of AH

The movement of the neurosecretory granules through the axons of the nervi corporum cardiacarum from the nsc of the pars intercerebralis protocerebri into cc is well known, but their role in cc and their passage from there into the haemolymph is yet unknown. Significant experimental evidence for the transport of AH from the cc to the pgl in Rhodnius by the haemocytes was produced by Wigglesworth (1956b). He showed that blocking the haemocytes by injecting Chinese ink, trypan blue or iron saccharate, the particles of which are phagocytosed, results in a significant delay of the next moult if the injection is carried out before the end of a specific critical period which corresponds approximately with that of AH. However, when at the same time he implanted active pgl or injected a sufficient amount of a solution of crystalline ecdysone, no such delay occurred. Wigglesworth concluded that under normal conditions the haemocytes transport AH from the cc to the pgl, or, it is possible, that they secrete a further substance for the activation of the pgl under the influence of AH.

As suggested by Wiggleswortha, other explanation are so possible, for example, that AH is adsorbed on to the injected material and removed with the material from the haemolymph by phagocytosis; or the haemocytes, damaged by phagocytosis, discharge some AH inactivating substance into the haemolymph. An alternative some AH inactivating substance into the haemolymph. An alternative explanation could be that, in normal insects, the role of the haemocytes is to phagocytose neurosecrtory granules, thus freeing the AH whilst digesting the carier substance. This would be in agreement with the observed occurrence of neurosecretory granules in the aorta dorsalis of various insects and with the observation of Hodgson and Geldiay (1959) who found that hyperactivity in *Blaberus cranifer* and, to a lesse extent, electrical shock treatments, resulted in bloods cells overflooding of all parts of the brain.

Hardly any information is one available regarding the mechanism of activation of pgl and other organs influenced by AH. But, it has been found by Williams (1952) that the same activating effect may by obtained by implanting another, active pgl. This is to be interpreted, together with Wigglesworth's (1957) conclusions, as meaning that continuous activity by AH is must for the pgl to provide an effective amount of moulting hormone. This is based on the discovery that the first change in the epidermis definitely attributed to pgl hormone commences about two days before the end of the critical period for AH, during which time the removal of the AH-source by decapitation stops the process of the moulting process.

The Moulting Hormone (Ecdysone) (MH)

The initial indication of the existence of a humoral factor controlling moulting can be found in the experiments of Kopec (1917, 1922), and later Wigglesworth (1934). During time of Wigglesworth, Fraenkel (1934, 1935) showed that similar hormonal factor, necessary for puparium formation, was present in the brain region of *Calliphora erythrocephala* larvae. Prior to this, however, Hachlow (1931), on the basis of his experiments with butterfly pupae (*Vanessa* io and *Aporia crataegi*), suggested the existence of a 'thoracic centre' which was necessary for development. Bodenstein (1933a, b, 1934) in his transplantation experiments with the legs of caterpillars got the same result.

The honour of being the first person to clearly distinguish between the hormone of the brain cells and that of a new hormonal source, the prothoracic glands, belongs to Fukuda (1944) who showed the importance of pgl for moulting by transplantation experiments in silkworms (*Bombyx mori*). After this, papers on the moulting hormone appeared at an increasing rate. The role of pgl, or the analogous peritracheal glands in lower Diptera, or the pericardial glands or ventral head glands in Hemimetabola, in the moulting processes has been demonstrated in all the chief groups of insects.

Earlier the research recognized and were mainly concerned with the importance of MH for inducing the moulting process, but most of the later investigators stressted on its indispensability for growth and morphogenesis and used for it the less suitable terms 'growth and differentiation hormone', or 'metamorphosis hormone' (see above). A certain amount of confusion appears to have been caused by the discovery of the effects of the ring gland in flies, which is a composite structure that contains the sources of all the three metamorphosis hormones also, the role of MH in diapause and imaginal differentiation in Cecropia pupae as resolved by Williams seemed at first to support the concept of the growth and differentiation concept.

The most significant stage in MH invitations, after its separation from AH by Fukuda, was its isolation in crystalline form by

Butenandt and Karlson (1954) from extracts of silkworm pupae. A significant requirement for this was the discovery by Becker and Plagge (1939) of a suitable test organism and a quantitative measure for judging concentration, the so-called Calliphora-unit. This method was improved by Butenandt and Karlson in their isolation experiments. While the specificity claimed by Williams (1951a, b) for the so-called spermatocyte-test appears to be questionable in the light of the findings of Laufer (1960). Becker and Plagge (1939) were also the first to show the broad inter -Order non-specificity of MH, which was confirmed by Wigglesworth for such widely separated orders as Hemiptera and Diptera. Another significant step was the successful extirpation of the ventral glands in migratory locusts by P. Joly, L. Joly and Halbwachs (1956).

The theory of the pgl hormone as a MH, which corroborates with the original findings of Kopec and Wigglesworth (1934) as well as with the gradient-factor theory of the author, has been approved by two independent pieces of work: Halbwachs and Joly (1957) showed, using Locusta migratoria, that transplantation of the pgl accelerates the moulting process without utilizing any positive effect on growth differentiation; Luscher and Karlso (1958) observed moulting but no and growth or imaginal differentiation following the injection of a large quantity of ecdysone into the nymph of Kalotermes flavicollis.

These discoveries have been confirmed by many other researchers. For instance, Zdarek and Slama (1972) proved that the injection of a large dose of ecdysterone at the outset of the last larval instar in *Calliphora* and *Sarcophaga*, inhibits morphogenesis and results in a supernumerary larva, whereas small amounts at later stages simply depress growth and result in small puparia. Administration at interval times produce forms between the larva and pupa.

The impact of ecdysone on colour change in *Cerura vinula caterpillars* (Buckmann, 110) also conforms with this theory; the greater the dose of hormone injected the quicker becomes the process of moulting whilst the effect on colour change decreases with an increase in the amount of hormone.

The function of MH (together with that of JH) at the cellular level was examined in detail by Wigglesworth (1963c) in the epidermal cells of Rhodnius prolixus. He came to the conclusion that MH is not a necessary ingredient for growth in insects in general, but it is necessary for the activation of the epidermal cells to produce their secretion (the moulting fluid and the chitinous cuticle) and to grow and divide. But precisely the same response is obtained by 'wound hormones' from injured tissue (cf. Wigglesworth, 1937). And the cells of the fat body and haemocytes do not need MH at all–their activation and mitotic activity is brought about by nutrition alone. The first discernible effects of MH on the necleolus of the epidermal cells is discussed and constrasted with the effects on the puffing patterns in the salivary glands in Chironomus.

Significant evidence supporting the view that MH only indirectly stimulates growth and morphogenesis is provided by the distribution of DNA synthesis during insect development. These authors conclude that ecdysone should be viewed primarily as a moulting hormone that initiates biosynthetic activities which caused moulting.

In some tissues, such as all chitinogenous epithelia, the Malpighian tubules and the nervous system, DNA synthesis occurred soon after MH secretion started. In others, e.g. pgl, midgut and haemocytes, it continued throughout the whole of the larval moulting cycle, but acquired a maximum at the peak of ecdysone production. In further tissues, e.g. the imaginal wing discs and muscles, no correlation between DNA and ecdysone production was discovered.

Schaller and Andries (1970a, b,) got the same results while studying regeneration and metamorphosis of the midgut cells in Aeschna cyanea and by Mouze and Schaller (1971) and Mouze (1971) in a study of development of the eyes in the same species, based on observations of mitotic activity. Agui et al. (1972) showed that an explanted pgl of *Mamestra brassicae* clearly stimulated the induction of moulting in the integument of the diapausing Stemborer (*Chilo suppressalis*) in vitro. On the other hand, MH failed to stimulate spermatogenesis in naked cysts of Cecropia pupae, but was active in a culture of intact testes.

Morohoshi and Iijima (1969) and Morohoshi et al. (1972) also reached the same conclusions with larvae and pupae of *Bombyx mori.* The only contradictory evidence is the report by Sondhi (1968), who claims to have found that in *Drosophila melanogaster* the injection of active ring glands into inbred larvae of the same sex and age did not affect either the time of puparium formation or of adult emergence, but raised the recipients' wet and dry weight by up to 10 per cent. The rate of the increase and the number of individuals are too few to allow definitive conclusions to be obtained from the results.

MH has a primary, direct effect on tissues of ectodermal origin only, i.e., the epidermis, stomodeum and proctodeum, the epithelia of the tracheal system, the ectodermal parts of the imaginal discs and the nervous system, etc. Its presence in the minimum active concentration, in the absence of JH, determines growth of the imaginal parts and degeneration of the larval parts according to the species specific morphogenetic pattern. This, together with induction of the moulting process, may per se be enough indirect stimulus to initiate metabolic and growth activity in other, MH-independnet, tissues also.

The Apolytic Glands (agl)

Many glands in the insect body have been considered to produce MH. they have some commonn characteristics. They are paired, laterally localized glands, mostly ribbon-like in form, and sometimes more or less branched. Their main component is large parenchymal cells with 'intricately interwoven cytoplasmic processes extending towards the glandular surface' (Scharrer, B., 1964), where their ultrastructure has been examined, different quantities of smooth tubuli of the endoplasmic reticulum has been found. These are supposed to be connected with steriod production . The glands also seem to have the same embryonic and phylogenetic origin.

There are six classifications of these glands have described in different parts of the body in various groups of insects.. Perhaps they can be extended to include the larval oenocytes. It was felt that a common term, applicable to all these glands, was needed and some authors suggsted the collective designation 'ecdysial glands'. Wigglesworth (1962) however, pointed out that this term has already been applied to a certain type of epidermal gland and might thus cause confusion. Another contrary reason is that ecdysis is precisely that part of the moulting process which is uncontrolled by the hormone from these glands. I would therefore suggest calling them 'apolytic glands' and have employed this common term in the present book. It is derived from 'apolysis', the first stage in the moulting process under the direct control of the glands in question.

Many researches has expressed their doubts (Locke, 1969; Romer, 1971, as to whether the agl actually produce MH. But these doubts seem to be vitiated by the findings of Agui et al. (1972) and Kambysellis and Williams (1972), who have demonstrated the activity of the glands in vitro. This discovery is substantiated by numerous authors in vast quantities of early and recent experimental data, whereas the contradictory observation can be explained by assuming that a primary MH source does exist.

1. The Prothoracic Glands (pgl)

Morphology

The pgl of the Cossus cossus caterpillar were almost accurately described by Lyonett as early as 1762. A more complete description along with data on embryonic development is given by Toyama (1902) who called them 'hypostigmatic glands'. Ke (1930) used the term pgl for the first time. Ever since because of the interest exicted

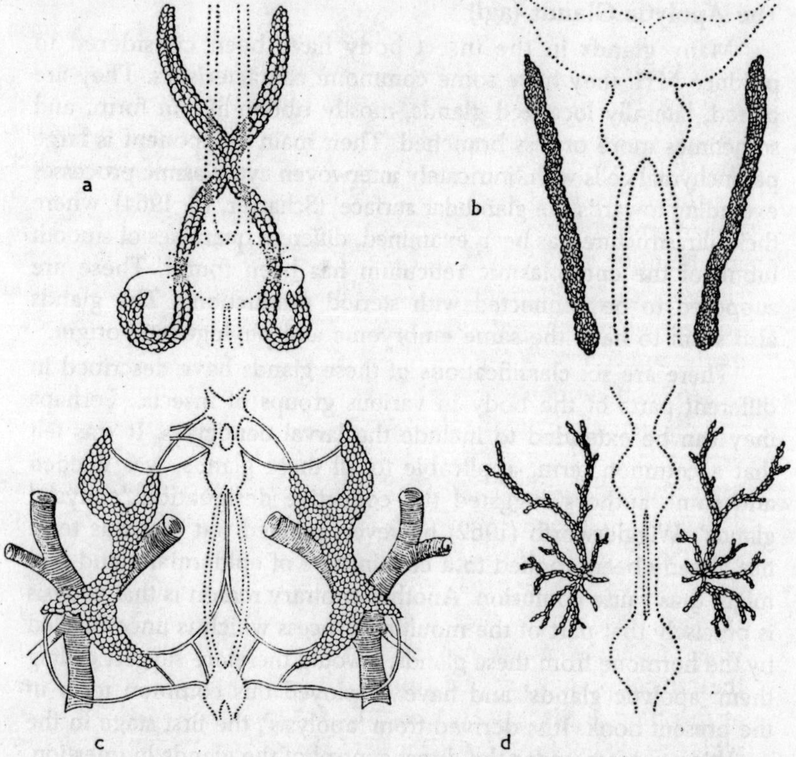

Fig. 7.8. Prothoracic glands in different insect orders: (a) Blattoptera, (b) Hemiptera, (c) Lepidoptera, (d) Hymenoptera.

by the discovery of their function, they have been described in detail for all the important groups in insects.

Many comparative morphological studies of the pgl and analogous organs in various groups of insects have also appeared. For example, they were described by Lee (1948) in many lepidopterous larvae, by Pflugfelder (1947d) in various orders of Hemimetabola, by M. Thomsen (1951) in Diptera, by Wells (1954) in Hemiptera, by Srivastava (1959) in Coleoptera, etc.

Although they change a great deal in shape, the pgl have certain features in common in all groups where they occur: they are paired glandular structures located in the ventrolateral areas of the prothorax (extending into the mesothorax in some species), and are sometimes partially, fused in the mid-line. They are generally closely linked with the main lateral tracheal branches near the prothoracic spiracles. In some bugs, like Rhodnius, they do not form independent compact organs, but chains of cells within the inner pair of fat body lobes. In cockroaches they are fixed in the body cavity by one or more muscles fibres and by a nerve ring from the prothoracic or, less often, the suboesophageal ganglion. Four types of pgl (apart from the ring gland) were differentiated on the basis by Joly (1968): the massive type (Apterygota, Ephemeroptera, Odonata), (2) the ribbon-like type (Orthoptera), (3) the blattoid type (Blattoidea), and (4) the diffuse type (some Heteroptera, Lepidoptera, Hymenoptera). They are supplied in great number with trachea in many groups they were till now unobserved by the researchers.

In each larval instar there is a distinct cyclicity in the secretory and mitotic activity of the pgl which agrees with the experimentally dertermined critical periods for MH activity. The secretion cycle in Pieris brassicae was described comprehensively by Kaiser (1949). The greatest size and adult insect they degenerate two to ten days after the imaginal moult. The changes in the pgl of Tenebrio molitor during development were critically studied by Srivastava (1960). In each larval instar they reach their maximum size at the time when feeding is interrupted, i.e., about two days before ecdysis (cf. the critical period for MH); they produce most of their secretion at this time and afterwards become reduced. A new cycle begins with the feeding of the next instar. The critical period occurs less than 24 hours after the interruption of feeding in the last larval instar. The glands are completely reduced at the commencement of eye pigmentation. However, they persist throughout the life of insect in Apterygota.

Herman and Gilbert (1966), accomplished a detailed anatomical and histological study of the pgl in *Hyalophora cecropia*, and found numerous, very diffusely arranged chain of cells, reaching to the posterior portion of the thorax, which arose from the four branches (anterior, dorsal, ventral and posterior) of the relatively compact tissue of the gland round the first spiracle. Each gland contained some 200 to 290 cells, which were larger and usually more in number in females and degenerated immediately after adult emergence

Histology

The pgl are usually formed of two strips of glandular tissue. The essential components of these are glandular cells equal to those of the corpora allata. Their cytoplasm is basophilic staining deeply with methylene blue, neutral red and other stains. Numerous deeply staining granules are found in the cytoplasm. Abundant black granules, probably lipoid in character, are found after fixation with osmium tetroxide. The glandular cells are joined by intercellular bridges which stain blue with azan and are connected by their anastomoses with the membrane envelopine the axial muscle fibres. a rich supply of thin tracheae was observed by Wigglesworth (1952a) in *Rhodnius prolixus* in contrast to the feeble tracheation of the surrounding fat body. In contrast to the pgl in cockroaches and butterflies, no nerve fibres were observed in the pgl of *Rhodnius*, either in dissected glands or histologically.

In *Hyalophora* each pgl cell consists of peripheral with striated border through which secretory substance is released. Cycles of activity are evident, characterized by nuclear changes followed by cytoplasmic vacuolization and correlated with the moulting cycle. A low level of seceroty activity is discovered in young pupae and a higher one in older pupae. Discernible variability of secretory activity could be seen among the cells of the same gland. a detailed histological study of the pgl of Galleria mellonella was carried out by Mala et al. (1972) with the aim of investigating JH effects at both the histological and the ultrastructural level. The more compact gland is composed of about 55 polyploid cells with giant nuclei and different layers of cytoplasm. The secretory cycles connected with the moulting process were examined comprehensively in the penultimate VIth and last VIIth larval instar and the pupal instar and differences between them were ascertained

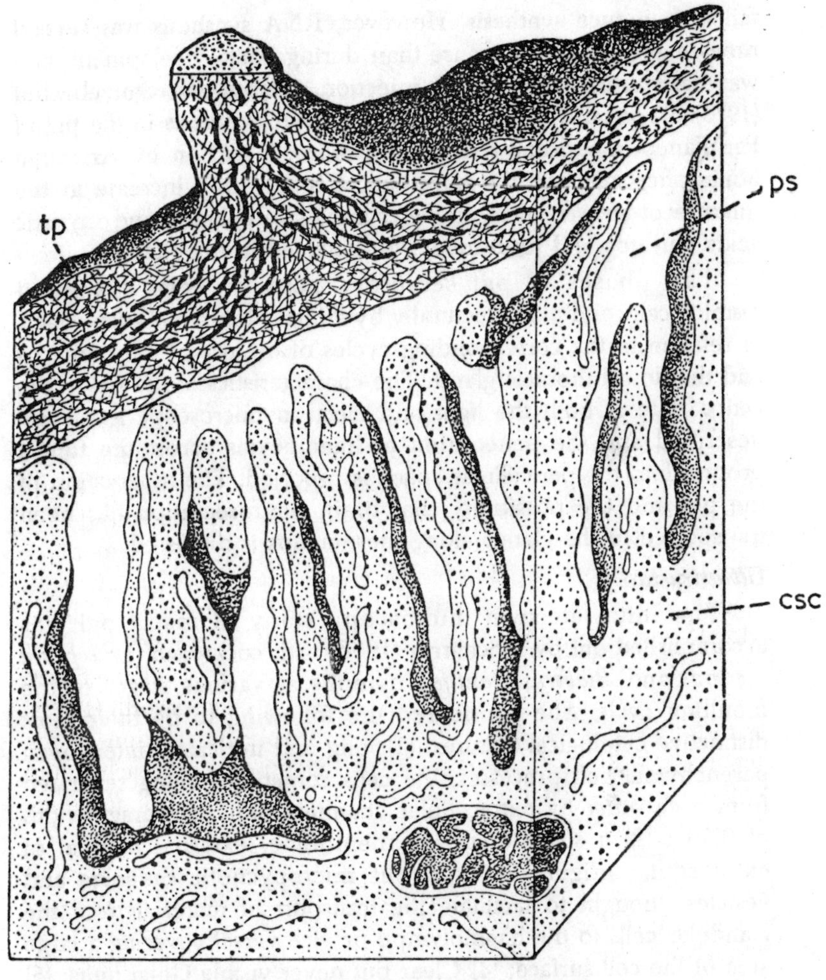

Fig. 7.9. Block diagram of the surface layer of a prothoracic gland cell of Antherea pernyi as seen under the electron microscope. tp—tunica propria, ps—pericellular space, csc—cytoplasm of secretory cell.

The amount and distribution of DNA and RNA in the pgl of Samia cynthia and several related species was considered by Oberlander et al. (1965) in the penultimate and the last larval instars and in pupae. DNA synthesis was found to be quite high in the IVth larval instar and lower in the last larval instars and pre-pupa. No DNA synthesis was observed in the pupa during diapause or during adult development. In these two stages, even JH injection failed to induce synthesis. However, RNA synthesis was several times higher during diapause than during adult decelopment, and was easily stimulated by JH injection. Gersch and Sturzebecher (1970) observed a similar increase in RNA synthesis in the pgl of Periplaneta americana nymphs following induction by AH.Eight hours after injecting the hormone, a noteworthy increase in the mixture of stearic, oleic, linoleic, linolenic, palmitic and myristic acids, but not MH (ecdysone) was detected in the glands.

The phases of pgl secretory activity were examined histologically and ultrastructurally by Hintze (1968) in *Cerura vinula*, in relation to the corresponding cycles of the nsc of the brain, ca and the dorsal thoracic gland. The characteristics of the pgl in the active state at both the light and electron microscope levels are described and the penetration of haemocytes across the tunica propria into the pericellular space is reported. The innervation of the pgl and the possibility that brain neurosecretion has been transported to the glands along these axons is discussed.

Ultrastructure

The inital electron microscopy study of insect pgl was accomplished out by B.Scharrer (1966) with cockroach (*Leucophaea maderae* and *Blaberus craniifer*) nymphs at various stages of the moulting cycle. She considered the following to be their most distinctive cytological features: (1) long and intricately interwoven parenchymal cell processes, the cells sometimes being separated from each other by extracellular channels with an average width of 0.5µ ; (2) a specific surface membrane (external lamina) of extracellular origin; (3) numerous micropycnotic caveolae and vesicles (thought to facilitate the transport of material from the glandular cells to the haemolymph or vice versa) covering a large area of the cell surface; (4) Clear but never visible Golgi units; (5) a small amount of smooth endoplasmic reticulum characteristic of steroid - producing cells, sometimes with a striking array of microtubules; (6) a varying incidence (according to the phase of the moulting process) of other structures, such as nucleoli, mitochondria and lysosomes. Degeneration of the glands started within a few days after adult ecdysis. Early signs of their destruction were observed in the nuclei, in the form of patches of contrasting electron density characteristic for pycnosis. later, large heterogeneous inclusion bodies, thought to be autophagic vacuoles, appeared in the cytoplasm. In the advanced stages of degeneration, the plasma membrane and nuclear membrane were slowly autolysed and the protoplasm was invaded by phagocytic haemocytes.

An exhaustive electron microscopial study of pgl in Lepidoptera (*Bombyx mori*, *Antherea pernyi* and certain other Saturniidae) was carried out by Beaulaton. The structute and origin of the tunica propria, the pgl surface membrane, was investigated. It is perhaps matches with the basement membrane of the epidermis, since it has a similar microfibrillar structure and several types of haemocytes take part in its formation. A pericellular space between the tunica propria and the surface of the glandular cells is explained. In agreement with Herman and Gilbert (1966), a neurohaemal control, different from AH control, was found to participate in the cyclical secretory function of the glands. Many kinds of cells are described in relation to these glands.

A remarkable increase in the external surface area of the cell membrane, caused by the projection of finger-like processesinto the pericellular space, a large quantity of free ribosomes and the presence of specific alveolar bodies within the nucleus of female glands are regarded as characteristic for the glandular cells. Rough endoplasmic reticulum and ergastoplasmic saccules are abundant, but smooth endoplasmic reticulum is somewhat rare and is regarded by the author as unimportant. The giant nuclei are differentiated by specific alveolar inclusions of a chromatin character (in females only) and by numerous polymorphous nucleoli consisting almost entirely of RNA. The distribution of phosphatase activities in the pgl of Antherea was examined by Beaulaton (1966). Their accumulation in the lysosomes (dense bodies) and vacuolated bodies are confirmed.

The histology and ultrastructure of the pgl in Tenebrio molitor and their changes within the moulting cycle were investigated comprehensively by Romer (1971). The use of ^{3}H-cholesterol showed that the glands absorb cholesterol from the haemolymph. Neurosecretory granules were discovered in the nerves innervating the pgl. Absorption of lipid granules by both rough and smooth tubules of the endoplasmic reticulum was observed. The extensive Golgi system was expected to be concerned with the production of ecdysone from cholesterol. Extracts of isolated pgl were shown to invite pupation in Calliphora.

Maia et al. (1974), Novak et. al. (1974) and Blazsek et al. (1975) made a detailed study of the structure and ultra-structure of the pgl in the last two larval instars and pupal instars of *Galleria mellonella*, for the purpose of determining JH effects. According to them all structural differences between the penultimate and the last larval instar as being due to JH deficiency in the latter. In consequence, the cytoplasm is vacuolated and less compact, and at first an increase in the amount of secretion is effected in the gland.

The impact of JH and MH on nuclear RNA synthesis in the pgl and corpora allata of saturniid pupae (*Philosamia cynthia* and *Hyalophora cecropia*) were examined in an autoradiographic study by Siew and Gilbert (1971). The pgl were stimulated by MH within three hours, ca three hours later. JH also stimulated both glands, but in this case the ca were activated sooner than the pgl. The administration of MH to pupae in which the adult moulting process had just started, resulted in drastic re duction of nuclear RNA synthesis. This is in agreement with the findings of Socha on hormonal interactions in various phases of the instar.

It has already been demonstrated by Toyama (1902) that pgl appear at a very early stage in embryonic development as an epidermal invagination of the lateral portion of maxillary segment and from here later extend into the prothorax and, where this is reduced, even into the mesothorax such result was also obtained by Wells (1954) for the bug Dysdercus cingulatus. Here, two pair of invaginations arise in the second maxillary segment; the anterior, larger, pair develops into labial glands whilst the posterior pair gives rise to the prothoractic glands. If, however, their supposed origin form the nephridian tubules and their innervation (in most insects by nerves from the prothoracic ganglion) is taken into account, it appers more probable that the invagination mentioned does not start in the second maxillary segment but in the most anterior portion of the prothoracic segment.

Phylogenesis

Pgl, or their equivalents are today known in practically all groups of metabolous insects (Pterygota). The conclusion, however, that their occurrence is not restricted to the Pterygota is well founded. Pfluggelder (1958) suggested the possibility of a homology between the ventral glands of Hemimetabola and the so-called head nephridia of Apterygota which are also developed from the second maxillary segment. The papers by Gabe (1953b, 1956) and Echalier (1955, 1956) suggest the possibility of homology between the pgl and the moulting gland ('laglande de mue') or organ Y in the Cruatacea Malacostraca. The view appears to be strangthened by the recent findings of Karlson (1957), who was able to induce puparion in ligated Calliphora larvae with a concentrated extract obtained from Cranton vulgaris. The earlier conclusions of Pfygfekder (1947, 1952) and B. Scgarrcr (1948), that both the pgl and vgi, as well as the ca, originate from the nephridia of the ancestral Annelids during the evolution of the Class Insecta are also in agreement with this. It may be expected that, if suitable techniques are used, the equivalents of the pgl will be found in all groups of arthropods such as the "Crustacea. Arachnida. Myriapoda, etc. If this is so, production of the moulting hormone may be looked upon as phylogentic adaption, which has developed in the close relationship with the chitinous cuticle as a mechanism to ensure the simultaneous moulting of the whole body surface, This is necessary to allow the organism to escape from the inextensible chitinous covering.

2. The Ventral Glands (vgl)

Morphology

Vgl, often called ventral head glands or tentorial glands, were first found by Pflugfelder (1947a), in the hind ventra region of head in Phasmidae. Boisson (1947) described corresponding structures in Bacillus rossii as corpora incerta and discovered a cuyclicity in their secretory activity, which was connected with moulting, Pflugfeelder (1947d) assumed that they were absent in Hemiptera and in all Holometanola.

An important feature of the vgl is that they degenetate at the end of metamorphosis or in the first few days after the imaginal moult in the same way as the prothoracic glands (= tentorial glands) of the worker and soldier castes of termites., which are less developed than in the sexuals but are active during the whole life period (Springhettin and Bernardini, (1955; Kaiser, 1956). This appears to be connected with the neotenic character of these castes insome lower termite groups, evidence by the underdeveloped state of their ovaries, in which they are maintained byinhibitory substances produced by the reproductives (cf. Pflugfelder, 1958). The influence of the vgl on the moulting process was shown by Strich-Halbwwachs (1954, 1958) and by Halbwachs, JOLY and Joly (1957) in migratory locusts. Tyhe implantation of an active gland

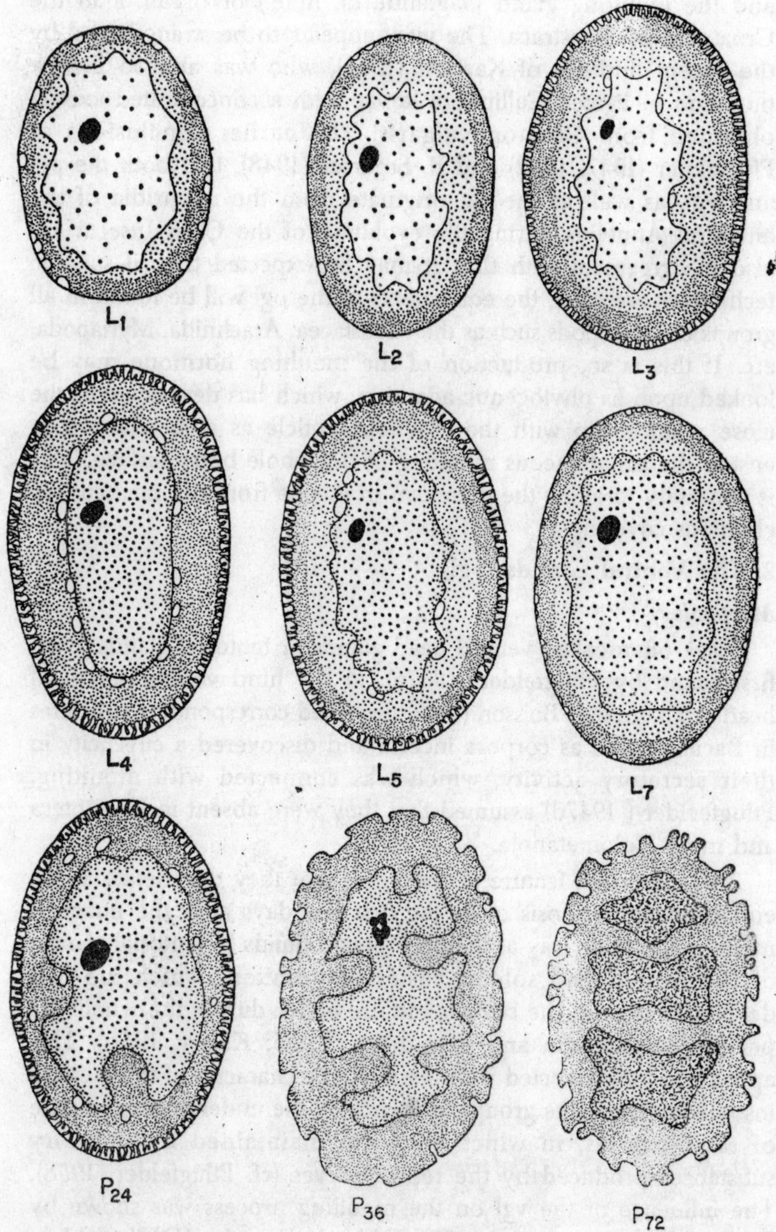

Fig. 7.10. Scheme of developmental changes in a prothoracic gland cell during the last larval and pupal instar of Galleria mellonella. L_1-L_7–Ist to 7th day of VIIth instar larva; P_{24}-P_{72}–24th to 72nd hour of pupal development.

at the geginning of the IV of V instar causes an acceleration of the moulting process with a feeble prothetlic effect. The so-called corpora adnexa in *Dinjapyx marcusi* which are epithelial in structure and situated caudally to the suboeso phageal ganglion are viewded by Marcus (1951) as a homologue of the vgl in Apterygota. Some authors suggested that the vgl can be uderstood as homologous with the pgl of Holometabola. But, it seems more probable that they are serial homologues of the pgl.

A regular cyclical secretory activity of vgl in *Carausius* was described by Pflugfelder (1958). The drops of secretion apper, closely connected with the nuclei, during and for a few days after ecdysis. This period of excessive secreetion is followed by a phase of exhaustion. Between 3rd and 7th day of the intermoult period both the nuclei and cytoplasm grow rapidly. On 8th and 9th days, there are many cell dimensions resulting in abundant cell aggregates between the 10th and 11th days. Ecdysis takes place on the 12th day relatted with the strong secretory activity. Similar observations have been made in Odonata, termites and other Hemimetabola.

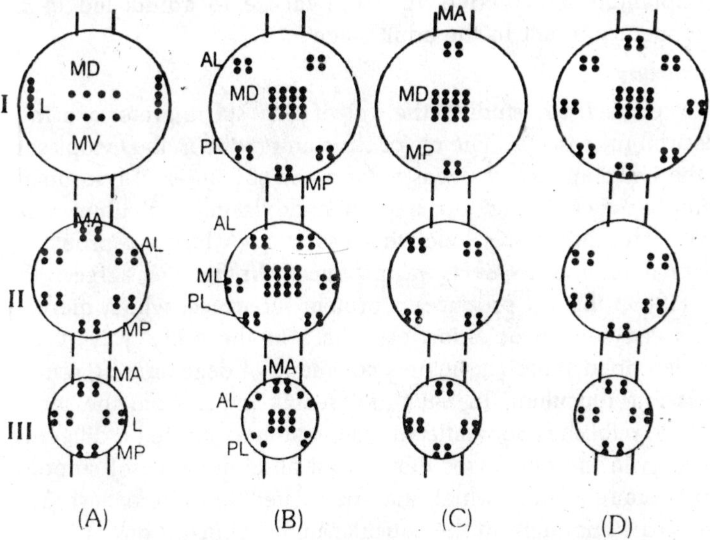

Fig. 7.11. Neurosecretory cells in the ventral ganglia. A–dragonfly, Orthetrun chrysis; B–cockroach, Periplaneta; C–locust, Schistocerca and D–moth, Othreis materna. I-suboesophageal, II-thoracic and III-abdominal ganglia. AL–anterolateral, PL–posterolateral, MA–mid anterior, MP–mid posterior, L–lateral, ML–mid lateral and MD–middorsal, and MV–midventral neurosecretory cells.

Histology and embryology

The histological structure of the vgl is quite similar to that of the pgl. Their embryogenesis was examined by Pflugfelder (1958) in *Carausius morosus.* They develop from epidermal proliferations in the ventrocaudal part of the head, but are soon separated from the epidermis and become bladder-like. Jones (1953, 1956) found vgl in *Locusta pardalina* and *Locusta migratoria* as invaginations in the head region of embryos at the end of the kataterpsis stage. He considered them to be homologous with the pgl of the other insects. He examinied their function in relation to the beginning of the secretory activity of neurosecretory brain cells and the first embroynig moulting process.

Degeneration of the vgl of the earwig Anisolabis maritima during metamorphosis was studied histologically and analysed experimentally by Ozeki (1968, 1970), who transplanted them to various other stages of the same species. He concluded that both the small amount of JH in the last larval instar and the latter's internal state were determining factors in the regression of the vgl. The implantation of active vgl from larvae to adults led to a supernumerary moult in the adult stage

Ultrastructure

Joly et al.(1969) studied the vgl of Locusta migratoria using the electron microscope. The major structures found in the cytoplasm were the Golgi apparatus, ergastoplasm and structures of lysosomal or autophagic character; no smooth endoplasmic reticulum was observed. The authors concluded that a substance which was probably protein nature, and not MH, was produced. In this they agree with Locke (1969,) the vgl produces a protein substance, which merely controls ecdysone production elsewhere in the body. Joly et al. (1969) described special structures consisting of degenerating rough endoplasmic reticulum. In another research paper from the same laboratory, published soon afterwards the authors formed a different opinion, as in the oenocytes they found an abundance of smooth endoplasmic reticulum, which may be related with the occurrence of a steroid component in the cuticular lipids. Unlike Locke (1969), they did not consider cytological structure alone as correct evidence for steroid production.

In an ultrastructural study of the secretory cycle in Locusta migratoria vgl, Cassier and Fain-Maurel (1968) based their conclusions on the belief that the glands participate in ecdysone production. They demonstrated that, during the vgl activity, cycles the tubules of the smooth endoplasmic reticulum were extruded from finger-like cytoplasmic evaginations into the intercellular spaces of the gland. They stressed that this extrusion coincides with maximum MH activity and concluded that it was the most effective way of discharging the hormone from the glandular cell. They suggested the use of isotopically-labelled cholesterol in order to solve this question.

Differences in the ultrastructure of vgl in the active state and during diapause were examined by Guellin (1971) in the grasshopper *Pyrgomorpha conica.* The ultrastructure of diapausing glands is indicative of remarkable depression of metabolism : the pericellular system is reduced. lysosomes are apparently absent and free ribosomes and eragastoplasm at the end of diapause, the author concludes that active vgl produce a protein type of substance, as did Joly et al. (1969) in Locusta.

Fain-Maurel and Cassier (1968, 1969) examined the involution of the vgl in the last larval instar of the gregarious phase of *Locusta migratoria migratorioides* (in adult, the glands persist in the solitary phase) and compared it with the secretory cycle in earlier instars. They claim ' an early and primordial role of Golgi apparatus; in the elaboration of the 'vacuolar bodies' of 'cytosomes' and participation of rough endoplasmic reticulum in the process of degeneration, through the formation of 'cytosegresomes' and 'particular microtubular areas'. The process begins with the autolysis of the named structures followed by fragmentation of the cells and nuclear lysis. Specific 'Golgian vacuoles' are formed in the Golgi apparatus, which afterwards undergo autolysis or are enclosed in autophagic vacuoles. degeneration can be avoided by the daily application of carbon dioxide from the IIIrd instar onwards.

The structure and development (degeneration) of the apolytic glands have been studied in several other insect groups and species, e.g., Mantis religiosa Coenagrion angulatum (Odonata), and Nauphoeta cinerea (Blattoidae) Hoffmann and Joly (1972) paid special attention to the physiology of the vgl in Locusta migrtatoria, and found complete inhibition of moulting following extirpation of the glands, but this was fully compensated by the implantation of an active ecdysone, however large the dose, was ineffective. The authors reached the conclusion that vgl produce a hormone which is different from ecdysone.

Phylogenesis

Rae and O'Farrell (1959) analyzed the formation of vgl and the retrocerebral complex in Grylloblatta campodeiformis, a represent- active of the primitive Orthopteroid Order Notoptera (= Grylloblattidae). They explained the structures obviously endocrine in function in the ventrocaudal region of the head and in the cervical area which resembled very closely the so-called 'head lobes' of Blatella germanica and the 'cervical glands' of Periplaneta americana. These researchers accepted that the coxal muscles they describe in Grylloblatta are the remnants of the degenerated pgl of cockroaches, and homologous with the axial muscle cord in Blattidae depicted as Scharrers' organ by Chadwick (1955).

The presence of both pgl and homologous of vgl in the cervical structures of cockroaches seems to be well accounted for by the morphological and experimental evidence of Rae (1955) and Chadwick (1955, 1956). Thus complete etirpation of the pgl of *Periplaneta americana* does not prevent moulting. Cyclical changes in the volume of the 'ventral lobes' in *Blatella* are related to the moulting process and regeneration

3. The Pericardial Glands

The initial indication of the existence of ductless glands associated with the dorsal vessel was made by Verson(1911a,b). The pericardial glands in various species of Phasmidae (Phyllium, Carausius etc. Were fully explained by Pflulgfelder (1938a-d, 1946b). Besides the vgl they occur in the form of paired glands, the histological structure and function of which, unlike origin, is equal to that of the pgl and vgl. They also decline in the adult.

4. The Peritracheal Glands

Matching structures were described by Possompes (1949 a-c, 1953a,b) and other researches in the Chironomidae, Simullidae, Tabanidae, and other groups of Diptera under the name of peritracheal glands. They have not been discovered in Tipulidae. Both reseatchers agree their homology with the pericardial glands of Phasmidae. The relation of the moulting process was demonstrated experimentally by Pflugfelder (1949b). No moulting was observed in specimens in which the glands were extripated.

Histology, embryology and phylogenesis

In their histological structure the pgl of both Phasmidae and Diptera correspond closely with the pgl and vgl of other groups of

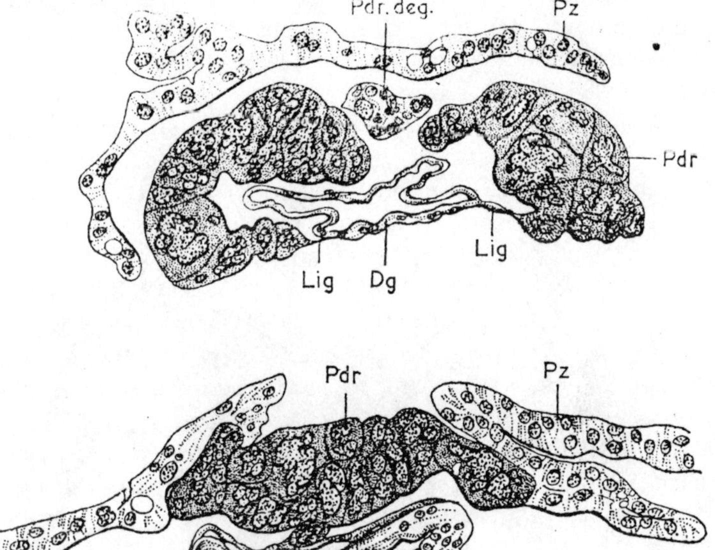

Fig. 7.12. Pericardial glands of Phyllium sp. in transverse sections. Pdr–pericardial glands; Pdr. deg.–degenerated part of a pericardial gland; Pz–pericardial cells; Dg–dorsal vessel; Cc–corpora cardiaca; Lig–Ligaments.

insects. The pericardial glands were initially accepted to be of mesodermal origin by Pflugfelder (1938). In a later work however, Pflugfelder (1958) mentions their ectodermal-like structure and emphasized the difficulty in distinguishing between mesenchymal and ectodermal cells in the early embryonic period of Dixippus morosus. This would remove the only possible objection to their identification with the peritracheal cells, which are of ectodermal origin according to Possompes. The problem of their homology has, however, been rather complex because of by their confusion with pericardial cells of obvious mesodermal origin, which is not at all simmilar to the glands.

5. The Ring Gland (rgl)

In higher diptera special endocrine organ occurs, which combines the source of all three metamorphosis hormones, the first signs of which can be seen in the Tipulidae. It was initially explained by Weismann (1846) in *Calliphora vomitoria* and is therefore referred to by some authors as 'Weismann's ring'. The first to suggest its

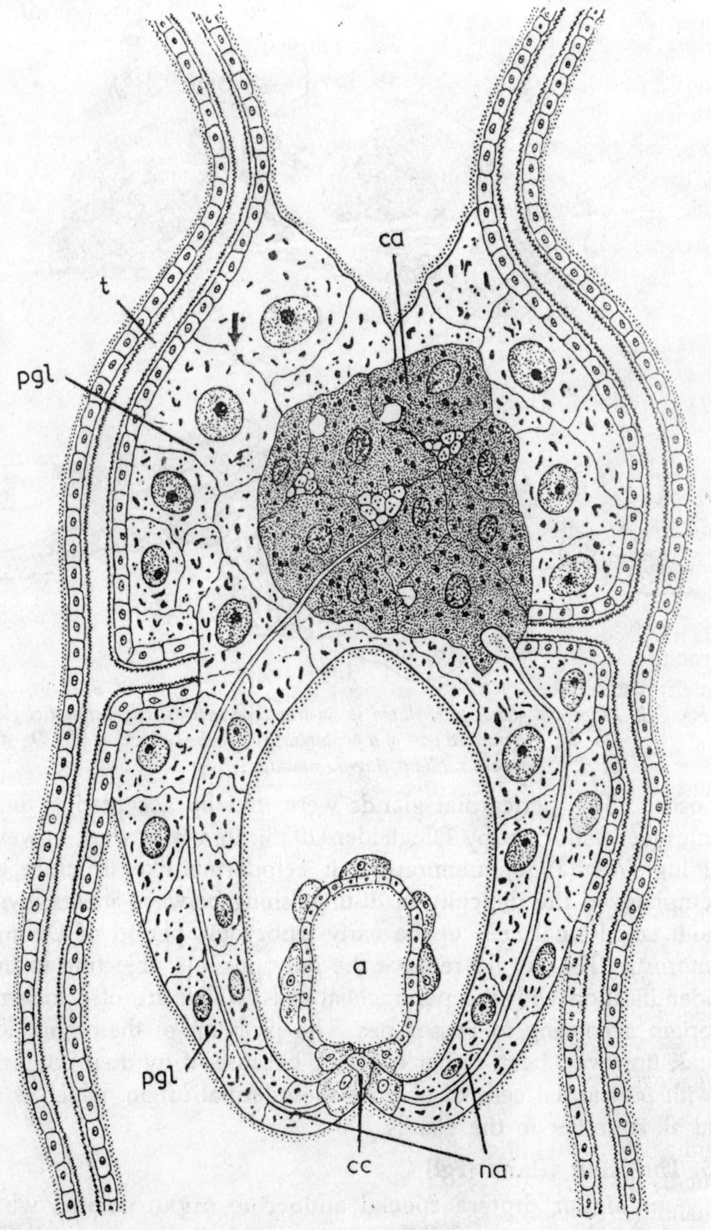

Fig. 7.13. Cross-section of ring gland of Drosophila melanogaster IIIrd instar larva. ca—corpus allatum, t—trachea, pgl—peritracheal gland cells, a—aorta, na—nervus allatus axons, cc—corpus cardiacum.

endocrine character was Hadorn (1937 a-c) although the experiments of Fraenkel (1934, 1935). In these papers the rgl was expected to exert a positive influence on metamorphosis. But Burtt (1937, 1938) postulated that the rgl produced an opposite, i.e., inhibitory, effect on metamorphosis, and he identified the rgl with the ca. Hanstrom, on the other hand, attempted to equate the rgl with the cc on the basis that nerve cells are found therein, which do not exist in other endocrine glands.

The first suggestion made by B. Scharrer and Hadorn (1938) who first suggested the ring gland as a composite structure, consisting of both ca and cc. The structure of the rgl during metamorphosis and the development of the adult cc and ca in *Calliphora* were thoroughly studied by E. Thomsen (1941, 1947). She recognized the lateral portions of the rgl, formed of large glandular cells, with the pericardial glands of Phasmidae by showing that they disintegrate during metamorphosis. She also found that the hypocerebral ganglion is associated with this composite structure. Her findings were confirmed and elaborated by Vogt (1941c, d) for Drosophila, M. Thomsen (1951) for many species of Diptera, and by Possompes (1953) in a monograph chiefly concerned with the structure and funcstion of the rgl in Calliphora erythrocephala based on the individual implantation of parts of the ring.

Four major components of the rgl may now be differentiated : (1) a corpus allatum of the small cell type situated in the dorso-median part of the ring, often covered by large cells of the lateral portions; (2) pericardial (peritracheal) glands formed by the large glandular cells of the lateral portions of the ring; (3) a corpus cardiacum in the ventromedian part of the ring, with only one a couple of nervi corporum cardiacarum formed by the coalescene of the nervi corporum cardiacarum interrni and externi before they enter the ring gland fusing with the nerve recurrens in some species; (4) the hypocerebral ganglion consists of transparent ganglionic cells and closely joined with corpus cardiacum.

Ultrastructure

An exhaustive study of the ultrastructure of the rgl of a normal and a 1/2/gl mutant strain of Drosophila melanogaster was carried out by King et al. (1966) and Aggarwall and King (1969). The quantity of smooth surface endoplasmic reticulum in the peritracheal gland cells increased about tenfold in the pre-pupal period in normal larvae, but did not exceed 1 percent in mutant larvae. Metamorphosis of homozygous larvae of the lethal mutant was impossible unless an active gland from a normal larva is impainted. The researchers reached the conclusion that the relevant gene (the + allele of the 1/2/gl mutant) was indirectly concerned with the synthesis of smooth endoplasmic reticulum, where cholesterol from the food is transformed to ecdysone. In the intracellular spaces between the gland cells they encountered numerous cytoplasmic vesicles, which were pinched off from the cell projections.

Willig et al. (1971) examined ecdysone and ecdysterone biosynthesis in 7-day Calliphora erythrocephala larvae by injecting the larvae with 14- C-Cholesterol. The same kind of experiment was performed with explanted brain-rgl complexes in vitro. In the latter case, no acative hormones were found, but occasionally a very small amount of active ecdysone glycosides or esters were produced, which exhibited lot of hormonal activity after enzymic hydrolysis with a-glucosidase of esterase. Therefore, the researcheres regard the inactive hormone glycosides as MH precursors produced by the gland.

The ultrastructure of the rgl of Drosophilia melanogaster was studied comprehensively by King et al. (1966). Typical smooth endoplasmic reticulum was found in the lateral branches, corresponding to peri-tracheal glands, which the researchers suggest as being involved in steroid hormone production.

Embryogenesis and phylogenesis

The embryogenesis of the rgl was examined in detail by Poulson (1954a). Amongst other things, the ectodermal character of the lateral, large gland cells, which correspond to the peritracheal glands of Possompes, was shown. There is not much information regarding the phylogenesis of the rgl. Possompes distinguishes four different types of development of the retrocerebral endocrine system in Diptera: (1) Tipula type : 2 ca, 2cc, (no equivalent of the pgl has so far been found here), (2) Chironomius type : 2ca, 2cc, and 2 peritracheal glands, each of which are connected with the cc and the ca. (3) Tabanus type : I ca, originationg by fusion of two original ones). (4) The true rgl of Calliphora with 1 ca (fusion of two), 1 cc and 2 peritracheal glands.

The Direct Effects of MH

The initiation of the moulting process

The implantation of active pgl on their equivalent results into moulting, not only where secretory activity is naturally suppressed, as in diapause, or where the glands have been experimentally removed, but also in cases where neither the pgl nor the moulting process normally occur, as in adault insects. The moulting process and consequent growth can be rosuced by active pgl along, without a simultancous impl antation of an active brain or cc along is not able to induce multing in the absence of an effective concentration of MH. The resufe was flly confirmed by experiments with isolated MH. the resuf was fully confirmed experiments with isolated MH prepared in crystal form by Butenandt and Karlson (1954).

The circulation of labelled ecdysone injected into the insect body was analyzed by Karlson and Sekeris (1963). They discovered that soon after injection the hormone accumulated in the epiderims and later in the fat body. The nuclei of the epidermal cells exhibited the greatest ratio activity. The authors realized that the sites of action of MH are the nuclei of the epidermal cells and that microsomes of the fat body eliminate the excess of the hormoes from the haemolymph.

The regulation of growth

The most important requirement for postembryonic growth in insects, as in any other arthorpod, is the loosening and subsequent freeing of the body from the inexpansible cuticle. MH is therefore indispensable, though indirectly so, for growth, as, for instance, the influence of Mh on the mitotic activity which precedes the detachment of the old cuticle. While, there is an increasing amount of evidence to the contrary: (a) Growth is possible without any moulting process and thus without MH (e.g., the growth of ovaries and other internal organs as well as of regenerating tissues). (b) The implantation of agl as well as the injection of the ecdysone induces moulting, often without any effect on growth or even with a negative one). (c) As distinct from JH, MH alone is not able to induce mitotic activity in larval parts of the body. Without further evidence the presence of MH is thus to be accepted as one of the conditions of growth but there is no cause for accepting it as a true growth hormone of the JH type.

The discovery, that the deposition of endocuticle and the secretion of wax also took place during the intermoulting period in the absence of any MH, led to the conclusion that these processes are not controlled solely by MH. MH action after the attainment of an active concentration of the hormone in the haemolymph is perhaps responsible only for a change in chitinase activity (or concentration), resulting in the chage from chitin synthesis to its decomposition, causing to moulting and subsequent gradual dissolution of the old endocutile.

Investigations by Morohoshi and Iijima (1969) were concerned with the effects of injection of phytoecdysoids (inokosterone, ecdysterone, cyasterone) into starved, ligated last instar Bombyx mori larvae and into pupae. In some cases supernumerary larval instars (a VIth instar) resulted. The researches confirmed that the MHd were concerned ony with the control of moulting and that the structure of the moulted insects depended on JH action. In another study, Morohoshi et al.(1972) injected Vth instar larvae of the same species with varying amounts of ecdysterone at different intervals after the preceding moult. They investigated the effect on the length of the intermoult period

Scrutiny of adaptive changes in larval behaviour prior to pupariation demonstarated interesting correlations with MH action. During 30 hours under dry conditions after the larvae have left the nutrient medium, they become increasingly sensitive to injected ecdysone. If the dry period is interrupted by a further exposure to moisture, however, their sensitivity diminishes. The researchers stated that this observation might be due to a 'convert' effect of ecdysone. It, seems practical enough however, to assume that leaving the moist enviornment induces MH production (undoubtedly by activating AH release) and as more endogenous MH is produced, less exogenous MH is needed to induce pupariation. Replacement in the wet medium inhibits AH production, and with it the amount of MH produced up to a given moment, sothat more has to be injected to induce puparium formation. The researchers concluded that this was an significant form of phylogenetic adaptation to the given conditions.

In another research paper, Zdarek and Fraenkel (1972) explained that injection of larvae with a large dose of ecdysone under wet conditions shortened the pupppariation time to 12 hours, as against 30 hours in the controls. Tanning occured at various stages of puparium formation (longer, less contracted body, and less smooth surface), again demonstrating the independence from MH and the moulting process of the individual process of puparium formation. Earlier experiments with the facefly (Musca autumnalis), in which the puparium undergoes calcification instead of tanning, proved that this process was equally independent of MH. MH is, of course, the essential releasing factor in this series of processes.

The evolution of typical puparia following the injection of ecdysone into Calliphora erythrocephala larvae deprived of their rgl confirms that MH controls contraction of the puparium as well as its tanning and other associated processes. During other set of experiments, Fourche (1969) starved Drosophila melanogaster larvae from the 48th hour of the last instar and then supplied them with ecdysone in pure gelation. If the concentration and time of action were adequate, the larvae formed puparia at the proper time. The relationship between endogenous MH and exogenous ecdysone and their correlation with pupariation were discussed. The above results prove that, as in morphogenesis, these processes are only indirectly dependent on MH.

The action on larval and pupal diapause

But it is quite clear that MH acts directly to break both larval and puppal diapause. Here, however, it must be noted that it is the absence of MH consequent upon the absence of AH which is the immediate internal cause of diapause. a state identical with diapause can be induced experimentally by removal of the source of AH.

The effects on the puffing patterns

The influence of the puffing patterns of the giant polytene chromosomes in the salivary gland cells and other tissues of Chironomous and Drosophila will be discussed later.

The action of MH

The impact of MH on the amounts of RNA and DNA during metamorphosis in Calliphora erythrocephala was studied by Berreur (1965a) in whole animals and in their separate organs and tissues. The resultant changes, which were tisuse specific for different parts of the body, were not discussed with reference to whether the action of the hormone is direct or indirect. In another investigation, using autoradiography of imaginal discs of the same species, Berreur (1965b) demonstrated that no DNA was synthesized in the absence of MH caused by removal of the ring gland, and that RNA synthesis was limited. Reintroduction of the hormone resulted in nuclear RNA synthesis and, somewhat later, in DNA replication.

Another group of researches, Sekeris et al. (1965) demonstrated that the injection of ecdysone raises the incorporation of radioactive orthophosphate (H^{33} $PO^{2}{}_{4}$) into nuclear ribosomal and messenger RNA in the epidermal cells of Calliphora erythrocephala. They accepted that MH stimulated DNA -dependent RNA synthesis.

Gillot and Daillie (1968) found a break in DNA production in the silk glands of the Bombyx mori at the beginning of each moulting period. They observed asynchrony of the DNA cycle and differences between the two parts of the silk glands, but the maximum amount of DNA per nucleus in the two parts at the beginning of the Vth instar was identical. Neufeld et al. (1968) analyzed the effects of injected ecdysterone on Calliphora stygia larvae and the conditions for its action on protein and RNA synthesis. Gorell and Gilbert (1969, 1971) demonstrated a similar positive effect by ecdysterone on RNA and protein synthesis in the hepatopancreas of the crayfish Orconectes viridis. The particularity of all these effects is yet to be throughly examined.

In Philosamia cynthia and Hyalophora cecropia a nuclear RNA synthesis in the pgl was activated 3 hours after injecting ecdysterone, whereas similar activity did not start in the ca until 6 hours after administration of the hormone. The administration of MH to pupae which had just began adult development culminated in a drastic decrease in nuclear RNA synthesis.

The action of enzyme activity

Various researchers have expressed that MH interferes with the reaction chain of some of the principal enzymes of the insect organism that regulate normal metabolism and growth, as for instance the cytochrome oxidase system. However interesting and important these findings may be in the study of the corresponding enzyme activity showing amongst other things the functional importance of insect hormones in research on any question of insect physiology and physiology in general–they do not bring any definite proof in favour of the assumption of a direct participation of MH in the corresponding reactions.

Marmaras et al. (1966) discovered an inverse correlation between 5-hydroxytryptophanase activity and the ecdysone titre during the pupal instar of Calliphora erythrocephala. Kobayashi and Kimura (1967) illustrated that labelled glucose was converted to haemolymph trehalose is decerebrated Bombyx mori pupae injected with ecdysone, but that in the controls (injected with water) it was converted to glycogen. Adenyl cyclase in the pupal wing epidermis of Hyalophora gloveri is stimulated after the injection of ecdysterone.

Injected ecdysone has a positive effect on different types of metabolism in ligated last instar larvae of Musca domestica. Time and tissue specific incorporatiion of tritiated thymidine was observed after the injection of ecdysone, but it was limited; however, tritiated uriding was incorporated intensively under the same conditions. Tritiated leucine had a different incorporation pattern. The incorporation pattern. The incorporation of labelled glucose was characterized, in various tissues, by a large increase 1 hour after the administration of ecdysone; it then fell rapidly, at 6-hour intervals until, 18 hours after injection, it reached zero.

The effects on metabolism

Bernardini and Laudini (1966) examined oxygen consumption by the prothoracic glands of young and last instar larvae of two species of cockroach (Leucophaea maderae and Periplaneta americana) and found that it has cyclical variations. Ultimate oxygen consumption was reached on day before, and minimum consumption one day after each ecdysis. The researchers discussed the possible significance of these changes. Fourche (1967) observed a typical U-shaped curve in Drosophila melanogaster pupae, with increases in respiratory metabolism accompanying pupariation, apolysis and ecdysis. In Thermobia domestica, Rhdendorf and Slama (1966) found specific oxygen consumption cycles associated with moulting and off-laying, which corresponded with those in pterygote insects. Oxygen consumption increased during actual ecdysis and fell about half way through the intermoult period.

Many researchers have examined the uptake, transport and utilization of cholesterol and other steroid derivatives by insects. Bergmann and Levinson (1966) tested 15 cholesterol derivatives and isomers in Musca vicina larvae and some of them in Dermestes maculatus larvae. Both the requirements and utilization of steroids in these insects was discussed. Chino and Gilbert (1971) fed silkworm (Philosamia cynthia) larvae on an artificial diet containing 14 C-cholesterol. Practically all the substance found in the haemolymph in vitro. It was finally accepted that the midgut is necessary for the incorporation of 14 C-cholesterol.

A qualitative and quantitative study of haemolymph proteins in *Dixippus morosus* demonstrated a remarkble relationship between them and the moulting process (or morphogenesis), and to yolk formation in adult females. In the final instar larvae, the haemolymph protein concentration reached the maximum on the 13th day after the preceding ecdysis, i.e., at about the time of the new adult apolysis. After that the protein concentration declined abruptly, largely owing to the decline of two glycoproteins, both of which practically disappeared during yolk formation in the adult females.

The effects on rhythmical activity of muscles

Roussel (1970,1971) explained specific acceleration of the cardiac rhythm in locusts following the implantation of vgl. In last instar nymphs, the implantation of active glands accelerated the cardiac rhythm only in the presence of the ca. In allatectomized adults the implantation of four vgl also acclerated the heart rate. While extripation of the vgl at the beginning of the last larval instar stabilized the cardiac= rhythm at a relatively high level compared with normal specimens. But even here the direct nature of this efect still requires verifacation.

The indirect effects of MH

The effect on differentiation

Including the effects of MH on growth, many researchers consider that it affects differentiatiion. While the question of the direct nature of an MH effect on growth may still be considered as a matter for debate, there is no longer any cause for assuming a direct influence of MH on differentiation. However, the dependence of any surface morphological change on the moulting process is unquestionable. The slightest morphological changes on the surface of the body can only be realized during the relatively short interval between the detachment of the old cuticle and the deposition of the new.

Any break in the process of moulting therefore necessarily means the inhibition growth and differentiation. But the initiation of the moulting process does not in any way mean a simultaneous initiation of differentiation. On the contrary, a premature introduction of MH, as in the ecdysone experiments, results not in more differentiation but in less. This was shown in Fukuda's experiments (1944) with prothoracic gland implantation, and in the experiments of Wigglesworth (1940a, b, 1948) with the parabiosis of moulting specimens of Rhodnium combines to those which has passed the critical period for the moulting hormone.

The same effect implies that the result necessarily results from the effects of juvenile hormone on growth and differentiation. By implanting active ca the gradient growth, resulting in differentiation, may be changed into an isometric one without change in form.

Such an inhibition of differentiation does not mean any inhibition of the moulting process which is distinctly accelerated. There is no basis for assuming any antagonism between MH and JH. Differentiation can be completely separated from both moulting and growth and there is no connection between the effects of Mh on one or the other. It is confirmed that the influence of MH on differentiation is therefore purely indirect.

Joly (1958) when analyzed the effects of the ventral gland hormone on the mitotic activity of the epidermal cells and an allometric growth in Locusta during metamorphosis, and found that MH only affects the moulting process.

The postulation that MH has only an indirect effect on growth and protein synthesis is supported by the following experiments. Chilled Antherea polyphemus about to start adult development and ligated abdomens of Hyalophora cecropia and Samia synthia adudlts were injected with crystalline ecdysone together with mitomycin C (which inhibits DNA replication). Ecdysone was found to cause apolysis in the pupae of these insects, in the presence of mitomycin, but growth, morphogenesis and the rest of the moulting process were inhibited. If ecdysone were administered after DNA synthesis, the epidermal cells also responded by depositing new cuticle. Similarly process took place in adult insects, in which no more DNA synthesis takes place, but the entire moulting process can be activated experimentallywith the same ecdysone.

Courgeon (1969, 1970) analyzed the effect of MH on the imaginal discs of eyes and antennae in tissue cultures. Generally mitotic activity stopped a few hours after explanation, but if an active ring gland was implanted at the same time it continued for up to five days. When the nervous system (the cerebral hemispheres and the ventral ganglion) was also added, there was a stronger effect.

The impact of MH on morphogenesis were examined in detail in ligated Galleria Mellonella larvae. Repeated dipping of isolated abdomens in methanol solutions of ecdysterone in different concentrations resulted in morphogenetic changes of varying degrees after ecdysis. Different types of cuticle were obtained in different changeing from a larval type, via a more or less normal pupal type, to an imaginal type with more or less developed scales. Taking into consideration the amount of hormone applied (the concentration and number of dippings) and the interval between treatment and subsequent moulting, the following explanation is submitted, in agreement with the gradient-factor theory. The degree of morphogenesis depends upon the length of the interecdysial period, or to be precise, upon the time between apolysis and deposition of the new cuticle. If this is shorter than normal (after a large dose of the hormone), the degree of differentiation is smaller and the cuticle retains a more or less larval of character. If the MH concentration is low, the interecdysial interval is longer and morphogenesis can progress further. This is the condition with omission of the pupal type of cuticle. Moulting is not caused by small does, but certain developmental changes in the internal organs do occur.

The action of ovarian development

The indirect character of the morphogenetical impact of MH is quite clear from its effect on the development of the ovaries. It was demonstrated by Strich - Halbawachs that the removal of the vgl at the beginning of the last larval instar in Locusta migratoria, which prevents moulting, also suppresses the development of ovaries but to a limited extent. The oocytes, which average 0.7 mm in length at the time of the operation, reached 1.8 mm within one month in the permanent larvae thus obtained, while they reached 7.2 mm in the metamorphosed controls at the moment of egg-laying. few ooctyes in the experimental series were abnormal or had even degenerated. It is evident that the same factor is afecting here, i.e., the inhibitory influence of the inextensible cuticle, the only difference being that its effect on the internal organs is less direct than on the epidermis. The relationship between the moulting process and the development of the ovaries also permits a rational explanation of the findings of Bodenstein (1953a-d) and Wigglesworth (1957) in which the induced nature of this type of MH effect is also evident.

The effects on ovarian development

The injection of ponasterone A affects the ovarian development (the weight of the ovaries and the number of eggs) in isolated pupal abdomens of *Bombyx mori.* It also had other morphogenetic effects. Ovarian weight and the number of eggs increased proportionally to the dosage up to an amount of 2.5 μg per specimen. No change was observed after larger doses, which resulted in the formation of a scaleless adult cuticle. The implantation of vgl into adult Locusta migratoria females, stimulated growth of the oocytes, but caused the more developed oocytes to degenerate..

The activity of the accessory glands and the genital apparatus was stimulated in allatectomized locustus of both sexes.

One may even include that the independence of the development of the ovaries from the moulting process and MH is quite evident from the fact that it normally takes place in adults where no prothoracic glands exist and no moulting occurs.

The action on the corpora allata and reproduction

The entomologist Englemann (1959) examined the effect on the function of the ca and the development of ovaries of implanting active larval pgl into adult females of the cockroach Leucophaea maderae. He obtained moulting of the adult in some cases and nearly always a partial inhibition of ca secretion. He describes this as a direct effect on the ca and an indirect one on the neurosecretory brain cells. Both these inhibitory effects were avoided by the simultaneous implantation of a brain and suboesophageal ganglion. An injection of ecdysone had an effect similar to that of the implanted prothoracic glands. In both cases cell division was found in the ca with a subsequent increase in tthe number of cells. As suggested by, Englemann, the results for a related species, Periplaneta americana to some extent contradict those obtained by Bodenstein (1953a, e). There is also a remarkable discrepancy in the conclusions if both these researches since the prothoracic gland secretion is produced during larval instars almost simultaneously with that of the ca without producing the slightest inhibitory effect. Some results in recent years have even shown that the ca may cause a direct positive effect on the induction of moulting which has no connection with JH Nevertheless, there is the possibility of a single simple explanation of all these facts provided in the 'law of correlation in consumption' by the different tissues inside the body. When comparing the findings of Englemann with those of Bodenstein, two major points emerge. Firstly that the moulting process, even when not accomplished, necessarily consumes part of the available body reserves. This deficiency becomes clear in the reduced activity of the corpora allata and in the restricted growth of the ovaries. Secondly, such a serious interference with the physiological processes of the body as that of allatectomy and the resulting regeneration processes may in itself produce an impulse strong enough to prolong the activity of the pg.

In Locusta migratoria, the disconnection of ncc and the resulting increase in the volume of the ca has no inhibitory effect on the moulting process. In this context it is interesting to note the effect of folic acid on diapausing pupae of Pieris' brassicae, where it increases the amounts of ribonucleic and desoxyribonucleic acids causing tumour growth in the absence of MH. The researcher attempts to explain these findings on the principle of artificial induction of endogenous viruses.

The action on regeneration

Other aspect of differentiation is also true for regeneration : that is, it is also independent of moulting. However, there is thought to be an indirect interdependence, at least as far as more extensive regenration is concerned. It can either be acclerated by the action of MH, or suppressed or greatly delayed by the removal of MH. Less extensive regeneration, including mitotic activity, may occur independently of MH.

A study of contrasting features of the activity of the pericardial glands in normal and regenerating (with the mesothoracic legs removed) Ist instar nymphs of Carausius morosus was made by Vietinghoff, Penzlin and Spannhof (1964). A statically significant difference in the volumes of nuclei was found between the glands of regenerating and normal specimens. The decrease in volume of the nuclei in normal specimens connected with the onset of the moulting process, was delayed by two days in the regenerating individuals.

Many researchers whve examined the effect of injury and wound healing on local cuticle formation on the general moulting process (Krishnakumaran, 1972). In both cases the effect was positive. The effect of injury on the general moulting process seems to be mediated by the brain, since brains from injured specimens, unlike those from the intact controls, displayed AH activity on both the pgl and ca. In the case of local cuticle formation, which was also observed in adults, the presence of MH does not seem to be absolutely necessary for the cuticulogenic activity of the epidermal cells. This is also pointedout by the findings of Gersch (1972), who observed chitin hypertrophy in the salivary glands of *Periplaneta americana* following injury in the thoracic region (pgl extirpation).

Schaller and Andries (1970a, b) and Schaller (1972a, b) analyzed the regeneration cells in Aeschna cyanea (Odonata) larvae deprived experimentally of their vgl resulted in an abrupt and significant increase in the activity of these cells. In another research paper, Schaller (1970) concluded that MH was responsible both for the activation of remaining regeneration cells and for regeneration of the midgut epithelium in the intermoult period, although these cells, like the cells of the eye imaginal discs, also displayed lower mitotic activity in the absence of the hormone. On the other hand, no mitosis was observed in the absence of MH in permanent larvae. A detailed study of the eyes and optic lobes in permanent larvae of the same species fully confirmed the above conclusion. Keeping in consideration the gradient-factor theory, it emerges as an interesting observation.

The influence on the morpholigical colour change

Remarkably visible morphlogical colour changes occur in various insect species during metamorphosis. Few exhibit this in the adult stage, but in the majority it is a transient phenomenon. These changes may be quite affected by the suppression of premature introduction of MH and can either be suppressed or prematurely induced. Because of this, many researchers have assumed that the moulting hormone exerts a direct effect on the chemical synthesis of the corresponding pigments. Buckmann (1956-1959), for example, in his extensive studies of morphological colour changes in *Cerura vinula* larvae during metamorphosis reached the conclusion that these changes are caused by low concentration of the MH, whilst large doses of the same hormone induce pupation. Generally the green caterpillars turn dark at the time the cocoon is made due to the production of a red ommochrome, which first appears in the epidermis and one day later in the fat body. The morphological process of the pupal moult does not start until five days later. The colour changes are completely halted by ligaturing the part of the body lacking the MH. When 66 *Calliphora* units were injected into such a permanently green abdomen, they cause reddening of the epidermis. Three hundred and thirty Calliphora-units caused reddening of the fat body only, whilst very large doses, 330 to 6600 units caused pupal moulting without any colour change. Buckmann (1947a) concluded that there the effect of ecdysone could be due to a simple reduction of the ommochrome pigment and he suggested that 'das ware ein idealer Modelfall einer chemisch einfach definierten Hormonwirkung'.

Even though an alternate, direct dependence of a pigment reaction on the hormone cannot be excluded, the following explanation seems to be the most probable : the effect of MH consists, as in other-metamorphosic processes, in the conditioning of species specific chemical and morphological changes, in the normal course of development. The discovery that removal of the ca (by decapitation) in the earlier, larval instars result in similar changes in the larval periodfullly accords with this.

Doubtlessly a similar dependence would be found in all other cases of morphological colour change described for various other insect species, with an experimental approach similr to that used by Buckmann with *Cerura.* It is confirmed from the experiments of Wigglesworth (1940) with *Rhodnius* that there is no direct influence of MH on pigment vanishes from in some places where it was earlier present.

The correlation between the prothoracic (= ventral) glands and phase development

This was examined in *Schistocera gregaria* and *Locusta migratoria,* The prothoracic glands which gemerally disappear in adults were found to sruvive for a longer period in the solitary than in the gregarious phase. Besides this the ca are larger in solitary than in gregarious nymphs. Extirpation of the greater part of one of the pair of the prothoracic glands in solitary nymphs ***resulted in the assumption*** of the gregarious habit. It is accepted that a complex interaction exists among all three metamorphosis hormones in fluenceing phase differences.

The action on the development of internal parasites

The group A virus *Sindlois arbovirus* is known to reproduce during the metamorphosis of *Calliphora erythrocephala.* In larvae deprived of their MH source by ring gland extirpation, no multiplication of the virus takes place.

The Control of MH Production

The secretion of the MH by the pgl and their equivalents is influenced by the presence of a certain effective concentration of AH in the hemolymph. It is therefore the control-system of AH which, initially, also controls the production of MH. It was shown by the experiments of Williams (1952) that pgl may be activated to produce MH even in the absence of AH by the introduction of MH. Thus, diapause in *cecropia* pupae may be broken by the implantation of active pgl or by the injection of ecdysone into pupae with inactive pgl, even in an amount which itself would not be sufficient to initiate the moulting process (in specimens with their pgl removed). One the basis of these experiments Williams suggests that co-ordination of secretory activity between the left and right pgl is definite. Furthermore, extensive or repeated injuries and the regeneration resulting from these injuries are brought about by the reactivation of pgl. They may even in some conditions cause supernumerary moulting in adult insects. But, such effect is probably restricted to the phylogenetically 'lowest' groups of Hemimetabola such as cockroaches, not yet every far remote from the Apterygota where moulting persists in adult insects.

New information on this subject is available in recent years. Zdarek and Fraenkel (1971) studied neurosecretory control of MH release during pupariation of *Sarcophaga argyrostoma*. In an earlier paper they demonstrated that there is strict environmental control of the initiation of the pupariation process in Cyclorrhapha. The larvae do not pupariate unless they leave their wet feeding medium and spend one or two days in a dry environment. However, implantation of the cerebral and ventral ganglion, together with the rgl, induced pupariation even in wet larvae. If the rgl alone was implanted, only larvae which has spent some time (insufficient by itself) under dry conditions pupariated. Neural connection with the cns proved to be essential for an independent pupariation effect. The activity of the rgl consistently increased in the pre-pupariation period in active flies.

Gersch and Sturzebecher (1968, 1970,) carried out a detailed examination of AH and its effect on the pgl in *Periplaneta americana*. The pgl were experimentally activated by one larval brain of *Antherea pernyi* of three brains of *Periplaneta americana*. After 8 hours' exposure to AH, a substance was produced in the pgl. Thin-layer chromatography and mass spectrometry have confirmed this substance to be a non-separated mixture of stearic, oleic, linoleic, linolenic, palmitic and myristic acid.

According to further discoveries on this subject, some juvenoids were found at activate diapausing or decerebrated *Antherea polyphemus* and Hyalophora cecropia pupae. Their effects on the pgl were proportional to their juvenilizing activity.

Sehnal and Edwards (1969) examined the inhibition of pgl activity body constraint in *Galleria mellonella* larvae. Constraint inhibited AH secretion and the authors interpreted this a nervous inhibition (possibly owing to the elimination of an environmental nervous impluse). MH production was also inhibited, although the epidermis remained capable of its moulting function (demonstrated by the administration of ecdysterone). This indicates that an appropriate external or internal stimulus is needed for activation of the neuroendocrine system.

Ohtaki *et al.*, (1968) examined the rapid decrease in MH activity in the insect organism after hormone administration. They concluded that a special inactivating mechanism was present in the tissues, but not in the haemolymph. When injected into mature *Sarcophaga peregrina* larvae, 1 Mg ecdysone lost 50 per cent of its activity in 1 hour and by 8 hours only 2 per cent remained. In the critical phase for pupariation, the larva contained only 7 per cent of one *Sarcophaga* unit. According to the researchers, this could have been because of accumulation in the tissue of convert effects determining the overt response. No accumulation was observed in the haemolymph. A more practicle explanation, however, is that early change in the target tissue (the ectodermal cells) are reversible up to a given point of time, until then the presence of MH is indispensable; but not later than that.

Chemical Characteristics of MH

Isolation of MH

The first prothoracic gland hormone of the insect metamorphosis hormones to be prepared was in crystalline form. Butenandt and Karlson (1954) used 500kg of silkworm pupae to prepare an extract using the following technique.

The pupae preserved in methanol were subjected to a pressure of 50atm and the resulting extract reduced by evaporation in *vacuo* . The 30liters remaining from about 650 liters of the original extracts were shaken several times with butanol. The separated butanol extract was then washed successively with 1per cent H_2SO_4, 10 per cent Na_2CO_3 and 5 percent acetic acid. It was evaporated and the remaining red-brown oily substance (79g) was redissolved in butanol, and after filtration through about 800g alluminium oxide, following Brockmann (activity V), re-evaporated and dried in *vacuo*.. It was then dissolved in water, shaken successively with ether and ethyl acetate and the water phase again evaporated. The remaining 1.15g of red-brown oil showed and activity of 1 *Calliphora* –unit in 0.5g. It was then purified by chromatography in aluminum oxide activity IV and, using ethyl acetatebutanol (3:1) and ethyl acetate-methanol (3:1), 167 mg of a highly active fraction (1 Calliphora-unit in 0.1µ 1) was obtained. After recrystallizing from ethyl acetate and methanol, 25mg of crystalline substance was obtained which depicted an activity of 1 C.u. in 0.0075 Ml. It is called endysone and has been identified with MH.

Chemical characteristics

This crystalline preparation formed silky fibrous needles. These change their appearance at 161-162°C probably because of the loss of water of crystalization, and melt at 235-237°C. Within these limits the substance is heat stable; it is soluble in water and very soluble in methyl alcohol; it shows distinct absorption bands in the ultraviolet and infrared parts of the spectrum. The ultraviolet spectrum reaches its maximum at 244nm (in ethyl alcohol) and at 249 nm (in water); and the infrared at 612 nm. It is optically active:

$$(d)\frac{20°}{D} = 58.5° \pm 2° \qquad (d)\frac{21}{546} = 82.3° \pm 2°$$

(at a concentration of 7 mg in 2ml ethyl alcohol).

Depending on these findings the presence in the molecule of α and β unsaturated keto groups was assumed. The infrared part of the spectrum also suggests the presence of at least two hydroxyl groups. It was further inferred that a $_2CH_5$ group and perhaps also a Ch_3 group are present. Both acids and bases easily inactivate the substance. There is another band in the spectrum at 265 nm which may be assumed to have originated by the splitting off of water which led to the formation of the double unsaturated system.

The hormone decomposes quickly in alkaline solutions, the decomposition being accompanied by a successive change of spectrum. The first analysis of the dried substance showed elements in the following ratio: $C_{4'},\ _4H_{7',3}O_1$. The molecular weight was originally shown to be about 300; and the emperical formula $C_{18}H_{31}O_4$. Concentrates from Crangon vulgaris which were found to be more active in crustaceans than in insects point to the existence of several different chemical and physical characteristics (e.g. increased polarity) in the crustacean hormone. Further research has confirmed that at least two differently substance are present with identical physiological activity, which are separable into α-ecdysone and β-ecdysone (now ecdysone and ecdysterone, by extracting with a cyclohexane-butanol-water mixture. The two substances may be separated by their melting points and solubilities though their spectra are very similar.

Karlson and Hoffmeister (1963a) worked incessantly on the chemical properties of ecdysone. 1000 kg of dried pupae of *Bombyx mori* removed from cocoons obtained from the silk factory were used. From these 250 mg of crystalline ecdysone with a specific activity of 100 000 *Calliphora*- units per mg was obtained.

With the help of X-ray analysis and mass-spectroscopy a molecular weight of 464 was confirmed. This corresponds to the empirical formula $C_{27}H_{44}0_{6.}$ Evidence was obtained of the steriod nature of ecdysone, one oxy-group and five hydroxyl-groups were found. On the basis of the empirical formula a relationship with cholesterol was suspected. Injection of tritium-labelled cholesterol into *Calliphora* larvae yielded crude radioactive ecdysone after extraction and purification. It was then confirmed that cholesterol is a precursor and ecdysone. A similar observation was made by Kanazawa and Teshima (1971) with C^{14}-labelled cholesterol in the spiny lobster *Panulivus japonicus.* The activation of the prothoracic glands in decerebrated *Bombyx* by ether of brains and by cholesterol, found by the Japanese authors, could probably be explained by this. It is no way contradicts the fact that cholesterol is inactive in diapausing sawfly larvae.

The substance proved to be active not only in various Lepidioptera but also in Hymenoptera, and Hemiptera Karlson (1957) later succeeded in isolating a similar substance from the shrimp *Crangon vulgaris,* using the same method. The *Calliphora*-test shows a distinct, but much lower activity in this substance (1 C.u. = 50 mg of the dried oil;. Similar preparations have been obtained from *Carcinus moenas* and *Portunus holstaus.*

Almost similar and yet more active MH preparation has recently been obtained by Stamm (1958) from the grasshopper *Dociostaurus maroccanus.* The researcher used 10 kg of adult specimens which yielded 11 mg ecdysone and 12 mg ecdysterone in crystalline form. The substances were shown to be chemically identical with the silkworm α-and β-ecdysone and were found to be more active in *Calliphora*–tests. The reduced activity of the silkworm preparations is assumed to be due to the presence of kynurenin and others substances harmful to *Calliphora.*

A number of research papers by the Karlson school deals with the question of the chemistry of the MH and the biochemistry of its production. On the basis of their work the definitive solution of the chemical structure of ecdysone was made by Karlson (1965). As stated earlier, it is a steroid molecule of the following structure:

It is a 2β, 3β, 22β F. 25=pentahydroxy-Δ7-5β-cholesteron-(6). The C_{20} has the normal steric configuration, i.e. 20β F-methyl 6 – keto-0 =0 –1; 14α - 0 = 0-2, 2β-0=0_3, 3b0=0-4,25-0=0-5,22-0 = 0-6.

Five various biologically active fractions of the ecdysone are assumed to have been obtained by chemical treatment of *Bombyx mori* extracts, three of them being undescribed and two known from the work of Karlson and his colleagues and earlier from that of Burdette and Bullock (1963). A positive effect on tissue growth in mammals is claimed together with stimulation of protein synthesis in the cytoplasm.

Immediately after Karlson's original formula for a-ecdysone was discovered, the compound was simultaneously synthesized in laboratories in West Germany and the U.S.A. (Furlenmayer *et al.*, 1966a, b, 1967; Siddall *et al.*, 1966a,b) and its structure fully confirmed. Furlenmayer *et al.* started with ergosterin and used 14 steps. The main steps were (20S)-2, 3β-diacetoxy-20-formyl-5α-pregn-7-ene-6-one and (22R)-2β, 3β-diacetoxy-14-22-dihydroxy-25-(tetrahydropyran-2-yloxy-)-5α-cholest-7-ene-6-one. Later ecdysterone was also synthesized.

As soon as the chemical structure of ecdysone and its artificial synthesis were resolved, the number of studies on its various chemical properties and those of its many natural and synthetic derivatives rapidly multiplied.. The most complete survey is the latest one by Slama *et al.* (1973), which is yet to be published.

The Mode of Action of MH

Inspite of the fact that the chemical composition of the moulting hormone has now been fully determined, the mode of action of this, the best known insect hormone, is still not properly understood. The way it affects the cytochrome oxidase system has already been mentioned, as has it action on the ommochrome changes. Its effect on development of the epidermal glands in *Ephestia kuehniella* has been examined experimentally by Hanser (1957), using the ecdysone preparation of Karlson and Butenandt. His paper shows : (1) The injection of MH at any time during the intermoult period induces the moulting process. This commences with the loosening of the old cuticle and the secretion of the moulting fluid, and continues with the formation of a new cuticle (pupal cuticle). (2) A single dose, even if rather strong (40 C.u. or more), is inadequate for the completion of the moulting process. If no further dose is supplied the moulting process stops in one of its later stages and can only persist when more ecdysone is injected.

It can be confirmed from the findings of Hanser (1957) and the results of earlier experiments with parabolses, and pgl transplantations that MH influences the activation of all the enzyme of the epidermal cells. The first phase of this activation consists of the swelling of the cells, which may or may not result in mitotic activity. It is followed by the second phase, the secretory activity of the cells, resulting in the production of moulting fluid, the initial effect of which is the loosening of the old cuticle by chitinase which subsequently dissolves the inner layers of the cuticle. In the third phase, the new cuticle is deposited and the secretory activity of the cells continues yo produce the first layers of the new cuticle, most probably under the action of the same enzyme (chitinase). The phases generally overlap one another. It seems that the presence of MH is imperative at least to begin all these phases.

The *consequence of thiourea injections* on moulting and pupation in silkworm (*Bombyx mori*) larvae were exmined by Chmurzynska and Wojtczak (1963). Injections up to 24 hours before larval moulting or cocoon spinning resulted in the retention of the old cuticle and accumulation of a haemolymph-like fluid between the old and the new cuticles. Partial inhibition and a delay in the sclerotization of pupal cuticle was observed in histological sections.

Advancement in the technique for isolating MH is small quantities have made it possible to compare the MH titre at various stages of an instar and in various periods of postembryonic development (Shaaya and Karlson, 1964; Shaaya and Sekeris, 1965; in *Calliphora erythrocephala*). There are two maxima-one just prior to pupation and the other on the 3rd to 5th day of the pupal instar. The first peak correlated with puparium formation, and the second with histogenetic processes. No ecdysone could be detected on the last two days of the pupal instar. The same method was employed to examine the activity of several enzymes was studied in a similar manner.

Many other research investigations were carried out to examine, the ecdysone titre in relation to body weight, in two lepidopteran species–*Bombyx mori* and *Cerura vinula*. In *Bombyx mori* a maximum was found in the IVth (penultimate) and Vth larval instar shortly before ecdysis; immediately before ecdysis occurred MH fell to zero. Two maxima were observed in the pupal instar. Some ecdysone was still present after adult ecdysis (about four times more in the female than in the male), but by the 8th day it had virtually disappeared. Two maxima were found in *Cerura vinula* – one at the beginning of the colour change and the other shortly before pupation. No ecdysone was detected in the pupal instar. But, the information may not corre;ate to the amount of active MH in the haemolymph. Adelung (1966) estimated MH titres similarly in the *crab Carcinus moenas*, and discussed their variation in the context of the crustacean endocrine system, i.e. the X organ-sinus gland complex and the rostral gland (Y organ).

The number of individual secretory and morphogenetic processes which take place in the epidermis during moulting cycle and their a hormonal dependencies were discussed by Locke (1965). From the results of experiments on the moth *Calpodes ethlius* (Hesperidae), Locke asserted that whereas the moulting process is controlled only by ecdysone, intermoult events such as wax secretion and endocuticle deposition need the assistance of a neuroendocrine factor from the brain.

A combination of recent views and findings on moulting control was published by Weir (1970), who on the basis of his experiments on *Calpodes ethlius* classified three phases of development in the last (Vth) larval instar. The first is the growth phase, from ecdysis up to 66 hours. After that the brain is no longer required for the induction of pupation, and the next phase, up to 156 hours, ends when the pgl are not longer necessary. Pupation takes place at 192 hours. From an analysis of the data on the pgl, the researcher concluded that certain abdominal tissues itneracted with the pgl in the induction of the moulting process and stated that the oenocytes might be invovled.

The Juvenile Hormone (Neotenin) (JH)

The metamorphosis-preventing activity of the corpus allatum was initially demonstrated by Wigglesworth in his classic parabiosis experiments with the blood-sucking bug *Rhodnius prolixus* in 1935. Wigglesworth (1936) demonstrated that joining a nymph in the penultimate or an earlier instar with a nymph at the beginning of the last larval instar results in the partial or total suppression of metamorphosis without preventing an increase in size. In this way VIth instar giant nymphs imagos. Conversely, when the ca of the younger nymph was removed by decapitation, a miniature precocious imago resulted.

Wigglesworth not only the source of this inhibitory effect but was also the first to provide definite evidence of its hormonal nature by demonstrated that the same results are produced when the specimens are connected by a glass tube so that only the haemolymph communicates. The original term 'inhibitory hormone' was soon changed to 'juvenile hormone' after the active role of the hormone in producing larval characters in adults was recognized. This discovery was confirmed by a number of other authors for various insect species, Bounhiol (1936, 1938a) for *Bombyx mori*; Pflugfelder (1937a, b) for *Carausius morosus*; Piepho (1938b-d) for *Galleria mellonella*; Bodenstein (1938b) and Weed-Pfeiffer (1938) for *Melanoplus differentialis*.

Contemporary researcher, Wigglesworth (1935, 1936) also discovered the effect of the ca hormone on ovarian development. This was later confirmed for a number of other species whilst in ohters, such as *Bombyx mori*, no ovarian control bya corresponding humoral factor was observed. The other JH effect were ascertained subsequently. The apparent contraction betwee the inhibitory, metabolism increasing influence, suggested to a number of researchers that there were two or more ca hormones having different effects. However, this hypothesis theory was not subsequent experiments, and it was soon recognized that all the available data could be explained by the operation of a single substance, JH, possessing the character of a growth hormone. Since then experimental observational evidence has accumulated in favour of this explanation. Interordinal non-specificity of JH was shown by the transplantation of ca between Blattoptera and Hemiptera, and hasmida and Lepidoptera. The first attempts at elucidating the chemical nature of JH were made by Williams (1956), Schneiderman and Gilbert (1958, cf. Schneiderman, 1960), using ether extracts of male abdomens of the Saturnid *P. cecropia*. Its chemical structure was first determined by Roller and his co-workers (1965a, b etc.). Many other chemically different substance, i.e. the juvenoids behaved in the same manner.

The Corpora Allata (ca)

The first description of the insect ca is probably that of Muller (1828), in *Blatta orientalis*; however, he expected the glands to be nervous in character, believing visceral ganglia supply the dorsal aorta and the anterior portion of the gut. This analysis was accepted by a number of other researchers. The first doubts regarding the nervous character of the ca were expressed in 1899 by Janet, from his detailed anatomical study of the female of the ant *Lasius niger* and to whom we owe the term corpora allata. In the same year, Heymons (1899) also concluded that the ca in *Carausius morosus* were non-nervous. The first systematic, comparative morphological study of the insect ca was carried out by Nabert (1913) the number of research papers dealing with their structure and function has been published in great number.

Morphology

The ca are found in all groups of winged insects in which they have been searched for, and in some Apterygota, where they are very primitive. Several different types of ca can be distinguished, on the basis of structure, position, relationships with other endocrine glands and phylogenetic considerations. The simplest and most primitive type known are the bladder-shaped ca of Thysanura and stick-insects, which in some species still keep their original position on the lateral epidermis. They are followed by the paired lateral ca with cc, as in Ephemeroptera, Odonata etc. The direction of development at this stage is through the approximation and later fusion of the ecc, as in Blattoptera, Orthroptera, Mecoptera and some of the Diptera, e.g., *Culicidae.* The other tendency was to fuse the ca which can take place either below the dorsal aorta as in Hemiptera, or, above it, as in Tabanidae. A special type with both cc and ca fused and connected by the prothoracic glands fused with the nervi allati to form a single structure, is represented by the ring gland of Diptera Cyclorrhapha.

Cyclical changes occur in the ca in each larval instar as well as in the adult. The ca in the plant-feeding bug *Oncopeltus fasciatus* have their smallest volume at the beginning of each larval instar and increase gradually in size to reach a maximum at about the middle of the intermoult period. Then they reduce in size with an associated decrease in the secretory activity before the next moult, the minimum size at the beginning of each instar being distinctly less than the maximum in the earlier one. The increase in size of the ca in each larval instar thus has two components–an increase due to secretory activity (i.e. the difference between the maximum volume during an instar and that the next ecdysis), and growth of the gland itself (i.e., the difference between its volume at one ecdysis compared with the previous one). In the last larval instar a distinct increase in the volume of the gland also occurs, but is comparatively less than in the earlier instars. In adult stage a clear increase in the size of the gland byins 3 to 4 days after imaginal moulting and continues for the whole period of egg-laying (from the 8th and 30th day), when the volume of the ca increases by about 30 times compared with that immediately after moulting. Legay (1950) found a distinct increase in the volume of the gland in *Bombyx* at the time of each larval ecdysis and a very small increase during the intermoult periods. This type of growth is probably connected with the displacement of the critical period for AH and thus, necessarily, also of that for JH towards the end of the intermoult period. It is a secondary adaptation in the phylogenetically 'hightest' insects, connected with the shortening of the moulting process.

The development of the ca in *Carausius morosus* and in both larvae and adults of the honey bee (*Apis mellifica*) was systematically examined by Pfulgfelder (1958). He found a remarkable increase in the volume of the ca of workers bees in the first days of imaginal life, at the height of activity of the pharyngeal glands (between the 6th and 12th day), in connection with the first orientation flight (9th and 10th day), and at the time of maximum secretion of the wax glands (14th and 15th day), followed by a slow decrease in volume. The ca of the queen larvae are larger than those of workers larvae during their entire process of development but especially during the pupal instar, which perhaps is connected with the development of the ovaries, as confirmed by Lukoschus (1955). In the adult queen their volume remains practically constant, distinctly below the maximum in the workers.

The increase in the volume of the ca between the first and last larval instars in 18 different species of both Holometabola and Heterometabola was examined by Novak (1954). It is comparitively lesser than the increase in volume of the body, but greater than the increase in volume of the brain and nervous ganglia. Kaiser (1949) analyzed the relationship between the growth of the ca and the prothoracic glands (pgl) during larval development in Lepidoptera (*Pieris*). Whilst the ratio was 1:2 in favour of the pgl at the moment of hatching it increased to 1:29 at the time of cocoon spinning. During the last larval instar the volume of the ca increased very little whilst that of the pgl was doubled. The conclusions of this researcher was based on the hypothesis of antagonism between MH and JH have found little support. We must also keep in mind that even if the volume of the pgl does increase, compared with that of the ca, it definitely does not do so compared with the body as a whole. It was recently shown by Mala et.al (1975) that the initial indication of pgl degeneration appear soon after the last larval ecdysis because of the non-existence of JH, and can be reversed by applying JH. The moulting process is distinctly accelerated in the presence of active ca therefore it is not possible that should be any antagonism between the two hormones.

In *Hyalophora cecropia* adults, and extraordinary sexual dimorphism in the size of the ca was discovered by Schneiderman and Gilbert (1959) in contrast to that in bugs and most other insects.

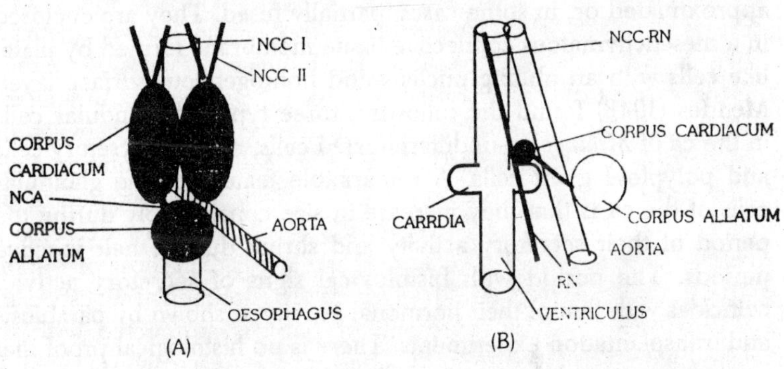

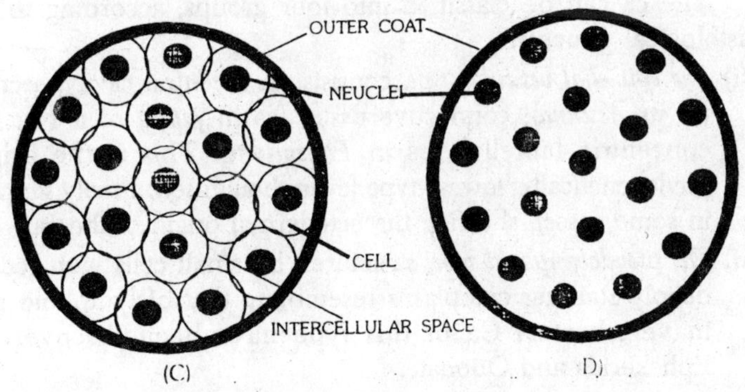

Fig. 7.14. Schematic representation of the corpora cardiaca–allata complex in–A–Hemiptera, B–adult housefly, C–cellular structure and D–syncytial structure of the corpora allata.

Whereas the pupal ca show the same weight of about 30 µg in each sex, those of adult males increase to about 400 µg whereas to those of the female adults only increase to about 65 µg. This is in agreement with the JH activity of ether extracts of male and female abdomens, that of the female beig only about 1/40th that of the male. A similar variation is observed in the ether extracted liquid (0.201 g in the male and 0.045 g in the female), which is correlated with the much greater flight activity of the male. Similarly, Rehm (1951) found a 20-fold increase in the ca adult male *Galleria mellonella*, whereas the increase in the female was only 3-fold. But the number of cells per ca is, almost equal in each sex.

Histology

The ca are sphericae, ovoid or pad-shaped bodies in most insects, connected with the cc by a nerve (nervus allatus) which may become reduced so that the two kinds of glands are closely approximated or, in some cases, partially fused. They are enclosed in a mesenchymatous connective tissue membrane formed by plate-like cells with an oblong nucleus and homogeneous surface layer. Mendes (1948) found the following three types of glandular cells in the ca of *Melanoplus*–undifferentiated cells, normal secretory cells and polyploid giant cells. A remarkable feature of the glandular cells of the ca is that they increase in size considerably during the period of their secretory activity and shrink during their inactive periods. The period with histological signs of secretory activity coincides with that of their hormonal activity as shown by parabiosis and transplantation experiments. There is no histological proof that more than one ca hormone is produced in any of the species studied.

The ca can be classified into four groups, according to their histological structure:

(i) *The epithelial vesicular type*, consists, in a central cavity, secretion (as in *Aeschna*), connective tissue (as in *japyx*) of a system of concentric lamellae (as in *Phasmidae*). This is the original phylogenetically 'lowest' type (even though secondarily simplified in some cases), showing the ectodermal origin of the ca.

(ii) *The pseudolymphoid type,* structured by small cells with reduced deeply staining cytoplasm resembling that of lymphoid tissue in vertebrates. Ca of this type have been discovered in Ephekerida and Odonate.

(iii) *The small cell type,* which is the most common, is found in Blattoptera, Orthoptera, Hemiptera, Coleoptera, etc. It is structured by numerous small irregularly shaped cells, in the intercellular lacuane of which, secretion accumulates during periods of activity. The nuclei are egg-shaped to oblong their cytoplasm being relatively clear.

(iv) *The large cell type,* phylogenetically the most evolved, and it is found in Panorpata, Hymenoptera, Trichoptera, Lepidoptera and many Diptera. The gland consists of few large cells with large, often labular nuclei and numerous uniformly distributed chromatin granules.

The histophysiological changes in the ca in adult *Leucophaea maderae* were examined compreyhensively by B. Scharrer and Harnack (1958) and Harnack (1958a, b). No great changes in the low secretory activity of the male ca were observed during the whole adult life except for a short 'activation period' just after imaginal moult. While in females significant changes occur in connection with the ovarian cycles. A conspicuous increase in the amount of cytoplasm, both absolute and relative (in relation to the size and number of nuclei), occurs during the activation period. The researchers describe the changes minutely. Harnack (1958a, b) also observed the effect of starvation on imaginal moulting. The ca volume decreased markedly shortly after a limited increase at the begining of the instar. Feeding following a long period of starvation results in a marked increase in the volume of the gland which greatly exceeds that in normally fed specmens, not only in the first ovarian cycle but also in the later ones. The researcher indicated tthe possibility of resemblance with the adenhypophysis in vertebrates.

B. Scharrer (1956, 1958) examined the impact of ovariectomy on the ca in *L. maderae,* and observed an increase in the volume of the glands, the number of nuclei, and an increase nucleocytoplasmic ratio following the operation. Similar changes follow denervation of the ca (interruption of the nervi corporum cardiacarum). The researcher calls attention to the analogy with the neuroendocrine cycle in vertebrates. The succession of developmental changes in the ca in the adult bug *Oncopeltus fasciatus* was analyzed by Johansson (1958a) and that in the pupae of *Mimas tiliae* (Lepidoptera) by Hughnam (1958b).

The disparity between ca from adult females which inhibited metamorphosis (induced 'supernumerary' larval moults) and those from adult males, which failed to produce this effect, was found by

Fukuda (1963) in the silkworm *(Bombyx mori).* This sexual difference in the activity of the ca persisted after gonadectomy, after both gonadectomy and reimplantation of the gonads of the opposite sex, and even in ca which were transplanted into another specimen after the biginning of the pupal period. The activity of the female ca was not affected by decerebration at the beginning of the pual period.

Akahira *et al.* (1967) examined the effect of the ca in 32 species of stingless bess (genera *Cephalotrigona, Trigona Scatura, Plebeia, Partamona Nannotrigona* and *Melipona)* and two subspecies of the honeybee *(Apis mellifera adamsonii* and *A.. m. ligustica)* and observed their biometry (size, number of nuclei, volume, etc). and histological characteristics. They agreed in many characteristics, but differed in others. In general, diversity was less in species of the same genus than in related genera.

Secretory activity of ca was observed histologically during sexual maturation and imaginal diapause in *Leptinotarsa decemlineata* and *Grullus domesticus.* Changes in the size of the ca were observed in association with different stages of glandular secretion. In both species the secretory cycle was further classified into three periods- a pre-secretion stage, a stage of nuclear activity and a stage of production and release of the secretory substance. The ca of constrated females displayed hypertrophy and accumulation of hormone, which in later stages could result in degeneration.

A process of the volumetric measurement of the ca was expained and used to analyse the relation between the size of theca and their activity, determined by the *Locusta* colour test. A close correlation was found. Lea and Thomsen (1969), however, found no correlation between the size of the ca, their activity and egg maturation. Activity was judged by the neurosecretory cell nuclear test.

Ultrastructure

Electron microscopic examination of the ca and its fuctional changes began with the work of B. Scharrer (1961, 1962) on adult females of *Leucophaea maderae.* The secretory cells interlock by mean of long cytoplasmic processes. Their cell membrances are folded in a chracteristic way in inactive glands. In active glands the cytoplasm of the cells becomes more abundant, the number of mitochondria and ribosomes increases, an endoplasmic reticulum and very numerous granular inclusions appear. The gland is encased by a thick membrance from which processes of varying thickness branch off and penetrate the spaces between the secretory cells. These processes form a 'stroma', containing denser bodies, through which the neurosecuretory axons of the nervus allatus penetrate the ca. the 'stroma' seems to be continuous with the adjacent cc. The researcher suggests that such process serves as a pathway through which food and other chemical are exchange between the gland and the haemolymph, including diffusion of the hormone.

Subsequently, electron microscopic studies of various insect species of different orders have been published with reference to the secretory activity of the ca, on the original elaboration, storage and release of the 'C body', a specific cellular product in the ca of *Leucophaea maderae*. 'C. bodies are found or hexagonal structures of varying size, composed of highly electron dense material. They orginate in the Golgi zone and are slowly transported to various intracellular and extracellular compartments affording them access to the haemolymph. From their incidence, distribution and fat in the insect organism, it was assumed that C bodies might have a bearing on the secretory function of the ca. Special attention was paid to the acellular stromal matrix, which seems to act as a vehicle facilitating the transport of C bodies to the haemolymph. C bodies might correspond to the 'osmiophilic bodies' described in the ca of the adult *Bombyx mori* female.

Number of research of papers on the ca of *Locusta migratoria,* their activity and the impact upon the glands of the A, B and C ncs of the pars intercerebralis was published by Joly and his co-workers. Abundant ergastroplasm was observed in actively secreting cells, which disappeared from cells in the non-active phase while the smooth endoplasmic reticulum showed conspicuous accumulation.

With reference to the previous discovery, Panov and Bassurmanova (1970) found an entirely different structure in the ca of the bug *Eurygaster integriceps*. Here, the smooth-surfaced endoplasmic reticulum was entirely absent in inactive cells, which were chracterized by diminished size, reduced cytoplasm and numerous ribosomes. The most pronounced feature of active cells was a large nucleolus formed of hollow spheres like the one observed by Blazek *et al.* (1973) in pgl cells of *Galleria mellonella,* while in inactive cells it was irregularly shaped with small vacuoles. the researcher discussed the possible fuctional significance of these changes in connection with similar observations in vertabrates (in the egg cells of *Protopterus ethipoicus* and in certain blood cells and liver cells of the rat). In addition, ultrastructural changes in the ca cells of specimens infected by larvae of the parasitic fly *Clitimyia helluo* were described (Panov *et al.*, 1972).

The ultrastructure of the ca was closely scrutinized by Deleurance and Charpin (1972) in cavernicolous beetles of the family Bathysciinae, by Odhiambo (1966a, b) in *Schistocerca gregaria*, by Waku and Gilbert (1964) in *Hyalophora cecropia* and by Fukuda *et al.* (1966) in *Bombyx mori*. The ca of *calliphora erythrocephala* female adults were examined ultrastructurally by E. and M. Thomsen (1969), with reference to hormone production. It was acceeeoted that the hormone or its precursoe originates in the smooth endoplasmic reticulum (which was found to be very abundant at the stage of incipient activity) and that it dissolves in lipids, finally forming fat droplete which are thought to be gradually extruded through the surface of the ca. The fat droplets seem to correspond to the osmiophilic bodies of Fukuda and the C bodies of B. Scharrer. An exhaustive comparative electron microscopic study of the ring gland in the larva and of the cc-ca complex in the adult female of *Drosophila mellanogaster* was carried out by King *et al.* (1966).

Embryogenesis

As opposed to earlier belief that the ca was of mesodermal or even of endodermal, origin (cf. Nussbaum, 1889), their ectodermal origin now seems to be confirmed. Mellanby (1936) proved that they originate as ectodermal invaginations in the area between the mandibular and maxillary pleurites in *Rhodnius prolixus*. The original paired rudiments fuse into a single median body in Hemiptera towards the end of he embronic period. Abnormalities due to incomplete fusion of the to embryonic invaginations, sometimes even showing two separate bodies, were often discovered by the resarcher in the plant-feeding lygaeid bug *Oncepelus fasciatus*. An increased number of such aberrations, amounting to 20 percent, was observed in animals kept at higher temperatures. It is suggested that they are the result of a selective influence on the process of fusion exerted by the increae in temperature. Similar changes have been observed in *Pyrrhocoris apterus*. The same origin was found by Pflugfelder (1937a) in Carausius morosus where the glands retain their original vesicular shae both in nymphs and adults. The embryogenesis of the ca as a part of the ring gland was elucidated by Poulson (1945a, b) for *Drosophila*. Their origin from ectodermal invaginations in bugs is also stated by Wells (1954).

Phylogenesis

Ca has been discovered in all groups of metabolous insects (Pterygoata) where they have so far been looked for. In Apterygota, they are fairly well developed in Japygidae and in a more primitie stage of development, in a lateral-position closely connected with the mandibulo-maxillar area of their origin, in *Ctenolepisma.* A distinct juvenilizing effect of the glands of *Ctenopisma* implanted into *Cecropia* pupae has been observed. Keeping in viw the earlier suggestions of the possible homology of the ca with the 'crops jugales' in Thysanura must be considered. It is probable that the original function of the ca was the activation of the ovarian follicles and that only gradually did they gain their function in larval development which led to the origin of Pterygote insects. The structure and development of the ca in the larvae of higher Diptera is quite complex. Here they form a part of the ring gland (1) the dorsal fusing of the originally paired ca; (2) the fusing of their nervi allati with the neighbouring pericardial glands; (3) the strengthening of the lateral arms and the dorsal part of the gland with tracheal trunks; (4) the bending of the ring with its upper part forward by a backwards shift of the brain. As pointed out by Cazai (1948) the formation of the retrocerebral endocrine system corresponds in a remarkable way to modern views on the phylogeny of metabolic insects based on the system of Martynov (1937) and Jeannel (1949).

The Direct Effect of JH

Ever since Wiggleworth (1935, 1936) discovered JH and demonstrated that it inhibited metamorphosis and activated the ovarian follicle cells, many rearcher papers have been written on various effects of the ca hormone and its different analogues on the most diverse structures and funcions of the insect organism, on its various biochemical processes and on its chemical composition. Actually taking the entire insect organism as a whole, and not its functions, structures, chemical processes or components has been found to be independent of JH.

The inhibition of metamorphosis

As it has been already stated, implantation of the active ca at the beginning of the last larval instar partially or completely prevents metamorphosis and results in adultoid forms or supernumerary larval instars. These extra instars (giant larvae) may actually change into giant adults at the next moult. The removal of the ca by decapitation or extirpation in earlier larval instars is followed by a precocious metamorphosis resulting in miniature adults, preceded (in Holomeabola) by miniature pupae. In most Hemimetabola only one instar occurs following allatectomy (e.g. Hemiptera), whilst in a number of the phylogenetically 'oldest' groups, such as Blattoptera and Phasmidae, two more moult appear following the operation, a special 'pre-adultoid' instar preceding adults.

The suppression of metamorphosis connected with the positive effect upon larval (isometric) growth is the quite obvious and important effect of JH one of the many features. Implantation of the c markedly accelerates the moulting process. For example, the intermoult period of the last larval instar in *Rhodnius* takes 24 days, whilst following the implantation of active ca at the beginning of the instar the animal moults in 17 day. On this Wigglesworth (1936) based is original concept of the JH effect, applying Goldschmidt's theory of differential velocities (1923). He finally came to the decision that morphogenesis could not be completed because of the acceleration to the moulting process. Such an effect really appears to exist but only till a limit in the case of the so-called progressive prothetely.

The effect of the JH on the growth and morphogenesis of internal organs

Sehnal (1965, 1968) in *Galleria mellonella*, fully accepted that JH affected the surface structures. The implantation of three complexes of brain-corpora cardiaca-corpora allata at the beginning of the last larval instar completely inhibited metamorphosis. Later implantation resulted in a series of transitions between the larval and imaginal shape of the brain and other organs.

Hlinak made the same discovery (1968) in the brains of a continuous series of transitions between a supernumerary larva and a normal adult in *Periplaneta americana*. Other similar results on the internal metamorphosis of the imaginal compound eyes were confirmed by Mouze and Schaller (1971) and Mouze (1970, 1971) who distinguished it from the metamorphosis of exteral structures,in which mitotic activity (and hence differentiation) depends on the moment at which an active MH concentration is reached. Only from that moment can JH action take effect, correlated with its particular concentration in the haemolymph. On the other hand, in the MH-independent organs, the moment at which only a minimum active concentration is reached is decisive. The results are also confirmed by Riddiford (1972) in *Hyalophora cecropia.*

Depending on corcumstances JH appears also to be capable of accelerating metamorphosis. At least this would be the most probable explanation of some of the observations of Wigglesworth (1948a), according to which the removal of the ca at the beginning of the last larval instar results in somewhat nymphoid adults. From this it would follow that the presence of JH in a concentration below the morphogenetical threshold is necessary for complete imaginal differentiation. On the other hand, the positive effect of the implantation of last instar ca on imaginal differentiation in earlier nymphs appears to have another reason.

The introduction of an active ca into an adult by implantation or parabiosis can result into a partial suppression of the imaginal characters, when an artificial imaginal moulting is induced simultaneously, either by parabiosis with the nymphs, or by implantation of active pgl, or by the injection of ecdysone or by exposing a part of the imaginal epidermis to the action of the larval hormone as in the epidermal bladder method.

The effects of JH on the indirect muscles of *Leptinotarsa* corroborated with concept of all the other effects of the hormone. There is, however, an exception–the experimental results of the action of JH and JHa on the indirect flight musculature of the cricket *Gryllus domesticus.* Here the impact of the hormone is the exact opposite of observations elsewhere: the muscles ripen without J.H. and its administration causes their degeneration. The flight musculature persists in allatectomized females while the reintroduction of JH or JHa, at any time, causes its rapid degeneration. Castration of females in the last larval instar affects neither maturation nor degeneration of the muscles. In the last larval instar the action of the hormone is the same as in all other insects. This is the only one of the yet expained paradoxical effects of JH. At present, the probable theory would be that the presence of the hormone acts as a signal for nervous action causing flight muscle degeneration. Nervous inervation is evident since removal, or even mereimmobilization of the wings also causes the flight muscle to degenerate or inhibits their maturation.

Recent data by many authors, supplemented by the study of JHa effects, show that JH can unquestionably affect morphogenesis at any point from the initiation of egg segmentation up to adulthood. This is demonstrated most convincingly by the action of JHa on embryonic development (Slama and Williams, 1966b; Novak, 1969; Riddiford, 1969-197ia, Matolin, 1970; Rohdendorf and Sehnal, 1973). Some papers describe the effects of JH on the pupal instar (Riddiford, 1972; Riddiford and Ajami, 1972) and its sensitivity to the administration of JH and JHa up to the next ecdysis. Various other authors, e.g. Sehnal and Meyer (1968), arrived at similar conclusions. Gametogenesis can likewise be inhibited by JH at any of its stages.

The influence on ovarian follicular cells

The effect of the ca hormone on the development of eggs in the ovaries was demonstrated in one of the initial researcher papers of Wigglesworth papers on JH. Wigglesworth (1935, 1936) ascertaned that the presence of an active ca is absolutely indispensable in *Rhodnius* for the ripening of eggs beyond a certain development stage, at which the ovarian follicle cells become active. Following allatectomy the growth of the eggs stops at a certain stage because of the degeneration of the follicle cells. This effect may be avoided the simultaneous implantation of an active ca from another specimen. Larval and imaginal ca show the same activity in this respect. Results similar to those with *Rhodnius* were obtained by Joly (1945a, b) with the beetle *Dytiscus,* by Weed-Pfeiffer (1945) with the grasshopper *Melanoplus differentialis,* by B. Scharrer (1946) with the cockroach *Leucophaea maderae,* by E. Thomsen (1952) with the blowfly *Calliphora erythrocephala* and by other researchers with other groups of insects. The research on *Calliphora* was further developed by Possompes (1955), according to whom extripation of the ca suppresses the ripening of the eggs only when performed in the adult. But it is still unknown whether or not there is at least a partial regeneration of the ca after its earlier removal. In the higher Diptera, the adult ca are quite different from those of the larvae (the dorsal part of the ring gland) so that their development may not be prevented even by the complete removal of the larval structures. The necessity of JH for ovarian development has been found in most species examined in this respect, although there are species where no JH control exist, such as *Carausius morosus. Bombyx mori,* etc. This independence of the development of the eggs seems to be a phylogenetically secondary adaptation connected with a particular type of egg-laying; for example, in *Bombyx* all the eggs

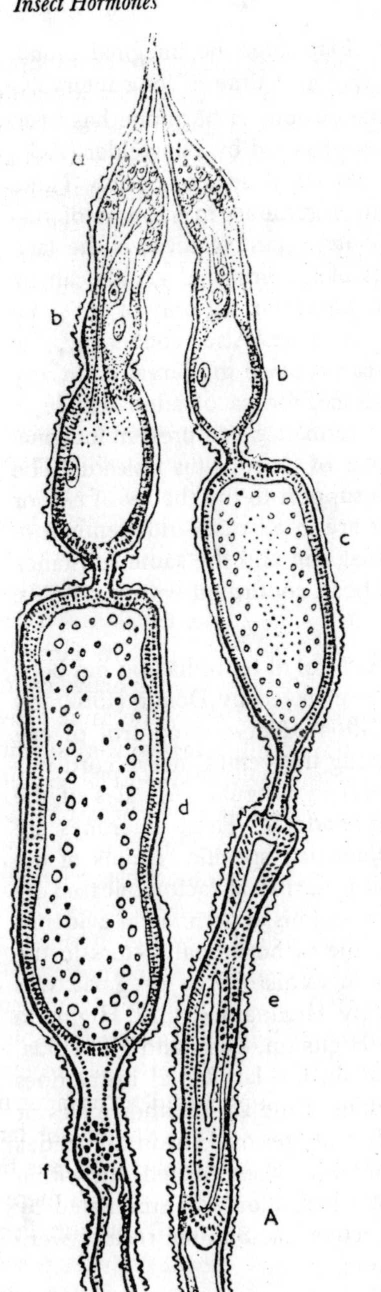

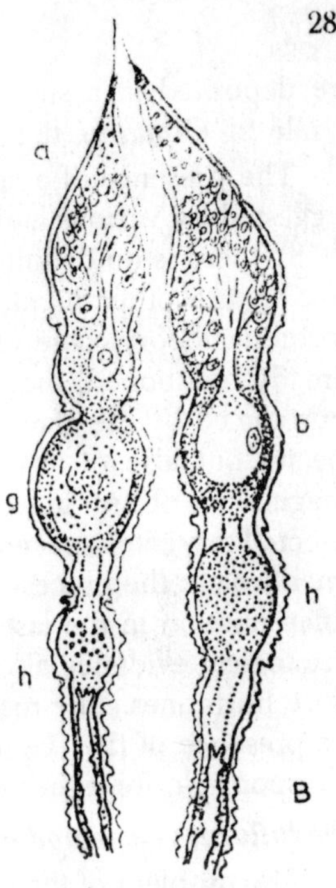

(a) germarium with nurse cells, (b) oocytes nourished via nutritive cords from nurse cells, (c) oocyte nourished solely by follicular epithelium, (d) egg almost fully developed, (e) empty follicle, (f) last remains of corpus luteum, (g) dead oocyte, (h) residues of dead oocytes and follicles undergoing autolysis

Fig. 7.15. Effect of JH on egg development in Rhodnius prolixus. A–Two ovarioles of normal female with active c. allatum; B–Two ovarioles of allatectomized female.

are deposited in a short time, not long after the imaginal moult. While in *Carausius*, they are laid one at a time at long intervals.

The presence of a special gonadotropic ca hormone has been suggested by various authors. As emphasized by Pflugfelder (1952, 1958), there is no definite proof in favour of this conclusion. There are, on the other hand, significant arguments in support of one hormone theory. One of the most important of these is the fact, already mentioned, that the effects of the imaginal glands can be reproduced by the glands of any larval instar not only of the same species but also of any other insect order. Also, conversely, the imaginal ca of any insect species can replace the larval ca in any insect (Pterygote) species. For instance, the ca of adult female P. americana in the process of ootheca formation can prevent imaginal differentiation in the last larval instar of Oncopeltus fasicatus. The histological evidence also lends no support to the theory of two or more hormones. The male gonads are in all cases independent of the presence of the JH. The fact that one and the same substance is responsible for either effect has been confirmed with juvenoids.

The influence on reproduction

The problem of the endocrine control of reproduction has been dealt with by many researcher and reviewed by Doane (1962), by Highnam (1963) and by Nayar (1964). Nervous control of egg maturation by specific neurones along the ventral nerve cord was demonstrated by Engelmann (1964) in pregnant females of the viviparous cockroach, Leucophaea maderae. These nuerones are supposed to inhibit the ca by influencing specific regions of the brain. The researcher assumes that a hurmoral factor released by the brood sac affects these neurones, but his experimental evidence does not seem to prove this. The role of hormonal factors in the control of ovarian development in Schistocerca gregaria was investigated in a series of papers by Highnam (1962), Highnam and Lusis (1962), Lusis (1963) and Highnam, Lusis and Hill (193a, b). In Locusta migratoria castration in the last larval instar does not seem to affect the cellular volume of the ca in either males or females. Allatectomy in a highly autogenous strain of Aedes taeniorhynchus inhibited egg maturation when carried out within one hour after adult moulting; yolk deposition was stimulated by the implantation of one pair of active ca. Similar results were obtained in other aedine mosquitoes.

Some researchers have recently reported quite a different type of JH and Jha action. JH administration in the early phases of

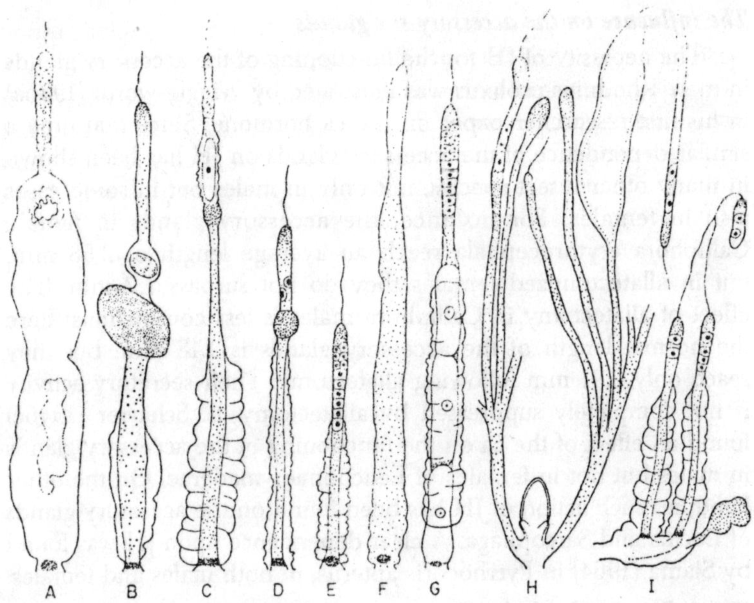

Fig. 7.16. Ovarioles of Thermobia domestica after juvenoid administration to preceding instar 20 to 50 days previously.

gametogenesis not only does not promote the process, but, on the contrary, inhibits it. The sensitive period for this inhibitory effect is, in general, earlier than for follicle cell stimulation. In some of the insects studied it comes at the beginning of the adult stage, in others in the pupal instar, or (in Exopterygota) in the last larval instar. It includes the whole period of gametogenesis, from the first oogonia and spermatogonia up to the mature sex cells. In principle, this is the same 'inhibition' of morphogenesis as observed in any other phase of ontogenesis. As with embryogenesis, the results of such inhibition of growth can differ considerably, depending on to the time of administration of the substance and the duration of its effect. Various kinds of deformation of the ovarioles mostly result.

Many researcher papers have been published on the ca and Jha endoerine control of oogenesis and of reproduction in general. Whatever the conclusions reached by individual researcher, the observed facts are all in agreement with the original explanation by Wigglesworth (1936) of the effect of JH on the ovarian follicles in Rhodnius prolixus.

The influence on the accessory sex glands

The necessity of JH for the functioning of the accessory glands in male Rhodnius prolixus was illustrated by Wigglesworth (1936a) in his first researcher paper on the ca hormone. Since that time a similar dependence of the accessory glands on JH has been shown in many other insect species, not only in males but in some cases also in females. For instance, the accessory glands in female Calliphora erythrocephala reach an average length of 2.58 mm, but in allatectomized females they do not surpass 1.7 mm. The effect of allatectomy in Calliphora males is less conspicuous: here the normal length of the accessory glands is 1.18 mm, but they reach only 0.91 mm following allatectomy. Their secretory activity is not completely suppressed by allatectomy. B. Scharrer (1946c) found an effect of the ca on the functioning of the accessory glands in males but not in females of Leucophaea maderae. On the other hand, no such action of JH has been found on the accessory glands of Lucilia and Sarcophaga. A clear dependence upon JH was found by Slama (1964) in Pyrrhocoris apterus, in both males and females.

The influence of polymorphism

Out of many types of polymorphism in insects the following have been found to be closely connected with JH: (a) The mechanism of caste polymorphism in social insects has been found to bear some relationship to JH activity, as for example in ants and in termites. The latter authors originally presumed the existence of a special, third ca hormone responsible for soldier production if administrated to larvae or pseudergates at the 'right time'. This possibility was ruled out by Luscher (1969), however, who showed that Roller's synthetic JH also induced the same effect. Soldier differentiation requires much larger JH does than those needed for the normal; i.e. juvenilizing effect. (b) Phase polymorphism in locusts. The influence of JH on the development of Locusta and Schistocerca, has been studied by Staal (1959, 1961), L. Joly (1960), Kennedy (1956, 1961), Loher (1960) and others. Their findings point to a complicated dependence of phase on JH activity. As confirmed by Staal, the implantation of active ca into young larval instars induces the green colouration typical of the solitary phase even in environmental conditions which would otherwise produce the gregarious phase. The adaptive value of either phase in an environment which induces this particular type of development has been emphasized by Kennedy (1961). The existence of a volatile factor acting through antennal chemoreceptors seems probably on the basis of the results of Loher (1960) who was the first entomologist to discover a substance accelerating sexual maturation in adult male locusts.

More information on this subject have been obtained by Joly and his co-workers. Seasonal polymorphism in aphids. A very complicated type of polymorphism is encountered in aphids. There is a strict seasonal alternation of forms with a pre-ponderance of parthenogenetic generations. The polymorphis forms differ with regard to the occurrence of wings, the presence or absence of vivipary and in a number of other, less marked, characters, both morphological and physialogica. Even though we are unfamiliar with the the various mechanisms involved, there are definite suggestions that the metamorphosis hormones and perhaps other chemical factors play a part in the determination of the character of the various forms. The apterous virginoparae of Megoura, for example, appear to be produced by the neotenic effect of increased JH activity. This activity, however, seems to take place during the embryonic period through the action of the maternal endocrine system on the ca of the older embryo. The still earlier activation of the embryonic ca by the maternal body appears to be the cause of differentiation of the oviparea. Similar conclusions were reached by Johnson and Birks (1960), with an Australian aphid species (Aphis craccivora Koch.), who assumed that all aphids started their development as presumptive winged forms which may be diverted from this path of development by various stimuli in the determination period between the late embryonic stage and the second larval instar. These researchers assumed that the results they obtained, could be applied to to other types of insect polymorphism.

The effect of JH on aphid morphogenesis was examined by Holman (1973). In a detailed analysis he found that the statistical regression expression the relationship between the length of individual parts of the body and total body length or the breadth of the head capsule differed relatively little in the 1st to Ivth instar, but displayed a significant shift between the Ivth instar and the adult. The regression curve corresponded to the total changes which occurs in any part of the body during metamorphosis. In most cases the absolute value of the shift was also dependent on body size. In supernumerary instars and juvenilized adults, this imaginal shift was reduced to various extends. The degree of JH effects can be determined from a comparison of these data with the average normal individuals. Mathematical formulae for the statistical evaluation of these effects are given. (d) Alary polymorphism in bugs. Alary polymorphism or the occurrence of divers degress of wing development in one and the same species is a common phenomenon in Heteropetera. A theory has been elaboratied by Southwood (1961) based in Wigglesworth's findings (1960) on the hormonal control of this type of polymorphism. Brachyptery is assumed to originate on the principle of metathetely in some species, particularly in colder, mountain conditions. It may be classified as a juvenile character in adults following the normal number of nstars (five in Heteroptera). In these insects (such as Gerris, Nabis, Bryocoris and various Tingidae) macropterous forms are found mostly in warm localities. On the other hand, brachyptery is assumed to originate in other species on the principle of prothetely (paedogenesis).

In species such as Dolichonabis limbatus and Microvelia, long-winged specimens appear in cold, mountain conditions whilst in the normal warm environment only brachypterous forms occur. This theory is yet unconfirmed as stated by its author. (e) Sexual alary dimorphism in Lampyridae. Davydova (1968) described the effects of ca implantation and extripation on wing development and metamorphosis in both males and females of Lampyris noctiluca (Coleoptera). Allatectomy often resulted in considerably prolongation of life without metamorphosis. Prothetelic males (regressive prothetely) with smaller wings were sometimes obtained. The implantation of several ca induced one or two supernumerary larval moults in the females, without inducing any wing development. The absence of wings in the females, together with other neotenic characters, is obviously of earlier phylogenetic origin and is incapable of phenotypic modification.

The effect on regeneration

A remarkable decrease in regeneration ability was noticed by Pfulgfelder (1939b) in Carausius morosus following extirpation of the ca. In some cases, particularly when an amputation was effected after allatectomy, there was a complete loss of regeneration ability. Generally the degeneration of whole groups of cells was observed in the regenerating epidermis whilst knotty proliferations appeared in other parts. But no difference was observed in Blattella germanica, either in the extent of regeneration or in its effect on the duration of the instar concerned, irrespective of whether the extirpation was carried out during the period of active JH concentration (Ist to Vth instar) or in its absence during metamorphosis. The relation between regeneration and the changes in volume of the ca were surutinizedin detail by O'Farrel et al. (1961) in Blattella germanica.

The duration of the moulting process was only affected when muscle tissue was involved and the greater the amount of tissue affected, the longer the intermoult period. But the moment the regeneration was restricted to epidermal structures, no changes could be seen in the moulting cycle. No evidence is available of any nervous influence on the regulation of the moulting process and all the effects observed appear to be the results of self-regulatory processes. A histological examination of the volume of the ca and that of the so-called headlobes carried out at twelve-hourly intervals showed a close connection between the volume changes of the two glands. Regeneration has a positive effect on the functioning of both pairs of glands, which is particularly interesting from the point of view of the experiments of Bodenstein (1953) and Engelman (1959) on the reciprocal effects of the ca and the pgl and their effect on the moulting process. Immediately after the moult following treatment which causes the regeneration, a hypertrophy of the ventral lobes and ca occurred. This hypertrophy, and the subsequent extensive regeneration seems to result in extra moults in the adult stage.

The delay caused by the regeneration observed in the moulting process in the Ist larval instar of Blatella germanica is followed by a shorterning of the subsequent instars so that the Ivth instar moult occurs at the same time as that of the controls. This can also be assumed to be as a of the positive effect of regeneration on the functioning of the gland concerned. The effect of metamorphosis hormones on the regeneration process in insects was explained in great detail by Bodenstein (1955, 1959) on the basis of his numerous experiments. He emphasized the need of MH for regeneration. It is only the degeneration of the prothoracic glands associated with metamorphosis which makes the adult metabolous insects unable to moult. All the tissues of the imago retain their full capacity for regeneration, which occurs in the extra imaginal moulting process brought about by the implantation of active prothoracic glands or by parabiosis of the adult with a nymph. It is yet unclear as to what extent the observed effect of MH on regeneration is direct or, perhaps more probably, whether it is only indirectly brought about by inducing the moulting process. According to bodenstein the results obtained by him contradict those of Pflugfelder with Carausius morosus, and attempts to find a common.

Penzline (1965a, b) has comprhersively analyzed, of regeneration and the ways in which it can be influenced hormonally. He illustrated that all three metamorphosis hormones were needed for normal regeneration and that it was inhibited by allatectomy. Another detailed survey was published by Needham (1965), who suggested that JH was necessary for regeneration processes to a limited extent in adult insects, e.g. the healing of integumental wounds and the regeneration of nerve fibres and internal organs. Thus capacity can be greatly enhanced by the transplantation of wing discs and leg fragments, etc., and by the implantation of active pgl. Rinterknecht (1964) found that allatectomy did not inhibit the reconstruction of the integument in Locusta migratoria, but caused a roughly two-fold delay and that it similarly slowed down the mitotic activity of the regenerating tissue. Normally allatectomy is supposed to speed up ageing.

The influence on tumour growth and other alterations in histogenesis

Including to the morphological effects earlier mentioned, JH has atremoendous impact on histogenesis. The first detailed analysis of the histological effects of ca extripation was carried out by Pfulgfelder with Carausius morosus. He noticed numerous changes of a pathological nature following alltectomy carried out in earlier instars. He distinguished tissue degenerations and tissue proliferations both of which often occurred in one and the same tissue. The most reoccuring pathological changes are: (a) Degeneration sometimes connected with a partial histolysis of fully developed tissue and often connected with a subsequent partial or complete phagocytosis. Simultaneously, proliferations usually appear as knotty structures scattered throughout the tissue, which sometimes contain single differentiated cells in the form of giant muscle cells. (b) Degeneration of the nervous system and the adjacent mesoderm and glial cells in which the formation of large irregular cysts often occurs. (c) Degeneration of the fat body during which vacuoles appear in fat cells, which resemble fat droplets but do not stain with osmium tetroxide and therefore probably do not contain either fat or any lipoid substance. It would be interesting to determine to what extentdefine the limit of JH deficiency syndromes agree with the alterations in the fat body of adults of other species, which are not connectred with patholotgical changes in other tissues. The prolonged effect of allatectomy is encapsulation of the pathologically changed cells by lymphocytes which often leads to the development of large cysts. Inside these cysts large concretions often appear which show concentrical layers. (d) Degeneration features are also, often observed in the Malpighian tubules in which cysts and concretions, similar to those in the fat body, appear. (e) Accumulations and proliferations of haemocytes (lymphocytes) often occur. These may give rise tonumerous knots of undifferentiated cells, which resemble those of spindle cell sarcoma in Man. (f) Proliferations of a carcinomatous nature are observed in the epidermis, gut and oviducts. Many changes occur following the implantation of extra ca (from adults into larvae), in, for example, the ovarioles, oviducts and other organs and tissues. But they are, certainly different from the changes produced by allatectomy.

Various modifications in tissue differentiation frequently occur during regeneration, particularly when a piece of embryonic tissue is implanted whilst there is disturbance in the hormonal balance, as shown by Pflugfelder (1948). He attempted verify this by the facf that young cells are more easily affected by external stimuli than older ones. Tumorous malignant growths have often been observed following the cutting of various nerves, such as the nervi corporum cardiacarum. Schneiderman and Gateff (1967) set up clones procured from a new mutant in Drosophila melanogaster, which they called lethal malignant brain tumour. The mutant imaginal discs grew considerably but did not respond to react the metamorphosis hormones. The mutant cells grew rapidly and invasively, killing the hosts, and behaved like true malignant tumours. They were cultivated for 35 transfer generations and further clones were secluded from single cells.

The Indirect Effects of JH

One must be careful to distinguish between the direct effects of the hormone as a chemical agent and its indirect effect, i.e. the consequences of change produced by the direct action of the hormone in various parts of the insect organism through the mediation of the circulatory and nervous system and other correlation mechanisms. Failure to note or respect this fact caused lot of confusion in the early days of insect hormone research. For example, JH was regarded by some researcher as a moulting hormone, because it induced extra larval moults, which MH was called the growth and differentiation hormone, etc. Since it is often difficult to identify the actual character of hormone activity, many authors are at variance as to whether specific hormones (e.g. JH) exert a direct or an indirect effect. There are, however, several cases where the indirect nature of the JH effect is confirmed.

The effect on moulting

The researchers, who were the first one to deal with the ca hormone assumed it to be a moulting hormone. Depending on the fact that, after the implantation of active ca into a last instar larva, one or several extra moults and supernumerary larval instars occurred. Further investigations have, however, shown without doubt that the moulting process is directly influenced by the prothoracic gland hormone (MH), and the remarkable effect of JH is actually the indirect effect of prolonging the action of the pgl which are normally histolysed during or immediately after metamorphosis. The seemingly contradictory observation of Bodenstein (1953a-e) that allatectomy towards the end of the last-larval instar in the cockroach Periplaneta americana may result in preservation of the pgl and thus cause supernumerary moults can also be explained on this basis. As demonstrated by O'Farrell et al. (1953, 1956), the amputation of legs or a more extensive injury to other parts of the body may have the same effect as ca extirpation. This, as noted above is an effect of the regeneration process, perhaps improved by selection into a self-regulatory mechanism, in the case of insects with a well-developed autotomy, and has nothing to do with the action or absence of ca. The indirect nature impact of JH on moulting is quite obvious from the fact, amongst other things, that neither the normal nor the experimentally induced action of the ca in adults causes any sign of moulting, whilst this may easily be induced in the adult, by the implantation of an active pgl.

The effect on the silk glands of caterpillars

It is generally accepted that in many Lepidopterous larvae the labial glands are almost inactive during the first top penultimate larval instar whereas they produce a large quantity of a silky material for cocoon spinning in the last larval instar. The situation is just the reverse in sawflies such as Cephalcia (Pamphilidae) where the silk glands are active throughout the larval instars producing the 'excrement webs' in which the larvae live gregraiously, and pupation takes place in cells in the soil without a cocoon. In each contact the action of the glands is connected with the presence of JH. The hormone, however, does not affects the silk glands directly, but affect their morphogenesis in a very general way. Whether its presence affects the function of the glands positively or negatively depends on genetic, species-specific factors.

The effect on instincts

The effect of JH on cocoon spinning in Galleria mellonella has been explained as a special case of affecting instinct. The larva of this species spins a cocoon before each moult. This cocoon is, however, different before the larval moults from that before the pupal. The larval cocoon is in the form of a tube open at each end whilst the pupal one is spindle-shaped and completely closed. It was shown by Piepho (1950b) that the implantations of ca at certain times before moulting caused the larva to produce cocoons intermediate between the larval and the pupal types, depending on the time of implantation. This effect cannot be expressed as a direct one on the instinct of cocoon spinning but, undoubtedly, as an indirect one through the morphegenesis of the brain. By preventing the morphogenesis of the imaginal brain at the required time JH determines whether, and to what extent, the material basis for a given instinct is developed, i.e., the corresponding nervous (cerebral ganglion) structure.

A detailed examination of this and similar effect will no doubt make possible an exact localization of various imaginal instincts in insects, especially when these differ from those in the larvae, as for instance in the case of insects with aquatic larvae. The question of hormonal influence on sexual behaviour in various species of the order Saltatoria has been described by Loher (1964). In some species, allatectomy influences the males, while in others the females are affected, and there are species which are not influenced by the ca at all (Gryllus campestris). No effect on sexual behaviour of either allatectomy or the implantation of additional ca was found by Roller, Piepho and Holz (1963) in Galleria mellonella. On the other hand, Beetsma, de Ruiter and de Wilde (1962) report that the injection of cecropia extract shifts the balance between the photopositive and photonegative orientation tendencies to larvae of Smerinthus ocellata towards a positive response. The implantation of active prothoracic glands resulted in the reverse effect.

Some JH effects are still debatable. However, only affects immediately connected with the morphogenetic processes of stimulation of larval growth, inhibited derepression of more advanced characters and suppressed degeneration can be regarded as direct. With the discovery of juvenoids (JHa), we have another type of effects which may not be the direct action of the hormone, but may be resulting from the side effects of molecular structure or an effect connected with experimental hormone supply (e.g. an effect on membrane conductivity, heart rhythm and also, perhaps, interruption of the diapause). More information on this group of effects is required before they can be properly difined. No difference has been discovered till now between the effects of the actual ca hormone and the various JHa, apart from their specificity form some groups of insects. This is not related to the mechanisms of action of the given JHa, but to their capacity for penetrating the integument, their resistance to various haemolymph enzymes, etc.

All JHa effects can thus also be ascribed to JH and vice versa. This is a specially significant in cases in which ca application is very difficult or even impossible, e.g. during early embryogenesis. It is often difficult to differntiate between the morphogenetical and the physiological effects of a hormone as well as between the direct and indirect effects. All morphogenetical effects are necessarily based on physiological (biochemical) ones, while most of the physiological effects can, sometimes, cause morphogenetical changes. In the earlier edition of this book (Novak, 1966) there were doublt about certain results of JH action among its indirect effects. Since then, however, we have acquired much information about the mode of action of this principleand my original assumpption on the primarily morphogenetic action of JH has been confirmed by dozens of new findings. Therefore, there is to be full justification for transferring most of the physiological and biochemical effects from the direct to the indirect group including, for example, all metabolic effects, together with oxygen consumption, the effect on fat metabolism and the fat body, etc.

The effect on the total metabolism (oxygen consumption)

It was indicated by Pflugfelder in as 1941 (on the basis of his work with ca implantations and extirpations in Carausius morosus) that JH is a factor which generally favours metabolism. This conclusion was later confirmed by the experiments of Weed-Pfeiffer (1945a). From her experiments on the effects of allatectomy in adult Melanoplus differentialis females, She discovered that the ca exert a positive effect on basal metabolism. She observed a connection between the presence of the ca and the consumption of food reserves in the fat body as well as an effect of JH on the increase in weight of the fat-free dry-matter and water. Further confirmation was obtained by E. Thomsen (1949), who found the allatectomized female adults of Calliphora erythrocephala show a 19 per cent decrease in oxygen consumption. A similar dependence on the presence of JH was discovered in males. Experiments carried out under the same conditions using castrated females. (Thomsen, E, and Hamburger, 1955) apparently suggest that this may be a direct, and not an indirect effect on metabolism and not, as earlier supposed by Pflugfelder (1952), caused by oxygen consumption of the developing ovaries. Pflugfelder himself (1952) observed an increase in oxygen consumption in Carausius morosus following the implantation of ca and a decrease following their extirpation. Both effects were, however, rather weak and only temporary. L'Helias (1956a) noticed a decrease in oxygen consumption in the same species following allatectomy. An even greater decrease was observed when the cc were removed at the same time.

L'Helias attempted to explain this as being connected with the accumulation of glycids and mineral phosphorus in the tissues and with the loss of energy caused by the breaking down of esters due to the increase in alkaline phosphatase. A progression of experiments on the oxygen consumption of normal and castrated adult females of Pyrrhocoris apterus, implanted with ca appears to indicate an indirect action of these glands on metabolism. It was noticed that castrated females do not show any increase in oxygen consumption compared with normal controls following ca implantation, very probably because of the increased share in the total metabolism of the ripening ovarian follicle cells which do not grow in the absence of the ca. Examination of the effect of ca implantation into the last larval instar of the same species gave corresponding results.

A direct connection was discovered between the extent of the induced morphological changes and therate of oxygen consumption. It was concluded from both series of experiments that the JH-effect on oxygen consumption is an indirect one, depending on the increase in the amount of metabolically active tissue. The direct effect of JH is probably concerned with protein synthesis. But, Sagessar (1960) claims to have theory an increase in oxygen consumption after ca implantation in the cockroach Leucophaea maderae even in castrated females and concludes; in agreement with E. Thomsen, that JH affects metabolism directly. His point of view based on the theory of two ac hormone appears tibe incorrect. Results which seem to contradict to some extent the above theory, are reported by Samuels (1956). He noticed a considerable increase in the oxygen consumption of the isolated thoracic musculature of Periplaneta americana adults of both sexes 2 to 3 months after allatectomy.

Samuels tried to explain the difference observed in oxygen consumption in the whole animal and in the isolated tissues by assuming that the glycide reserves, accumulated in the tissues because of allatectomy, and thus prevented from being used inside the organism, are freed and used in the isolated muscle, thus increasing its oxygen consumption. Slama (1960) examined the correlation between oxygen consumption and postembryonic evolution in various groups of insects; he found a typical increase at the beginning of each instar and a decrease in the second half of each larval instar. The period of metamorphosis was characterized by a typical U-shaped curve in Exopterygota also.

A similar correction between oxygen consumption and ovarian development was found in adult Pyrrhocoris females, in which typical O_2 consumption cycles were observed in connection with the ovarian cycles. The cycles were controlled in a typical manner by AH and JH. The correlation was less obvious in males, in which metabolism is low and displays no clear cyclicity (Slama, 1964b). Experiments with isolated body fragments showed that JH regulated oxygen consumption indirectly, by stimulating the growth, function and metabolism of specific parts of the body (Slama, 1964c). A similar close correlation was shown to exist between oxygen consumption and the degree of morphogenetic changes in Galleria mellonella. Sehnal (1966) performed a systematic study of the effect of JH on oxygen metabolism during larval and pupal development in the same species. No effect of ca implantation was observed in the penultimate larval instar, but implantation in the last larval instar was followed by a definite increase in metabolism in specimens which transformed to supernumerary larvae.

Effects on protein and nitrogen metabolism

The first research work on the ca effect on nitrogen metabolism was published by L'Helias (1956a), who observed the accumulation of amino acids in the tissues of Dixippus morosus following allatectomy. Janda and Slama (1964) studied the effects of JH on the body nitrogen, glycogen, fat and uric acid content in Pyrrhocoris. They were all proved to be indirect and the outcome of changes produced by the hormone on the growth process in particular parts of the body. Allatectomy, inhibiting ovarian development (and hence the utilization of reserve materials) resulted in enormous accumulation of these substance in the body. If cardiacectomy was performed at the same moment the production of reserve materials likewise stopped, so that compared with the controls no important differences were observed.

In another study, Slama (1964) observed excessive accumulation of the haemolymph proteins following allatectomy or castration, which was diminshed by ca reimplantation. Minks (1965) found differences between various protein fractions and free amino acid concentrations in the haemolymph of normal and allatectomized specimens of Locusta migratoria. The two entomologists Bentz (1969), and Bentz Girardie (1969) observed changes in ovarian protein composition during egg maturation after both allatectomy and nsc electrocoagulation. Engelmann (1969) detered the de novo production of a specific protein in the serum of adult Leucophaea maderae females treated with JH.

The protein was absent from the in nymphs of either sex or in adult males, and it disappeared in allatectomized specimens. Engelmann, Hill and Wilkens (1971), using various JHa, found dependence of a similar female specific protein in Schistocerca gregaria and Sacrophaga bullata. Engelmann (1971) further analysed these effects in the original speices (Leucophaea), using different JHa, and obtained largely the same results. Since then, female specific proteins have been found in representatives of several insect orders. Engelmann (1972) later found that synthesis of this protein was blocked by actinomycin D. In agreement with this finding, JH also increased the synthesis of total sodium dodecyl sulphate phenol-extractable RNA, which stimulated specific protein synthesis in a completed cell-free system from egg- maturing females, whereas RNA from males or allatectomized females had no such effect. As stated by the researcher the information seems to support Karlson's hypothesis of hormonal control of the transcription of specific messenger RNA.

Effects on protein metabolism

The influence of JH and the ovaries on the amount of pupal and adult body far was examined in Musca domestica by Adams and Nelson (1969), who observed a decrease in the rate of disappearance of the pupal fat body following both ovariectomy and allatectomy. The size of the adult fat body was not affected by allatectomy, but in ovariectomized specimens it was larger than normal. Baumann (1969) investigated the effect of several JHa he detectedgreater stability of the membrances, which varied with the kind of juvenoids and the type of membrane.

The influence on the fat body

It has been ascertained by Weed-Pfeiffer (1945a) that alltectomy causes an increase in the fat content which is only slightly affected by castration. Similar results were obtained by E. Thomsen (1942) with Calliphora erythrocephala and by Vogt with Drosophila. The latter researcher also observed reduction in size of the nuclei and nucleoli of the fat body cells which she explained as being due to the cytoplasm of these cells becoming filled with fat droplets. Day (1943a) also mentioned conspicuous histological changes in the fat body following extirpation of the ring gland in the flies Lucilia and Sarcophaga. It is not, however, very obvious from his experiments to what extent these changes were because of other components of the ring gland. Gilbert (1966) examined the variation in the lipid content of the ovaries and fat body of Leucophaea maderae and found that lipid synthesis by a maturing ovary was enhanced by JH application in vitro. He gathered that JH might depress lipid synthesis in the fat body, with resultant mobilization of the reserve substance on behalf of the developing oocytes. There are no proof against the opposite possibility, however, which would fit everything else we know about the action of JH, i.e. that the action of the hormone consists in activiation of the ovarian follicles.

The metabolism of the ovarian follicles is much greater than that of the fat body, so they may attract reserve substances not only from the haemolymph but from the fat body also. Stephen and Gilbert (1970) also noticed that fatty acid composition in Hyalophora cecropia changed during metamorphosis and that it was affected by JH. Here again, the effects of JH are clearly indirect and caused by the different requirements of the differentiating tissue. Morophoshi and Kiguchi (1969), Morohoshi and Fugo (1972) and Morohoshi et al. (1972) studied various aspects of the effect of JH on lipid and other forms of metabolism in Bombyx (1969) and attempted to explain their findings on the lines of Morohoshi's hypothesis. Again, however, there is no information which could not be discribed on the basis of the much simpler hypothesis formulated by Wiggleworth (1936) nearly 40 years ago.

The influence on the oenocytes

Reduction in the volume of the oenocytes and in the size of their nuclei with associated pycnosis was observed by Day (1943a) in Lucilia and Sarcophaga and by Vogt (1947) in Drosophila, following ring gland extirpation. It is difficult to decide the of changes resulting from JH deficiency and the absence of the other two metamorphosis hormones (AH and MH), which also originate in the ring gland. The possibility that the effect observed may be merely the result of inhibition of the moulting process must also be considered.

The effect on the water balance

Including the neurosecretory cells of the pars intercerebralis and the cc, the ca are also said to produce an effect on water balance. Allatectomy results in a reduction of the water content of the body, including that of the haemolymph. The effect of JH is particularly evident in castrated females where retention of the ca results in a considerable increase in the quantity of haemolymph with an associated conspicuous distension of the abdomen. In such individuals the amount of dry matter also increases. However, it is not clear exactly what effect is due to JH itself and what is because of the neurosecretion stored in the ca.

The effect on colour change

Quite obvious changes have been detected in the colouration of both larval and adult insects following experimental interference with JH production. Some of these are obviously of morphogenetical origin, such as the green colouration in the solitary phase of the migratory locust. Others are, undoubtedly because of to the brain neuro-hormone. There are, however, other changes which can be assumed to be due to JH, such as those causing the colouration of allatectomized stick-insects (Carausius morosus). The colour changes are assumed to be caused by an accumulation of uric acid in the tissue together with a decrease in the purine, pterine and melanin contents. Changes in colouration were also observed in the same species following implantation of active ca. A marked morphogenetical effect of the ca hormone on the black pattern in the lygaeid bug Oncopeltus fasciatus has been written about.

Conforming the earlier findings on the colouration of Gryllus bimaculatus (1960), Roussel (1967) obtained several interesting results on the action of ca and the control of pigmentation in this species. Roussel reported very high individual variability in the response of females to the action of JH and found agreement between the control of melanin pigmentation in this species and the gregarious Locusta migratoria. The absence of the ca leads to an increase in black pigmentation, while only a small fragment of ca induced orange pigmentation. The researcher confirmed that the ca had a direct effect on the integument.

The cuticle colouration of the tobacco hornworm, Manduca sexta, was sanalyzed by Truman et al. (1973) as the basis for an ultrasensitive JH test. The presence of JH inhibited melanization of the cuticle in the larvae of this species. If the hormone was administered up to a given moment (between the 17th and 25th hour of the last larval instar), pigment formation depended on the absence of JH. This test has the same sensitivity as the wax test in Galleria, but can be experimented upon only two days after administration.

The effect on the mitochondria

An increase in the oxygen intake is stated by Clarke and Baldwin (1960) in mitochondriae preparations from Locusta migratoria- L. on a sodium succinate substrate, after the addition of ca. The increase was, however, merely detected in mitochondria prepared from Locusta migratoria adults, those from Vth instar nymphs. Of Schistocerca gregaria showing a decrease in oxygen consumption. The addition of 2 : 4 dinitrophenol in physicological amounts, on the other hand, resulted in an opposite effect : it decreased the oxygen consumtion of the mitochondria of adult Locusta and increased that of mitochondria of Vth instar Schistocerca. The addition of both dinitrophenol and ca at the same time increased the oxygen consumption of the preparations from both species. The researcher discuss the possibility of the ca controlling metabolism by their action on the mitochondria and thè possible relationship between the effect of JH and those of the dinitrophenol. But it is possible that these are indirect effects, and that they depend on the type of mitochondria on which JH acts. JH has no effect on imaginal (JH-independent) cells, but may act on the structures of larval (i.e. JH-dependent) cells. This also seems to be indicated by the lack of differences between the alary muscle mitochondria of normal and allatectomized Locusta migratoria (Joly et al., 1969), findings by Novak and Slama (1962).

The effect on pheromone production

It has been confirmed by Loher (1961, 1962) that JH controls sexual maturation and the corresponding changes in colouration and mating behaviour in the locust Schistocerca gregaria males in a very interesting way. The maturation of the adult males of the gregarious phase in this species is marked by three distinctive features : colour change from the light-brownish grey and white of the freshly moulted adult to the yellow of the mature specimen; a specific sexual behaviour pattern; and the production by the epidermal cells of pheromone which acclerates the maturation process in young adult males in the immediate vicinity of the mature specimen by an olfactory effect on their chemoreceptors. None of these indications is demonstrated allatectomized males. They are however, induced by the implantation of active ca from a mature specimen. The same type of pheromone was discovered by Barth (1962) in the cockroach Byrsotria furnigata.

The virgin females of this species produces a volatile sex attractant which stimulates the courtship behaviour of the male. The production of this pheromone is also controlled by JH. It is not manufactured by allatectomized specimens, but its production can be induced by the implantation of active ca from another specimen. In this species no other role of JH in the reproductive behaviour of either of the sex was found. More particular JH (or ca implantation) effects in various indectswere observed by researchers. All of them were definitely indirect, or were perhaps, in some cases, caused by extrinsic neurosecretion contained in the ca. For example Baumann (1968) found that JH clearly increased cell membrane conductivity in the salivary glands of Galleria mellonella, causing strong depolarization of the membrane. The source of the hormone was cecropia oil. The same oil extracted from allatectomized males was inactive.

Here again, the action of JH can be taken to affect prolongation of the activity of similar larval (JH-dependent) cells about to degenerate, or which have already actually started to degenerate, and that this also indirectly influences membrane potential. This theory is yet to be tested in both larval and imaginal cells. Shaaya and Sekeris (1970) examined the effects of JH on protocatechuic acid-4-0, β-glucoside synthesis in Periplaneta americana. They found that JH acted in the biosynthetic chain leading from tyrosine to protocatechuic acid and that the primary site of biosynthesis of the substance might be the integument. Specific local effects of JH on the integument were described by Levinson et al. (1966). Roussel (1970) described the influence of the ca on the heart beat of last instar larvae and adults of Locusta migratoria. Allatectomy in the last larval instar slowed down the heart rate, while ca implantation accelerated it. In adults, alltectomy caused a remarkable decrease only if performed after the 8th day. The implantation of a pair of ca did not have any discernible effect on heart-beat. The researcher suggests two possible methods to bring about this effect - the induced release of neurohormones from the cc, or a direct JH action. A third possibility would be that the ca themselves contain an active neurosecretion from the brain. Extirpation of the ca in the last larval instar of the red locust, Nomadacris septemfasciata, produces a state comparable to adult diapause, which can be reversed by reimplanting several pairs of ca from normal adults.

Is There More Than One Corpus Allatum Hormone?

The seemingly opposing effects of JH, such as the negative effect on metamorphosis and the positive effect on growth, the activation of larval tissues and, on the other hand, its necessity for the development of such typically imaginal structures as the ovarian follicle cells, led to the hypothesis of two or more ca hormones. A number of indications in favor of this theory have recently been collected. It was debated by Luscher and Springhetti (1960) that the ca in Kalotermes flavicollis F. produce two different hormones which have different effects in caste differentiation, a 'juvenile hormone' and 'gonadotropic hormone.' The first of these is supposed to increase the capacity for supplementary reproductive differentiation and the other to initiate pre-soldier differentiation.

There is, however, no indication in other insects for the absence of one of the two effects in a gland which both would be present. The practice of this theory, therefore, results in complex contradictions. Sagessar supported the theory of two ca hormones in his finding that there is no qualitative difference in the oxygen consumption curve between the penultimate and the last larval instar in the cockroach Leucophaea madera. Actually, there is only a quantitative difference in the supposed production of JH, and the qualitative changes in its effects on morphogenesis can be completely explained on this basis.

Moreover, there are many other processes acting during each instar, e.g. all those connected with moulting which are practically equivalent in both the penultimate and the last larval instar and which more or less mask the changes because of differences in endocrine balance. All other experimental findings indicate towards the fact that there is only one ca hormone. According to Schneiderman uses the fact that the extract of the cecropia male abdomens does not induce the development of ovaries favours the theory of two ca hormones. This, however, is much more an argument against the identifying the extract with the ca hormone. This however, is much more an argument against the identifying the extract with the ca hormone. On the other hand, Wigglesworth (1961) found another morphogenetically active lipoid substance, farnesol, with a positive effect in preventing metamorphosis and in activating overian development.

Wigglesworth interprets his results as a 'support for the belief that the yolk formation hormone and the juvenile hormone are likely to prove identical.' There are various arguments in favour of this identity: (1) As confirmed by the histological investigation of the ca during the secretory cycle, there is no evidence for more than one type of secretion, there being only one type of secretory cell in the ca. (2) There is no difference in the endocrine activity of the ca in either direction during the whole life cycle: the ca of larvae from the first to the penultimate instar affect both morphegenesis and ovarian development in the same way as those of the adults. (3) As illustrated elsewhere, there is no major difference in the process of on growth and function of the larval parts of the body and that on the ovarian follicle cells, and both of these effects necessarily result in an increase in oxygen consumption. Even if none of these arguments can be taken as definite evidence of the existence of only one ca hormone, they are, nevertheless, much more relevant than the opposing arguments. There is therefore no evidence for the hypothesis of two ca hormones.

We get the same result from the fact that all these effects can be reporduced by a single juvenoid (JHa), i.e. by a single chemical substance. Pener (1967) arrived at the same conclusion regarding the identical nature of the ca secretion of both male and female Schistocerca gregaria adults. The minute differences which to have been noticed by this author were related to differences between the target organs in the two sexes rather than to differences in actual ca activity.

The Control of JH Production

Criteria of corpora allata activity

The activity of the ca may be understood either indirectly, by the size and the histological appearance of the gland, or directly, by its effects on the recipient after transplantation. Each of these methods has both advantages as well as disadvantages. There has been a great deal of discussion regarding the reliability of estimating the function of the ca by the increase in the volume of the gland. It must be considered, that there are various instances of increase in volume, both normal and pathological, which are not connected with a corresponding increase in hormone production. For example, the gland increase in size from one instar to the next and this increase is not related with an increase in activity per volume unit. It may be estimated by comparing the size of the gland at two consecutive ecdyses. Similarly, the increase in volume observed following castration in adult females is probably not because of an increase in the secretory activity of the gland but due to restriction in the removal of the secretion by the haemolymph.

The increase in volume owing to the secretory activity of the ca can be calcuted approximately as the difference between the total increase in the volume of the gland and the increase due to growth of the ca up to a given moment. It may be accepted that under normal conditions in comparable developmental periods or during short periods, an increase in volume is a definite sign of an increase in the secretory activity of the gland.

Histologically, the active stage is characterized by the swollen cytoplasm which stains more deeply with acidophil (nuclear) stains. The amount of cytoplasm increases in proportion to the nucleus even if the nuclei also become swollen. The local or general inhibitory effect on metamophosis following the implantation of a ca into the last larval instar is usually taken as experimental evidence of JH production. The defect of to this method is that the effect cannot be calculated until after the next ecdysis. During this time the activity of the implanted gland may change drastically. The probability of a quantitative test for ca activity based on its metabolic effects has been stated by Slama (1963). The implantation of an active gland into an allatectomized adult female of Pyrrhocoris apterus causes a specific increase in the oxygen consumption of the recipient. This increase appears to be directly proportional to the activity of the gland. Before it can be adopted for general use, however, detailed elaboration of the criteria for the test is required, together with the determination of the time limits of its applicability and the experimental conditions for its full acceptability.

The changes in corpora allata activity during the instar

The confirmation all the experimental (Wigglesworth, 1936, 1940a), histological (Mendes, 1948) and morphological (Novak, 1951b) information agree that the ca are inactive at the beginning of each instar and that the haemolymph does not contain an effective quantity of JH at that time. The amount of hormone gradually increases and the maximum volume of the ca in usually observed in the second half of the intermoult period. The production of the hormone ceases as the moulting process progresses and, as a result, its concentration in the haemolymph decreases. The main reason for this decrease is, undoubtedly, the excretion of JH by the Malpighian tubules as shown by the experiments of Bounhiol (1953). He discovered that ligaturing the Malpighian tubules of the silkworm at the beginning of the last larval instar has an effect similar to ca implantation, i.e. is suppresses metamorphosis to a greater or lesser extent. The operation is effective during a period which corresponds to that for an effective ca implantation. It is longer than the critical periods for AH and MH. The length of this period may, however, vary noticeably in different species.

In Lepidoptera, for example, the critical periods for all these hormones seem to be greatly prolonged as a result of the long feeding period and short moulting process. This may explain the discovery by Legay (1948) that the ca of silkworm larvae reach their maximum volume at about the time of each ecdysis. The selective value of the deviation, together with extreme abbreviation of the actual moulting process in many holometabolous insects, evidently depends on the prolongation of the feeding period; this is particularly significantin the last larval instar, where it enables the accumulation of reserve substances for metamorphosis and eventually for ovarian development.

Corpora allata activity during postembryonic development

The production of JH begins in the final stag of embroyogensis after the ca have been fully formed. The secretory activity of the gland continues during larval development with a temporary interruption at the time of each ecdysis, as mentioned above. A particular time is therefore necessary at the beginning of each instar before the active concentration of the hormone is again reached. It is not reached at all in the last larval instar in Hemimetabola, or in the larval and pupal instars in Holometabola. The production of the hormone continues in adult insects. It is clearly cylical in the females of those species which lay eggs in several separate batches, such as cockroaches and many beetles and bugs. No such cycles are noticed in males and in females of those species which lay only one batch of eggs (e.g. Bombyx mori) or which lay one egg at a time with wide gapof time in between the process. (e.g. Carausius morosus).

The secretory activity of the corpora allata in the last larval instar

Wigglesworth (1940, 1948) initially believed that JH production ceases completely in the last larval instar or even that the ca actively eliminate JH from the haemolymph during this period. While Piepho (1951), condidered the hypothesis that the presence of some JH in the last larval instar of Galleria (Holometabola) was a necessary condition for the development of the pupa : in the complete absence of JH, adult differentiation should occur. This theory was later confirmed by Wigglesworth (1954), Karlson (1956), Schneiderman (1960) and a number of other workers. That the ca continue to produce JH in the last larval instar was shown by Novak and Cervenkova (1959). They implanted ca from last instar nymphs of Pyrrhocoris apterus into other last instar nymphs immediately after moulting and at different times during the intermoult period and in most cases they obtained juvenilizing effects of varying extent. This does not prove, however, that the ca were producing a morphogenetically active concentration of JH as assumed by the above hypothesis. There is definite evidence against thisassumption : when the ca are removed from the last larval instar immediately after moulting no morphological change is observed. While, the implantation of another last instar larva ca which itself is inactive in the donor, is sufficient to produce a juvenilizing effect.

Hypetrophy of the corpora allata

Castration of Calliphora erythrocephala females immediately after the imaginal moult results in hypertrophy of their ca (Thomsen, E., 1946). The most possible elucidation this phenomenon is that in the absence of the ovaries JH is not removed from the haemolymph. When the concentration of the hormone here, exceeds that inside the gland cells, further diffusion of the hormone into the haemolymph is inhibited. It Therefore, it collect in the gland causing its excess growth and a subsequent decrease in secretory activity. This seems to be a more acceptable explanation than the theory suggested by E. Thomsen (1946, 1947) of a specific hormonal influence of the ovaries which suppresses the activity of the ca. The hypertrophy of the ca observed in functional termite females appears, however, to be of quite a different type, being connected with hyperfunction of the glands.

Effect of an implanted corpus allatum on that of the recipient

The first person to demonstrate at was Pflugfelder (1939a) that the implantation of an active ca causes a more or less complete degeneration of the recipient's gland. A detailed biometrical analysis of this relationship was carried out by Novak and Rohdendorf (1961) using adult females of Pyrrhocoris apterus. They fixed active ca into freshly moulted adults and measured the volume of the recipient's glands on the 3rd, 6th and 9th days after implantation and compared it with that of controls implanted with a piece of muscle of similar size. The ca of the recipients were distinctly smaller than those of the controls. No such effect on the recipient's gland was observed if the implantation was made later in the instar (6th day after moulting), when the recipient's gland had reached its full activity. The explanation of the effect of the implanted gland followes. The follow of the hormone from the gland into the haemolymph can take occur only by diffusion through the surface of the gland as there is no particular mechanism (e.g. muscular) for this.

But diffusion can only take place particular when the concentration of the hormone inside the gland is higher than that in the haemolymph. This condition is not maintained when an active gland is implanted before the recipient;s gland has reached its full activity. Under such circumstances the concentration of the hormone in the haemolymph, produced by the implanted gland, exceeds, that inside the recipient's gland and so diffusion and thus also the secretion of the hormone is avoided of course, if the implantation is effected later in the instar, when the ca of therecipient has reached its full activity, there is already a high concentration of hormone in the haemolymph. This goes beyonds the concentration in the implanted gland so that the position is now reversed and the secretion of the implanted gland is stopped.

As mentioned by Engelmann (1965) the ca was first activated by an increase in the blood protein level (to a small degree) and then by maturation. The decrease in ca activity associated with egg maturation can be attributed to a decrease in the haemolymph JH level caused by reduced food intake at this time. In other insects, however, such as the viviparous cockroach Leucophaea maderae, it was shown to be due to a hormonal factor produced by the brood sac. JH secretion in Dermaptera was probnable first examined by Ozeki (1965).

External factors controlling the activity of the corpora allata

The direct influence of various factors on JH productio is yet unknown. The only paper analysis the effects of external factors on development is that by Wigglesworth (1952a) on Rhodnius prolixus. He discovered that subjecting Ivth instar nymphs to a high temperature approaching the living maximum (c. 35°C in Rhodnius) causes an extension of the intermoult period and the development of somewhat adultoid Vth instars (regressive prothetely). A low temperature approaching the developmental minimum (below 20°C in Rhodnius) prolongs the intermoult period more than the high temperature, but the resulting Vth instar nymphs are somewhat juvenile (progressive metathetely). The results may be explained as effects on the action of JH in the tissues. A reduced oxygen concentration (below 5°C) produces exactly the same effect even at of high temperature.

A comprehensive morphological investivgation of the shape and size of the ca in the bug Eurygaster intergriceps during postembryonic development and in adults in the different seasons of the year was made by Teplakova (1947). She noticed an increase in the size of the gland in adult females in the spring, associated with the development of the ovaries. The ca of bugs parasitized by larvae of flies of the family Tachinidae were very small. In the summer period of increased activity, the ca of some of the females with functional ovaries were extraodinarily large, whereas in most others they were of medium size. Variations in the size of the gland were also found among females from different heights above sea-level.

Internal factors controlling the activity of the corpora allata

Various explanations have been given for the what has been called by Wigglesworth (1948) the 'counting of instars'. Wigglesworth stated on the basis of his experiments with Rhodnius prolixus first that the decrease in JH activity in the last larval instar is not connected with the age of the gland. This is obvious from the fact that the ca of the IVth (penultimate) larval instar when implanted into the Vth (last) instar is able to induce juvenilizing effects, i.e. it retains its activity even when its age is equivalent to that of the last larval instar when it would normally cease to function. From this, Wigglesworth deduced that 'it is not the corpus allatum itself that counts the instars' and he suggested the possibility of a nervous stimulus being the cause of this decline in ca activity in the last larval instar. The good innervation of the gland by the nervus allatus appears to indicate nervous control of the ca.

The theory of nervous control was confirmed by Engelmann and Luscher (1956, 1957) on the basis of their experiments with adult females of Leucophaea maderae and further developed by Engelmann (1965). They examined the stimulating effects of an interruption of the nervi corporum cardiacarum on the swelling of the ca previously observed by B. Scharrer (1952). They discovered that a similar result could be obtained by destroying specific part of the protocerebrum. On the while, no changes were observed after electrocoagulation or surgical removal of the neurosecretory cells of the pars intercerebralis. The treatment induced egg formation in adult females and several extra larval instars in last instar nymphs. The researchers assumed their results as evidence for nervous control of the ca. They believed that the absence of JH both in females carrying oothecae and in last instar larvae is the result of by nervous inhibition. But this is not the only explanation of their results. Deducing from with other known facts, it seems probably that the observed swelling and activation of the ca is because stimulation by the nervous irritation and regeneration process induced by the operation.

The fact that no effect was noticed after the removal of the neurosecretory cells may be due to neurosecretory material accumulated in the cc, as ca are incapable of either secretory activity or growth in the absence of AH. A nervous stimulus inhibiting AH production and in this way, undoubtedly, JH production might, however, be the mediator of the inhibitory effect of the oothecae on the activity of the ca, in the same way as the distension of the abdomen has a positive effect on AH production in Rhodnius. A nervous control of the ca was even accepted by Johansson (1958) in his work with adult Oncopeltus fasciatus. He realized that interruption of ca innervation in a starving female can induce egg-laying in the same way as the implantation of an active ca. This treatment, however, was infective when a brain with unbroken nervicorporum cardiacarum was implanted at the same time as the ca.further evidence nervous control of ca activity was provided in the research papers by joly (1945a, b) on Dytiscus marginalis, by Detinova (1954) on Anopheles messae, by de Wilde (1958) on Leptinatarsa decemlineata, etc.

The control of the function of the corpora allata

Over the years there has be lot of research on the question of the innervation of the ca and the form of the supposed nervous and neurosecretory control of their secretory activity. The relevant papers were reviewed by Highnam (1963). The existence of afine nerve connecting the ca with the suboesophageal gang lion, demonstrated by Engelmann and Luscher (1956) in Leucophaea maderae, was also found in periplaneta in schistocerca and in Locusta. Neurosecretory granules were detected in this branch of the nerves allatus. When this nerve is ligatured, neursecretory material accumulates on the ca side of the ligature, neurosecretory material accumulates on the ca side of the ligature. The different forms of control of the activity of the ca have been put forward.

Nervous control, was considered to be of an inhibitory character in some insects, e.g. Leucophaea while possibly having a stimulatory effect in others, e.g. in Oncopeltus or Schistocerca in which removal of the brain neurosecretory cells by cautery or extirpation prevents the ca from attaining their full size. The possible cause of this contradiction was pointed out by strong (see below). The non-specific effect of regeneration of the operation injury, must also be taken in to account. It was discovered, for instance, in Schistocera, that the development of the eggs can be greatly advanced by wounding, as shown by the experiments of Norris (1954) and Highnaman (1961b, 1962a, b). The function of the nervous system in the activation of the ca was examined by removing parts of the brain and by sectioning the nerves of the glands infemale adults of the Central American locust? (Schistocerca paranensis Burm), and by a histological investigation of neurosecretory activity with the following results striking changes in the volume and appearance of the ca were found during maturation, where as only slight changes occurred in the amounts of neurosecretion at the same time; no material from the median neuroseretory cells was found in the nerves which innervate the ca, but granules from the lateral cells with different histochemical properties from those of the median cells were found jin these nerves; extirpation of the cerebral neurosecretory cells resulted In reduced volume and inhibited the activation of the ca. Unlike other experiments, the implantation of ca did not restore oogenesis in allatectomized females.

The implanted glands could only become and remain active when they retained intact connections with the central nervous system. Never sections and cautery of the neurosecretory centres of the brain showed that the activation of the ca was mediated through the lateral neurosecretory cells in the ipsilateral half of the brain. Transsection of the nervi corporum cardiacarum II resulted in inactivity and reduction of the volume of the ca on the corresponding side. The contradiction between the supposed inhibitory effect of the neurosecretory cells of the brain and their stimulatory effects seems to be due to a disregard of the possible breaking of this connection. It is yet unknown whether the effects of the lateral neurosecretory complex upon the ca are mediated through nervous or neurosecretory stimuli, or, perhaps more probably, through the humoral action of AH., No function could be attributed to the nervous connections between the ca and the suboesophagela ganglioin. Local effect of neurosecretory granules from the neurosecretory cells of the brain reaching the ca via the nervous allatus.

Granules of Gomori-positive materials were found in the ca of various species by a number of authors. B.scharrer (1958) assumed that 'the restraining influence on the ca takes place by nervous activity, whereas the stimulatory affect is brought about by a substance present in the neuro secretory material'. Khan and Fraser (1962), who found neurosecretory granules in the ca of newly moulted adults of Periplaneta americana, believed, contrary of B.scharrer (1958) that they were responsible for the straining influence there. Mordue(1963) observed neuro secretory material in the ca of female pupae of Tenebrio moliitor before imaginal moulting; after mating, it disappeared from the ca, remaining, however, in the allatic nerves in close proximity to the gland. The ca were found to grow more rapidly in mated than in unmated females. Here again there is no unequivocal evidence of a local physiological function of the neurosecretion in the gland, and on based on the information of neuro hormones and their way of reaching the target its explanation is improbable.

Humoral activation by the AH of the brain, corresponding to the activation of the pgl. It seems that activation or inhibition of the production of AH by nervous stimuli from other parts of the body (e.g. the ovaries, ootheca etc.) or by various external factors (e.g. photoperiodism) might be the only way to control ca activity. Among other findings , this is shown by the fac that the imagical dispause in Leptinotarsa, aphids and Pyrrhocoris which is an AH-deficiency syndrome, can be completely abolished by implanting active brains and/or cc, whereas the implantation of active as is not effective in the absence of AH (in brain and cardiacectomized specimen), as demonstrated by Slama (1964) for Pyrrhocoris.

The following result from the available experimental evidence seems to be the most probable. The brain controls the activity of the ca both by neurosecretion (AH) and by nervous activity. The latter may produce either a stimulatory (distension of the abdomen in blood sucking insects) or inhibitory effect (carrying of oothecae) depending on to the biolohy of the species concerned. Neither of these factors, however, seems to be responsible for the decrease in activity of the ca in the last larval instar which is the immediate cause of metamorphosis. The nervous system here is merely a mediator of the external stimuli which on average remain unchanged in each instar.

The dependence of JH production on gland volume

A different view towards the problem of the control of metamorphosis was stated by Kaiser (1949) who stressed on the increasing disproportion between the size of the ca and the pgl and concluded that the amount of JH produced, rapidly decreases compared with that of MH, and the surplus of MH causes metamorphosis. This idea was further developed by L'Helias (1956). He based his views on to suppositions which have been disproved : that of a direct positive effect of MH on metamoirphosis; and that of an antogonism between MH and JH. In addition, the actual amount of MH as estimated by the volume of the pgl in releation to the volume of the body, and thus also the concentration of MH in the haemolymph, remains static.

Another solution which takes into account the surface area and volume of the gland in relation to the volume of the body as responsible for the time of reaching the active concentration of the hormone, was used by the author. Thwe concentration of the hormone in the haemolymph depends on the following factors: (1)

the quantity of hormone produced in a unit of time; (2) the volume of the haemolymph and, in direct proportion to this, the volume of the body; (3) the quantity of hormone consumed in the body; (4) the quantity of hormore removed from the haemolymph by the Malpighian tubules (and perhaps other organs) in the unit of time; (5) Another factor controling the rate of release of the hormone is the surface-volume relationship of the gland.

Observation of the changes which these factors undergo during postembryonic development confirms that JH concentration in the haemolymh necessarily decreases with each subsequent instar. Sooner or later, therefore, there comes an instar (which varies with the species) in which an effective minimum JH concentration is not attained before the advancing moulting process inhibits further growth and differentiation. This decides the last larval instar and the beginning of metamorphosis.

The relationship between the surface area and volume of the ca and that of the body, during development

To investigate the above theory, measurements were made of the surface area and volume of the ca and the volume of the body in freshly hatched larvae (the beginning of the first larval instar) and freshly moulted larvae int he last instar. The result obtained in 18 differnt speices of both Hemimetabola and Holometabola show that there is a marked decrease in tghe volume of a ca relative to that of the body, in addition to a very great decreaase in the surface area of the ca compared with the volume of the body. This comparative decrease in the volume of the ca is because of the very slow growth of the gland, which during the whole larval periods is slightly greater than that of various parts of the nervous system and much less than the growth of the most other parts of the body, such as the breadth of the head, breadth of the pronotum, length of the tibia, etc. It is confirmed that the relative decrease in the surface area and volume of the ca in all the species studies is too great not to affect the production of JH, and is large enough to explain the lack of an effective concentration of JH in the last larval instar and thus to induce the onset of metamorphosis.

The Mode of Action of the Juenile Hormone

Many theories have been forwarded to explain the mode of action of JH, some obviously contradictory. According to Wigglesworth's original idea (1936), JH affects morphogenesis on the principle of Goldschmidt's hypothesis of differnt reaction velocities. Wigglesworth believe that the degree of differentiation attained by a given specimen after ca implantation is the product of competition between two simultaneous processes : the differentiation of the imaginal structures, and moulting. Variation is only possible in the period between the moment of detachment of the old cuticle and that of the deposition of the new. The effect of JH was assumed to depend on accelerating the moulting process so that the permaturely deposited new cuticle would inhibit further differentiation. On this basis, Wigglesworth initially used the term 'inhibitory hormone' for the secretion of the ca. Wigglesworth (1940) himself replaced this theory with the hypothesis of two alternative enzyme systems inside each epidermal cell, larval and imaginal, The larval system was supposed to be dependent upon the presence

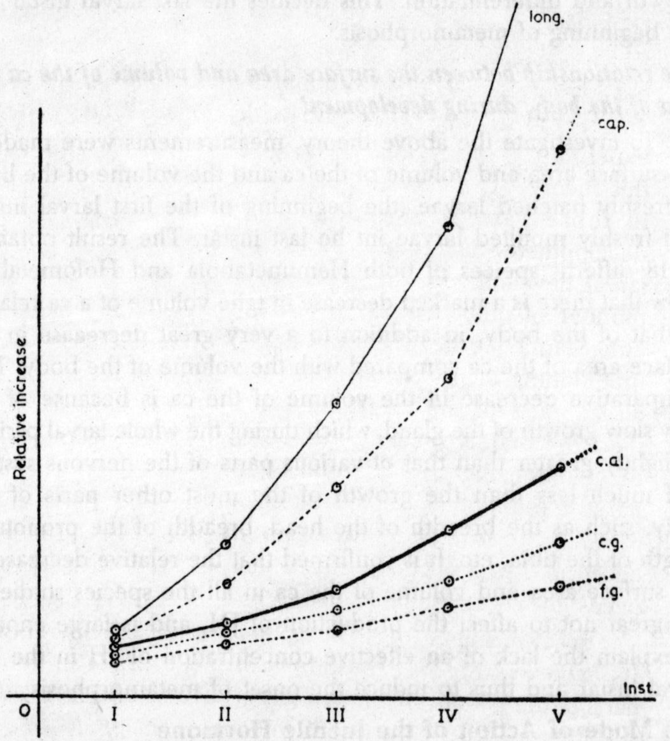

Fig. 7.17. Growth of the corpora allata compared with other parts of the body during the five larval instars of Bombyx mori. Abscissa-larval instars, ordinate–relative increase; f.g.–frontal ganglion; c.g.–cerebral ganglion; cap.–breath of the head; long.–length of the body.

of JH and the imaginal system in order to operate in the absence of JH. The terms 'inhibitory hormone' was therefore changed to 'juvenile hormone' or to 'neotenin'. The hypothesis of polymorphism which has been described earlier was based on this concept.

While Pflughfelder (1939, 1941) and Weed-Pfeiffer (1945a, b), stressed on the positive effects of JH on total metabolism as shown by their experiments. To avoid the obvious contradictions, the hypothesis of two differnt ca hormones was created; one of them being identified with the JH of Wigglesworth, and supposed to affect only larval development. The other, a 'gonadotropic' hormone, was assumed to activate the ovarian follicles and affect metabolism. However no cogent evidence against the first concept Wigglesworth on a single a hormone has been produced.

On the basis of the gradient-factor as a factor conditioning the JH-independent growth of the imaginal parts of the body, a theory which is discussed later, a hypothesis on the mode of action of JH has been suggested by the author (Novak, 1951a,b, 1956; cf.p.258). This seems to resolve the above-mentioned contradictions in the experimental results and to synthesize earlier views. According to this theory, JH produces its effect by taking the place of the GF in those parts of the body which lose it in the course of development. Such parts are the lateral parts of the body during larval development, the ovarian follicles in adult females, and a number of other tissues at particular times during development and Slama (1963), the effect of JH seems to depend on the moditioning of protein synthesis as well as other functions in the larval parts of the body. Its effect on oxygen consumption is therefore marely indirect, arising from an increase in the amount of metabolizing tissue (larval structures, ovarian follicles). The negative result on oxygen consumption produced by the implantation of can into castrated females seems to show that JH has no direct effect on the immaginal parts of the body. This question, however, needs further experimental evidence.

Put it in nutshell JH is principally a morphogenetic hormone. Its action consists in restitution of the capability for growth of parts of the body which have lost it in the course of morphogenesis. All other changes observed in the body after the administration of JH are indirect consequences of this action and are due to the altered biochemical balance produced in the body by the changes proportion of metabolically active tissue.

Chemical Characteristics of JH

Previously nothing was known of the chemical character of JH, largely because all attempts to extract it from ca failed. The situation changes when Williams (1966a) found that a lipid extract from the adbomen of cecropia males produced JH effects in Coleoptera (Tenebrio molitor, Schmialek and Wigglesworth, 1958, 1961) and in bugs and cockroaches, etc., as well as in cecropia and other moths. This discovered facilitated chemical identificaion of the active principle. Since then many papers on the principle of action of substances of this type have been published and many different chemicals, both natural and synthetic, have been found to produce similar effects. Of the many worker investigated the nature of this 'cecropia oil', the most successful were Roller and his co-workers who succeeded in isolating the active principle by means of gas chromatography. Later they identified the chemical formula of male cecropia extract, which was found to be methyl- 10- epoxy-7ethyl-3, 11-methyl-236- tridecadienoate. A number of related substances are also active.

The occurrence of the active principle in the body of the cecropia silkworm (Hyalophora cecropia) during ontogeny was studied in detail by Gilbert and Schneiderman (1960). As illustrated in the following table, the cecropia JH content is fairly high in unfertilized eggs, during the embryonic period and in the freshly hatched larvae. It decreases at the end of larval development and in the pupa, diasppears when development recommences after the pupal diapause and does not reappers until just before adult emergence, when it increases slowly in the female, but at faster rate in the male.

Roller and Dahm (1970) demonstrated that the cecropia factor was identical with the actual JH produced by the ca of this species. A number of brain-cc-ca complexes dissected out from pupae two to three days before adult emergence were cultivated in a tissue culture medium for six to seven days. The ability of the complexes to produce JH after this time was illustrated by reimplanting them into last instar larvae of Galleria mellonella. The culture medium was then extracted three times with diethyl ether. With 50 brain-cc-ca complexes, an extract containing 10 000 to 15 000 Tenebrio units was obtained. This is evidence that the active principle of cecropia oil is actually JH produced by the ca is quite convincing.

Table 7.1. Juvenile hormone content of cecropia during development

Stage of development	*Hormone content per gram fresh weight compared with adult male (%)*
Unfertilized eggs	4.3
7-day-old embryos with yolk	3.7
Ist instar larvae (freshly hatched)	6.4
Vth instar larvae (mixed ages)	0.50
Freshly moulted pupae	0.75
Diapausing paupae (1 month old)	0.55
Chilled pupae (6 months old)	0.0
Pupae 2 days of adult development	0.0
Pupae 11 days of adult development	0.0
Pupae 17 days of adult development	0.0
Pupae 20 days of adult development	0.5
Pupae 22 days of adult development (males)	50.0
Adult males, 2 days	100.0
Adult females, 2 days	3.2

8

INSECT PHEROMONES

Taste and smell, which can be also termed as chemical sences are much more important to insects than to humans. Insects conntinuously receive with meaningful chemical signal that tell of food, of nest sites, of prey, of predators, or of a suitable mate. Many of these signals produce a rapid and particular behavioral response. The study of the chemical signals that act between organisms constitutes a new disciplines called chemical ecology. Much study in this field has resulted into several recent books and innumerable research articles, and is one of the most exciting frontiers in entomology. Chemical signals include hormones, pheromones, and allomones.

Hormones are internal signals that are produced by an endocrine gland and that are carried through the blood to act on target tissues within the animal. Chemical ecologists study external signals that are secreted into the environment by exocrine glands and that act between different organisms. Pheromones act between individuals of the same species. Pheromones attracts the sexes, aid in courtship, announce the presence of danger, mark trails and home territories, and control many other intraspecific-interactions. E. O. Wilson and W. H. Bossert, of Harvard University, divided pheromones into two general subclasses: (1) *releasers*, which act rapidly to produce a behavioral response, and (2) *primers*, which act to modify the physiological state. *Allomones* act between different species to benefit the producer. Allomones repel predators, confuse prey, and mediate symbiotic interactions. Inspite of the fact that the terms *pheromone* and *allomone* are of recently coined, our knowledge of chemical signals began in the nineteenth century for pheromones and even earlier for defensive allomones. Yet it is only since 1950 that the analytical power of organic chemistry has developed to the point that identification of the molecules can be accomplish repeatedly from very small samples. With the resurgence of information on the molecules themselves, is now possible to ask precise questions about their impact on behavior and physiology.

The molecular structures are interesting in themselves but the most fascinating aspect of chemical ecology is the delicate intercourse among organisms that is mediated by the chemical signals. The factor which are necessary for chemical communication are a source of signal molecules, the chemical signal itself, and a receiver. The source is most often an *exocrine gland*, which opens on the surface of an insect. Glands may be associated with mouthparts, intersegmental membranes, sclerites, legs, and other body parts. Somehow the products of these glands must cross through the impervious insect exoskeleton. Type I secretory cells lie well below the epidermis itself and pass their secre- tions to the outside through elongate ductules that are invaginations of the epicuticle. Type II secretory cells have cuticles that are filled tiny perforations so that the secretion percolates up from the secretory cell to the outside. The molecular signals and their receptors will be discussed below.

Pheromones

From among a noisy jumble of competing chemical stimuli, a pheromone must become conspicuous and convey a clear message. Not all molecules make good pheromones. The signal must be unique and in a terrestrial environment it must be volatile. Glucose and glycogen, for example, are poor candidates on both counts. In theory, the larger the molecule, the greater are the possibilities for a unique structure that conveys a clear message. However, in actuality the requirement for volatility restricts pheromones to molecular skeletons of 20 carbons or less. Most are derived from common biochemicals like fatty acids or amino acids. In the entire class of Insecta, the molecular structure of most known hormones varies little.

In contrast, many thousands of different molecules have been identified as pheromone components. Their diversity among species helps to avoid signal confusion. To increase the information content even more, most pheromones are mixtures of several chemical substances so that in the actual signal several components are mixed together.

Active Spaces

Pheromone signals occupy space and persist for some time. Many of the special features of the communication systems are related to the expansion of the signal and to its fade-out in space and time. Imagine an insect sitting in the middle of an empty basketball court and emitting pheromone. Assume that there are no breezes. As the pheromone begins to evaporate from the insect, its concentration rises in the air nearby. As times goes on, the molecules diffuse outward to the edge of the center circle and beyond. There will be a hemispheric could of pheromone, with its concentration highest at the source and spreading around it. To begin a chain of reaction the pheromone must be perceived by another insect.

In most cases the response is fairly stereotyped, when the pheromone has exceeded a threshold concentration. Far away from the source, pheromone concentrations are extremely low, and no insects respond. As one moves closer to the pheromone source, concentration in the hemispheric cloud reaches threshold. And at lesser distances, concentration is above threshold. Thus there is a zone within which other insects respond and a surrounding environment where they do not. Wilson and Bossert called this zone of response the active space. In Wilson's words, "The signal is the active space." The size of an active space is determined by both sender and receiver. When an animal has given off pheromone for a long period, the active space is likely to be at its maximum size.

An active space will be larger for a receiver with a low threshold of response than for one with a high threshold. When the emission frequency of the pheromone is high the active, space will expand faster and achieve a larger volume than when the emission rate is low. By altering the rate of emission, the threshold, and the chemical signal itself, the communication system can be tuned to optimum performance for a particular task. The active space for a sex pheromone may be large, but that for an alarm substance should be quite small. A basketball court is an artificially created environment, but the lessons learned there about out hypothetical insect can be applied to the natural world. The major differences between the gymnasium and the forest are winds and obstacles. In nature, breezes act on odor clouds. The cloud of pheromone drifts downwind and forms a plume, or aerial odor trial. The plume will rarely be perfectly symmetrical, since breezes increase and decrease unpredictably. The resulting turbulence gives the plume ragged edges, and these are further distorted by flowing over hills and into valleys , by going around and through trees, and by changes in temperature and air pressure. Every plume is a little different from all the others. Although there remains much debate about the detailed shapes of these clouds and about the effects of wind and weather, the general concept of an odor plume serves well to give us some clue for the chemical signal.

Sex Pheromones

Some pheromones attract one sex to the other over long distances, while others act at close range to actival a sequence of the elements is a chain of courtship behaviors. These Pheromones are not produced continuously by all members of a species, but only by animals that are reproductively mature and often only at a particular time of day. Because disruption of reproduction is an appealing strategy for control of insect pests, a large number of species have been studied in university laboratories and agricultural research stations throughout the world.

The research began as a descriptive search for the molecules, and as more and more have been identified, the field is shifting toward precise study ofhow pheromones function. During the period around 1960, it was thought that one sex of each species might produce a single, unique substance that automatically triggered a rigid sequence of stereotyped behavior in the opposite sex. That view of pheromones was attractive because of its simplicity. However, natural selection has left much more interesting and complex phenomena for the entomologist to analyze. Most pheromones are complex medleys of scents, and these odor blends have many subtle shades of meaning. Rather than list all the detail the biology and chemistry of the sex pheromones which affect the reproduction of two groups of insects: moths and queen butterflies.

Moth Attractants

The French naturalist J. H. Fiber a nineth century entomologist demonstrated that male peacock moths are attracted to a paper that previously held a female but pay no attention to a female under a glass bell jar. He correctly concluded that a residual sex scent on the paper was responsible. But it was not until 1959 that Adolf Butenandt, of the University of Munich, Germany, became the first to determine the chemical structure of a pheromone. That was the culmination of a 30 year effort with Bombyx mori, the oriental silkworm moth, which has been cultured for many centuries in Japan and China. As soon as a female Bombyx moth comes out from her pupal cocoon, nearby male become highly excited and will copulate immediately with her if given the opportunity. Males do not even need to see the female. They exhibit characteristic excitement (wing fluttering, flying upwind, abdominal movements, attempting copulation) when exposed to female scent, generally, the scent is emanated from two glands that open in the intersegmental membrane at the tip of the female abdomen. Butenandt and his colleagues painstakingly snipped off the tips of the abdomens of 500,000 virgin female *Bombyx.* These tips were then extracted with ethanol-ether, and the active components were carefully purified from the crude mixture and then characterized by brilliant microchemical techniques that were devised by Butenandt for this project. He obtained about 12 mg of pure attract ant which is termed as bombykol and identified it as a 16-carbon alcohol.

Male *Bombyx* are extremely responsive to female scent, and the male fluttering response is used as a bioassay to compare different concentrations of pheromone. For each test a sample of dilute pheromone is placed on a filter paper cartridge and blown by an air-stream to resting males. A typical does-response curve for pure bombykol is shown in Figure. If carefully examined such curves may seem dry an abstract but they nearly summarize the data and show how they fall into a pattern. Below 10^{-6} µg the response is no different from background activity. Between 10^{-6} µg and 10^{-4} µg there is an abrupt rise as an increasing percentage of males respond. Above 10^{-4} µg nearly all the males respond. For a considerably period of time, it was wrongly presumed that bombykol is the only component emitted by female Bombyx. But in 1978 K.E. Kaissling and his colleagues at the Max Planck Institute in Sewiesen, Germany, found that the female also produces the corresponding aldehyde in the ratio (bombykol:bombykol) of 10: 1. When males are stimulated with pure bombykol at moderate doses, there is no behavioral effect. But when the alcohol and aldehyde are blended together as in the natural pheromone, bombykol partially inhibits the response of the males to bombykol. The dose-response curve for this mixture (which presumably approximates the natural pheromone). Note that relative to pure bombykol, the curve is shifted to right, and its rise is much more gradual as the dose increases. The curve shown for the blend reflects that the use of pheromone concentration results in a more graded response.

The part of male Bombyx which detects pheromone are the antennae. The large, feathery antennae are studded with about 64,000 sense hairs, of which about 80% are specialists that respond only to sex pheromone. One sensitive to bombykol, the other to bombykol, and each sends a signal independently to the brain. The brain computes the relative concentrations and decides the physical responses. On any summer evening many different species of moths emit pheromones simultaneously, and somehow males must distinguish conspecific female scents from all the others. Wendell Roelofs and his co-workers at Cornell University have shown that the pheromone blends emitted by some species are carefully times. For example, two species of the genus *Archips* (leafroller moths) share the four components in their pheromones. But the two species differ in the relative proportions of the components-60:40:4:200 for *A. argyrospilus* and 90:10:1:200 for *A. mortuanus.* Males of each of species have specialist receptors for each of these molecules. Their sensilla have at least four neurons, and apparently each neurons recognizes one of the components. Signals from these sense cells are sent in parallel to the brain, where the relative proportion of the components must be computed. In this way males are able to discern the particular type of scent emitted by females of their own species.

Now the question arise that how well can moths discriminate between a signal, such as bombykol, and other molecules of similar size and with similar properties? The fluttering response can be used to compare similar molecules, such as geometric isomers of bombykol. Bombykol has two double bonds and thus four geometric isomers, Butenandt synthesized all four isomers, and experiments showed that the threshold for bombykol itself is 100 to 1000 times lower than for any other isomer. Most of the experiments have resulted into the belief that precise molecular shape is recognized by pheromone receptors.

Another question which is arised about the number of male each female can attract. The pheromone gland of each female *Bombyx* contains about 164 mg of bombykol, in theory enough molecules to excite 10^{12} males if each received a threshold dose. But such calculations ignore the impact of bombykol in the blend, and even if so adjusted, they would not be biologically meaningful. The important advertising signal is the active space. Production of large amounts of pheromone is largely an adaptation to maximize the volume of active space. By spreading wide the area of active space, the female increases the likelihood of attracting males. As soon as attracted by pheromone within the active space, a male requires orientation cues in order to find the female. For long-range orientation, he appears to use mostly *anemotaxis*-movement based on the direction of wind. The upwind movement of insects takes them toward the source of wind-borne scent.

It is difficult to observe sustained oriented movement in free-flying moths, but Iise Schwinck, of the University of Munich, Germany, found a clever alternative system. She used a flightless *Bombyx* mutant. When excited by female pheromone, flightless males wave their useless wings and try to run to the female. Schwinck watched the males as they ran along a table in the laboratory. When exposed to female scent, males ran upwind in a *zig-zag* course irrespective of the actual origin of scent. Further research has proved that the "Schwinck effect" also applies to free-flying moths. J. S. Kennedy and D. S. Marsh, of Imperial College, London, confined male moths to a special wind tunnel and placed a "calling" female upwind. Males flew upwind in a series of diminishing, irregular *zig-zags*. But at the sudden removal of the calling female, males flew back and forth across the wind with frequent left-right reversals. In some other field experiments, the males actually blundered out of the plume. To find the female, they flew black downwind to reenter the odor trail and began anemotaxis again.

Orientation of moths can also be done by *chemotaxis*, detecting slight differen- ces in pheromone concentration that would permit them to follow a chemical gradient upstream toward its highest concentration where the female sits. Some evidence suggests that moths can detect and follow chemical gradients in still air, but it has yet to be shown that they can tell upstream from downstream. Chemotaxis certainly is used for short-range orientation in insects, but as an important mechanism in long-distance orientation of moths is yet to be experimented. As soon as males get close to a female, their behavior changes. They hover or land and run until collide with her. There are many ways by which males locate the calling female. In some species it appears that the male sees a silhouette against the sky or sees the female herself and then alights.

In other species a hierarchy of active spaces may be involved. In the redbanded leafroller moth, Roelofs and his colleagues found that the pheromonal blend included components that acted at a distance as well as one, dodecyl acetate, that served in close-range orientation and woke her responses of landing. Unique feature of some arctiid moth is their use of pulselike signal William Conner and Thomas Eisner, of Cornell University, have recently shown that a female arctiid moth emits its sex lure in short pluses (100 times per minute). At distances of several meters from the source, the pulses would probably be smeared together by turbulence, but close in, they remain discrete. When males are but a few centimeters downwind, their antennal receptors show corresponding pulsations in electrical potential. It is possible that when the continuous signal transforms into an on/off pattern of pulses, males switch from upwind anemotaxis to the research in shorter range.

Queen Butterfly

Visual signal attract males to females in most diurnal butterflies, such as the queen (*Danaus gilippus berenice*). The courtship sequence, a chain of signals between male and female, was described in detail by Lincoln and Jane Van Zandt Brower while they were at Amherst College in Massachusett.

Male queen butterflies have a pair of large organs, called *Hair pencils*, which are really tufts of fine hairlike processes of cuticle. As soon as a male queen sees a female in flight, he quickly overtakes her. As he passes over her dorsal surface, he extrudes large hair pencils from the tip of his abdomen and bobs up and down over the female's head and antennae. As demonstrated by Thomas Pliske and Thomas Eisner, of Cornell University, the hairs are covered with dustlike particles that are brushed onto the antennae of the female. These numerous particles are actually microsopic spherical bits of epicuticle that were formed during pupal development. The particles clump together and stick to the hairs and to the antennae because they are coated with an oil film. The oil contains the real signal, and the dust is just a means for its transfer. The oil is apparently produced by a secretory cell at the base of each hair. Jerrold Meinwald and his team at Cornell identified two major components in the secretion: a ketone and a viscous terpenoid diol. As a result of to the hair penciling, the female lands. The male continues to hair pencil vigorously, and the female folds her wings. The male then alights and attempts copulation. Pliske and Eisner deprived males of their hair pencils and found that, although they courted actively, the courtship very rarely led to copulation. Males raised indoors on cuttings of the major food plants failed to mate successfully, although males raised outdoors, mostly on the same intact food plant, courted successfully.

Comparison of these two male groups showed that the indoor males had the diol but lacked the ketone, whereas the outdoor males had both components. The outdoor males obtained the chemical precursor of the ketone from another plant on which they spend very little time and of which they eat little. Therefore it is clear that the intake of ketone by males is vital in the courtship procedure. When artificial ketone in an oily vector was applied to the deficient males they began to court successfully. The behaviorally active aphrodisiac is clearly the ketone. Which the diol provides the oily vehicle for its spread and adherence.

Aggregation Pheromones

Many types of insects form aggregations for feeding, mating, shelter, oviposition, or some other purpose. The gregarious behavior may wax and wane with the passing seasons, perhaps occurring only in the cold months for hibernation. It may be limited to a certain time of the day, as in sleeping aggregations of butterflies. It may be associated with a developmental stage, as in tent caterpillars, which are gregarious as larvae but not as adults. It is generally understood as opportunistic behavior, as when lycid beetles form large aggregations to feed on sweet clover and to mate. Once the temporary or restricted resource is exhausted, the aggregation breaks up. The majoirty of insect group form in response to pheromones. The most intensively studied species are bark beetles and ambrosia beetles, both members of the family Scolytidae.

Bark beetles and ambrosia beetles join together for feeding, mating, and oviposition. Because the beetles and the fungi that they carry from tree to tree do tremendous damage to pines of our forests and to the American elm, his aggregation behavior has been exhaustively studied. Bark beetles' pheromones explain the complexity and natural history of a chemical signal system.

Immediately after their final ecdysis, the adult bark beetles emerge from their host tree and begin a dispersal flight to find a new host. Depending on the species of beetle, either males or females (but never both) are the pioneers. When they reach a new host, the pioneers burrow through the bark and construct galleries. As they chew their way along, the beetles seed the galleries with fungal spores, which germinate to form the mycelial mass which become the food of beetle larvae. A new tree is colonized by pioneer beetles that orient to a particular host tree by several cues, such as its shape and odor. In the genus *Dendroctonus*, females are pioneers; in the genus *Ips*, males are the first to colonize. As soon as they land, the beetles burrow into the bark.

Healthy trees produce a flow of resin or pitch that may eject the burrowing beetle from its partially completed hold or may drown it. But when a few pioneers manage to complete the entrance tunnel and to establish themselves, they release aggregation pheromone. More beetles are attracted, and the balance begins to shift toward the intruders. The mass attack causes damage to the phloem, and fungal hyphae block the conducting vessels of the tree. The tree gradually begins to die. Since dead timber dries rapidly, a quick explaitations of the resources become must for beetles. The bark beetle pheromones which have been till now identified are monoterpenes, structurally similar to those produced by the host tree. The aggregation is a response to a medley of molecules that enhance each other's effectiveness. Such molecular components are described as *synergistic*. Each of the synergists may be somewhat attractive when tested alone, but the attractiveness of the blend is much greater than we would expect the sum of all the contributions from each separate component is combined.

The very first bark beetle pheromone to be identified was that of *Ips confusus*, the California five-spined ips. As the beetles chew their way into ponderosa pine trees, the wood passes through their gut and is deposited pellets called frass. David Wood, an entomologist at the University of California, Berkeley, and Robert Silverstein, and organic chemist now at Syracuse University, isolated three terpene alcohols, ipsensol ipsdienol, and cis-verbenol, from the frass of attacking male engraver beetles. In a laboratory bioassy with walking female beetles, ipsenol alone was excitatory as a single component, and it was synergized by mixing with one or both of the others. In field tests only the combination of all three synthetic compounds proved attractive. When the synthetic mixture was later compared with "natural" scent from male infested pines, that natural sent was much more effective than a synthetic blend of ipsenol, ipsdienol, and cis-verbenol. Clearly the natural pheromone contains components which are till now unidentified.

Near the male-infested tree, a female runs about, searching for entrances to a male burrow. Outside its entrance, she stridulates by rubbing the serrated edge of her elytra against a scraper on her abdomen. In response to the female chirps, the male may allow her to enter his burrow; mating follows. One male will allow only three females to enter the nuptial chamber, and after mating, each female constructs an egg gallery in the phloem and outer xylem. When more females stridulate or try to intrude the burrow, the male uses his body to block the intrance.

Many species of Ips have common geographic distributions. Both *Ips paraconfusus* and *Ips pini* attack ponderosa pine in California, but the two species rarely attack exactly the same tree. Martin Birch and his colleagues at the University of California, Davis, have shown that resource competition is minimized because each species responds only to its own attractant blend and because each blend includes an inhibitory scent for the other species. The pheromones emitted by males is composed:

Ips paraconfusus	Isp pini
100% S-ipsenol	
90% S-/10% R-ipsdienol	100% R-ipsdienol
IS,4S,-5S-cis-verbenol	

S-ipsenol inhibits females of I. pini, and pure R- ipsdienol inhibits the attack by females of *I. paraconfusus.* One component of the aggregation pheromone of each species is an inhibitory allomone.

As it has been stated earlier, in bark beetles of the genus *Dendroctonus* it is the females that are pioneers. This genus includes many serious pests of coniferous forests, such as the southern pine beetle and the western pine beetle. From the frass of the latter species, Wood and Silverstein isolated and identified the unusual terpene *exo*-brevicomin. Pure exo-brevicomin is quite attractive to males, but many more males were attracted to synthetic mixture of exo-brevicomin and myrcene, a terpene produced by wounding the host ponderosa pine. After responding to these substances, males at a tree laden with female and gove off a pheromone of their own, frontalin, which attracts other females. Like exo-brevicomin, frontalin is synergized by myrcene. The combination of these substances is cause for the mass attack of both sexes. After mating, both sexes begin to release two additional pheromones, *trans*-verbenol and verbenone, which stops the arrival of other beetles.

Inhibitory pheromones are one device to control density so that the tree is not overcrowded with larvae. Another device to control density is high attractant concentration, as demonstrated in trapping experiments. Beetles are readily caught in artificial traps baited with a mixture of brevicomin, frontalin, and myrcene. However, when pheromone concentration gets extremely high, the catch in the trap declines. At very high concentrations of pheromone the incoming beetles turn away from the baited tree and shift their attention to the nearby trees. Consequently a cluster of infected trees is produced, which becomes a "hot spot" of brown foliage in the forest.

It is remarkable that the of beetle receptors to recognize the differences between the pheromone components and similar molecules. Geometric isomers are easily discriminated; thus transverbenol is an effective inhibitor for attacking western pine beetles, while cis-verbenol is not. It is yet unknown that how the chemicals in bark beetle aggregation pheromones are manufactured. No discrete exocrine glands seem to be responsible. The aggregation scent is not produced if the host terpenes are absent, and many of the pheromones seem to be oxidized derivatives of the terpenes.

For instance, male Ips paraconfuses convert myrcene to ipsenol and ipsdienol. David Wood and his accociate, John Byers, have proved that the oxidation of Myrcene to ipsenol and ipsdienol is inhibited by the antibiotic streptomycin. Then suggest that symbiotic bacteria, restricted to the male, are responsible for myrcene oxidation. If bacteria play such a role, the beetle depends on the tree for the starting material and bacteria for the metabolic conversion in order to produce an insect aggregation pheromone. Initiating an aggregation carries a risk. Pradaceous carabid beetles, which feed on the bark beetles, are also attracted by the bark beetle aggregation pheromones. When the carabid arrives at the infested tree, it feeds on the beetle which emits the signals.

Alarm Pheromones

The major disadvantage of such mass gathering of insects is that a predator has an easier time finding prey. Upon approach of the predator, The group may disperse or it may invoke a colonial defence response. In some hemipterans and in social insects, an alarm pheromone results in the similar behavioural pattern.

Aphids gather in large number to feed. When a predator attacks, the victim secretes a droplet of fluid from a large gland opening called a *cornicle.* Within that fluid is an alarm pheromone, of which the most common is farnesene. Nearby aphids respond by falling to the ground, walking away, or leaping from the plant.

Alarm Pheromones of Ants

Anyone who has stepped on an ant nest has seen the outrush of defenders, which open their mandibles and swarm over the attacking foot. Such alarm behavior is caused by a pheromone emitted by the mandibular glands and by glands associated with the sting. It has been predicted by Wilson and Bossert that an alarm signal should be localized, blatant, and brief. The pheromone should spread rapidly to coordinate colonial defense and should fade out rapidly to avoid prolonged false alarms. For the site of the disturbance to be easily located, the active space should remain small. To prove there findings they investigated the size and time-course of the active space for alarm in a harvester ant.

The major component of this ant's Mandibular secretion is 4-methy1-3- heptanone. Wilson and bossert qucikly crushed the head of a worker ant to release all of its mandibular gland contents at once. They watched the response of nearby workers, and by repeating the experiment many times, they were able to deduce the properties of the active space. The active space expanded to its maximum radius (6 cm) in 13 seconds, and as the pheromone continued to diffuse, the active zone shrank to nothing by 35 seconds. In a natural situation with a serious invasion of the nest, other workers, excited by the first secretion, would have encountered intruders and added their own pheromone. This added pheromone would amplify and prolong the signal and expand the defensive frenzy. The extent of the group response decide to the severity of the attack.

Alarm pheromones are very volatile, and most are of low molecular weight, containing 12 carbons or fewer. Those from the mandibular glands usually have a keto-or aldehyde group, while those from Dufour's gland (near the sting) are most often hydrocarbons. Many are mildly toxic and distasteful; thus they function as alarm defensive devices. In case of the african weaver ants construct an elaborate nest by sewing leaves to one another with thread from the silk glands of their own larvae. Both the nest itself and the territory surrounding it are vigorously defended by the major workers. Any insect that happens to wander into the territory is rapidly attacked and often becomes prey for the ants. The defensive behavior consists of four stages. (1) alerting, in which ants raise their heads and open their jaws; (2) attraction toward the disturbance; (3) stopping near the source of the disturbance; and (4) bitting and holding for several minutes.

The distress signal follows the expulsion of the fragrant but musty of or from the mandibular glands of a major worker. J. W. S. Bradshow, R. Baker, and P. E. Howse of the university of Southhampton, England, have shown that at least 33 volatile components are released by crushing the head of a major worker. Four of these seem to trigger specific aspects of the alarm response. These molecules differ in their volatility from hexanal (most volatile) to 2-butyl1-2-octenal (least volatile). Their defense mechanism can be easily apprehended if we imagine that an excited ant deposits a droplet of mandibular gland secretion at the center of the concentric circles. In still air each component diffuses away from the site of deposition. Because they differ in volatility, some fill space more quickly than others. After a short time, hexenal defines the largest active space, indicated by the outermost circle. When ants enter this space, the alerting behaviour is characteristic. As they move into the next zone (hexanol), the major workers are attracted toward the site of its deposition. In the innermost zone 2-butyl-2-octenal is the biting marker, and 3-undecanone contributes to biting and is also a short-range orientation signal. When the secretion has been allowed to evaporate for some minutes, no effects of hexanal or hexanal are detectable, but as soon as the workers come close to the deposition site, biting behavior is released. Consequently a chain of behaviors is sequentially triggered by the discrete chemical components of the alarm pheromone that are localized on the intruding object.

As opposed to the major worker and the minor workers of the weaver ant emit a considerably different set of components from their mandibular glands when they are disturbed. If other minors are exposed to that secretion, they do not defend the nest but instead flee from the site of disturbance. In contrast, major workers are attracted and arrested by high concentration of atleast one component in the minor's perfume. Thus major workers come to the defense of the distressed minor worker and of the nest.

How much different is one alarm signal to the other? Are they specific for each ant species or for each colony? Apparently they are much less so than for sex pheromones. Often the same components are shared along many species within a genus and among genera within a subfamily. However, there are subtle and biologically important differences between similar species, as shown by studies in the ant genus Myrmica. Four species of this genus that coexist in similar habitats in England and Belgium have been found to have similar blends of several components in their alarm pheromones, but their relative proportions differ. In effect, each species has a different form of communication.

Trail and Territorial Markings

Both food and habitat are restricted. Territorial and trail pheromones enable efficient resource exploitation. Many foods, like blue-berries or caterpillars, act as the container of insects eggs. Female insects that place their eggs in such organic matter and mark them with a pheromone that is like a sign that says "Occupied." Some parasitic wasps include such a territorial pheromone with each egg injected into an insect host. Female apple maggots mark each fruit after ovipositing in it. After she has laid an egg, the female drags her ovipositor over the surface of the fruit and thus deposits pheromone. The size of the mark depends on the future needs of one larva from hatching to maturity. A small fruit, like a cherry, is completely coated after one oviposition and thus will be avoided by subsequent females. A large fruit, like an apple, is marked on only a portion of its surface. It will go on collecting several eggs until its surface coat of pheromone reveals that it is loaded to capacity with larvae. Effective utilization of the resource material requires choices among alternative.

Many social insects use the guiding trails to direct nestmates to the richest food sources. Recruitment trails are continuous directional signals that precisely guide the foragers to abundant goodies. Stingless bees use secretions of their mandibular glands to lay recruitment trails. Like honey bees, stingless bees have large colonies with many workers, use wax as a building material, and collect and store honey and pollen. Martin Lindauer, of the

University of Munich, showed that stingless bee foragers guide each other efficiently. When a pioneer forager finds a new nectar source, she makes a direct flight back to the nest with food. If the nectar source is still copious after several trips, she returns to the nest in a different manner. Instead of a quick flight home, she flies in small stages-stopping repeatedly to wipe secretions from her mandibular glands onto rocks, plants, or dirt along the route. At the nest she arouses a number of her sisters, and with the pioneer as leader and pilot, the group flies along the trail of scent marks. The foragers stop along the way to check the marks, and finally arrive at the source of rich nectar.

The tactics which guide these stingless bees require both the pilot bee and an odor trail. In many ant and termite species an odor trail alone is sufficient to guide a follower to a food source. In the case of ants and termites the trails are laid by dragging their abdomens along the substrate and depositing streaks of chemicals. The active space of the trail is a narrow corridor and extends upwards in a half-cylinder. Followers run along with in this active space, weaving from side to side and turning back to the center as their waving antennae break out of the active space into an usecented zone. The followers do not have to touch the odor marks but have only to stay within the active space. They can even follow a trial drawn overhead on the ceiling if ceiling and floor are not too much apart.

The function of these trails differ according to the biology of each species. In harvester ants, Bert Holldobler, of Harvard University, has shown that some are fixed trunk trails while others are more temporary recruitment trails. Trunk trails, laid by secretions of Dufour's gland, are like spokes radiating out from the nest entrance into foraging territory. Trunk trails are fairly permanent and are used mostly by ants that are returning to the nest. Recruitment trails, laid by the poison gland, branch off from trunk trails. They are laid by a successful forager as she returns from a food source. These trails entice other foragers to follow the chemical signals and thus to help retrieve the food. As food-laden foragers come back to the nest, they lay recruitment trails that reinforce the previous signals. Foragers that do not find food and return hungry to the nest lay no recruitment trails. The recruitment trail has a short half-life, it gradually fades away along with the finishing of the source of the food.

It is apparently easy to say that a given species of ant or termite follows a chemical trail, but the molecular components of the trail signal have proved difficult to isolate in most cases. To demonstrate the presence of a chemical trail, a researcher need only paints line of a chemical sample along a substrate, such as a piece of filter paper, and then watch to see whether foraging ants or termites follow the line when they chance to cross it . Extracts of whole insects or their glands or chemical composition can be easily analysid.

Not more than a dozen trail Pheromone components have been identified, and most of these are not very volatile. Neocambrene A, a complex terpene, is the trail substance of nasute termites, and a pyrazine is the trail substance is several species of myrmicine ants. At least in these ants, the true trail pheromone is not species specific. This pyrazine is found in all seven species of British *Myrmica*, and it is found also in related genus of tropical leaf-cutter ants. The power of modern microchemistry is such that only 50 Myrmica workers provided enough pyrazine for the adequate examination of the chemical composition of the trail.

The poison gland produce trail pheromone of Myrmica. When Marie-Claire Cammaerts, of the University of Brussels, Belgium, drew lines with poison gland extracts, workers that blundered across the lines turned to follow them, but foragers were not attracted to the trail. Further experiments revealed that the volatile components of Dufour's gland are attractants and excitants. Both glands have a common opening at the tip of the abdomen and both are responsible in forming the trail.

In the natural habit trails are not public thoroughfares that are shared commonly by all species. Each species, and usually each nest, follows its own trails. An artificial trail, laid with a combination of the pyrazine and volatile components of the Dufour's secretion of one species of Myrmica, is followed quite well by foragers of the same species and equally well by foragers of other species. There is no obvious species indentification. However, when Cammaerts added the remainder of the Dufour's secretion, which is an oily mixture of low volatility, she produced trails that were species specific. The oils differ among the species and allow each species to identify its own unique trail. The secreted oils from Dufour's glands are *territorial pheromone* for Myrmica. Pioneer foragers, seeking new food sources far from the nest, move forward slowly and tentatively. As they walk ahead onto new ground, each forager repeatedly marks the ground with her Dufour's gland. The oils persits for long periods, and subsequent foragers will run quickly across the same ground that was so slowly crossed the first time. The oily marks differentiate home territory from foreign territory.

The natural foraging trails of *Myrmica* have three kinds of components, each of which serves a distinct purpose. Secretions of the poison gland define the trail itself. Volatile substances from Dufour's gland attract and increase running speed, but these signals are short-lived. Finally, oily substances from Dufour's gland survive and mark the home territory.

Pheromones and Individual Differences

A few members of species do not emit the same complement of pheromones. Differences between life stages (Larva versus adult) and between sexes are obvious. Also, genetic strain, physiological state, and nest and caste identification is social insects are identified by pheromones. For example, for successful mating of the fruitfly, *Drosophila melanogaster*, the female must emit pheromones that stimulate the male to court, and the male must emit pheromones that cause the female to accept. Both sexes prefer to mate with "strangers" that are genetically quite different, and they recognize the differences between related and unrelated partners by their differing pheromones. Males and females from the same inbreed line do not spontaneously court one another unless they are foled by being brought together in an environment filled with alien pheromones.

Most of the halictine bees are colonial and nest together in burrows. At the entrance to each colony, a guard bee inspects newcomers and rejects those that are not residents. Charles Michener and his associates at the University of Kansas wondered how the guards of one species. *Dialictus zephyrus*, distinguished nestmates from aliens. Bees were brought to the nest opening in a piece of plastic tubing and porked out into the entrance with a pipe cleaner. Of the 228 nonresident bees introduced into other nests, 50% were attacked by guards before contact and 92% on contact. Fifty residents were introduced in a similar manner and all were accepted. Then, Michener killed 20 residents and 32 nonresidents by freezing and presented them, after thawing, one by one to the guard bees. Merely 10% of the frozen residents were rejected, while 97% of the nonresidents were attacked. Nonresidents were also rejected when the room was illuminated with a red light to which bees are blind. Thus, it is confirmed by these experiments and others established that touch, vision, and hearing are not used for rejection. Odor must account for recognition of nestmates. Does the guard remember the scent of a nestmate, or does she merely compare the odor of the entering bee with that of the nest? When guards are placed on duty in a foreign nest, they accept only their own nestmates and reject the owners of that nest. It is the odor of the bee, not that of the nest, that these guards use to detect the identity of the intruder.

The scent if the colony has a strong genetic component. Nestmates are sisters and are genetically similar. Guard bees of *Dialictus zephyrus* can recognize individual odors that are variants on the pheromone blends produced by their nestmates, and they will thus admit genetically related bees that were reared in other nests. In honey bees. Even in the case of honey bees, colony odor has a genetic basis. When a honey bee queen is suddenly presented to workers of another colony, she is usually attacked. But she is much more likely to be accepted by these workers if she is closely related to their own queen. Michael Breed, of the University of Colorado, Boulder has shown that inbred sister queens are accepted in 35% of the exchanges, while the queen bees which are unrelated are definitely not accepted.

Conditioning plays a part in the recognition of individuals or classes of indviduals. In laboratory colonies, an halictine guard bee, placed in a "foreign" nest, can habituate to the odors of that nest and then will accept a bee with that odor, but such a situation is probably rare in nature. The response of male halictines to females is more meaningful. Upon first scent, a male responds avidly to the novel odor of a new female. But if she is unwilling to be courted, he becomes unresponsive to her now familiar odor. When exposed to a second new female with yet a different odor, he immediately attempts to court again. It is clear that the male bee learns and accepts quickly the response of the female bee. Probably the glands which emit a distinct body odour are many. The surface layers of insect cuticles contain odorous molecules that originate in the underlying epidermal cells. Other contaminants are left on the surfaces around the exocrine glands and dissolve into the coat of epicuticular and surface lipids. Each exocrine gland has many components, and their proportions vary slightly between individuals. The rich mixture of secretions from many glands, the scents from the environment, food odors, and the intrinsic odors of the epicuticle combine to produce pheromonal polymorphism enable each insect with unique identity.

Primer Pheromones

Primers bring changes in the physiological state of the recipients. It is usually assumed that primer effects are mediated by the endocrine system. Thus pheromones influence endocrine glands, and hormones directly regulate the physiological state. Primers are found in subsocial and social situations where the persistence of the group allow stimuli produced by one insect to influence another. Most primers influence reproductive capacity. A pheromone from male desert locusts increases the rate of development in immatures of both sexes. The scents of male and female mealworm beetles accelerate egg development and other aspects of reproductive maturation in young female beetles. Honey bee queen substances stop the growth of ovaries in the workers.

It is quiet interesting to note that the primers that switch development from one end point to an alternative. Desert locusts have such a primer, locustol, which triggers development from the solitary form to the gregarious form. The gregarious form is responsible for the devastating locust swarms that have revaged crops in Africa since biblical days. As more and more locusts feed together, they begin to produce locustol in their faeces. Locustol causes a gradual transformation of the light-colored solitary types into much darker gregarious forms that gather together and migrate, devuring all that comes in their way.

Generally the colonies of termite have only two functionally reproductive indviduals at any time. If an accident or an entomologist removes the male or female reproductive, one of the nymphs of the same sex develops into a "supplementary reproductive". The same sex develops into a "supplementary reproductive". The presence of a functional reproductive inhibits the maturation of others of the same sex. It is presumebly each reproductive insect produces an inhibitory pheromone, which is passes around by mutual grooming to workers and among workers. In at least one species by termite, it is possible to inhibit sexual maturation of nymphs by introducing extracts of the functional reproductives. Soldier production is probably limited by an analogous soldier-specific inhibitory primer, but there is no experimental evidence to confirm this.

Defensive Allomones

Most insects prevent the possibility of being eaten because they are distasteful. Foul-smelling and eviltasting chemicals in both animals and plants confer protection from attackers. Probably the easiest chemical defensive behavior for an insect in regurgitation-vomiting gut contents at an attacker, as do many grasshoppers when they "spit tobacco." The regurgitate may be enriched an unplesant chemicals derived from the host plant. For instance, sawfly larvae that feed on pine trees sequester terpenoid resins from the needles in two pouches of the foregut. The purches, like the rest of the foregut, are lined with a cuticle that prevents diffusion of the resins into the blood of the insect. When attacked by an ant or a spider, the sawfly rears up its front, regurgitates a droplet of resinous fluid, and wipes the regurgitate onto the attacker. The regurgitate is rich in the resins of the ine tree, which are repulsive and distasteful to birds and so arthropod predators such as ants and spiders.

Blood-Borne Allomones

Most of the unpalatable insects are aposematic (warnings colored), such that predators can readily associate the appearance of the animal with its unpalatability. For example, monarch butterflies are strongly aposematic with their striking orange and black wing patterns. When blue jays try to eat monarch butterflies, they vomit up the partially eaten carcass. Linclon and Jane Van Zandt Brower, them at Amherst College, Massachusetts, offered moncarch and other butterflies to captive jays. The birds quickly learned to associate the monarch color pattern with vomiting and thereafter the butterflies were refuced. Monarch butterflies are unpalatable because their blood and tissues contain cardiac glycosides. These glycosides were sequestered by the caterpillars from milkweed in their diet.

The insects which carry the defensive allomone in blood, it may be liberate a special localized bleeding, or autohemorrhage. Blood bubbles out from weakened sites at leg joints, antennal joints, or between sclerites when the insect is attacked. Rough handling of a blister beetle will cause it to release clear droplets of blood from its leg joints, and lady beetles exude yellow drops of blood from similar sites. The blood of the blister beetle contains cantharidin (Spanish fly), While that of the lady beetle carries the alkaloid precoccinelline. In contrast to the cardiac glycosides used by monarch butterflies, both cantharidin and precoccinelline are manufactures by the insect itself rather than extracted from a food plant. The substance such as, cantharidin and precoccinelline are distasteful to many vertebrate predators, but not to all. Thomas Eisner reports that toads will repeatedly consume blister beetles. Small predators are further deterred by the mechanical properties of the droplet. Ants become entangled in the sticky blood. As it coagulates, one leg is glued to another, antennae become stuck to mouthparts and ants adhere to each other. Ants retreat off the attack and go through a frenzy of cleaning activity.

Allomones Produced in Exocrine Glands

Defensive allomones are frequently found in cuticular reservoirs (as were those of the sawfly). Most often the reservoirs are storage compartments for secretions made within exocrine glands. The reservoir may be eoerted and wipied on the site of attack, as seen in the larvae of sawllowtail butterflies. The defensive fluid of these caterpillars cantains isobutyric and 2-methyl butyric acid. Alternatively, the secretion may ooze from the gland opening, much as blood oozes from the knee joints of blister beetles. Finally, the secretion may be ejected in a jet of acrid mist. The spraying glands are dramatic in their impact and often extremely accurate. The defensive fluid of a Florida walking stick is aimed directly at the site of attack and effectively repels both vertebrate and invertebrate predators.

Cuticular reservoirs increases the chances of potential defensive allomones, for within a sac of impervious cuticle, small reactive molecules can be safely stored. In contrast, these toxins cannot be stored in the blood. The chemical substances which protect the insects operate in a very different context from that of chemical signals such as pheromones. The potential prey is at an advantage when the signal is so blatant that the predator cannot ignore it. The predator is at an advantage when it can ignore the blatant insult or when it is insensitive to the defensive substance. Any insect has a host of potential predators, and one might expect natural selection to favour defensive allomones that are nonspecific. That, are the repellant to many species. The ofensive secretions of many insects contain small reaction molecules that are broadly toxic to most living tissue. Small acids, like formic or acetic, readily coagulate protein and irritate excitable tissues.

The chemical substances, such as formic and acetic acids are used in the defenses mechanism of ants, beetles, and caterpillars. Reactive aldehydes, which bind readily to proteins, are painful tear gases to humans and animals. Molecules such as these are found in the defensive secretion of walking sticks (*Anisomorpha*), cockroaches, beetles, and true bugs. Benzoquinos, which also bind proteins and "tan" them, are common in the defensive secretions of beeltes and millipedes. Many predators find such small reactive substances unpleasent and split out the prey that sprays them. In addition, these defensive allomones are effective as antibiotics against microorganisms. Storing the reactive defensive molecules could pose a problem for the insect producing them.

There are two ways by which an animal avoids being poisoned by its own secretions. First, the secretions are stored within reservoirs of cuticle that have special permeability properties that prevent any leakage of the toxic fluids. Effectively the secretion is stored outside the animal. Second, many of the secretions are produced in a stepwise manner, such that the cytoplasm of the secretory cells contains only harmless precursors rather than the toxic end product. The progression of the production machinery is best seen in the adaptation for quinone production by bombardier beetles and darkling beetles. The term bombardier beetles is in keeping with the defensive behaviour of this particular species of beetles. When disturbed, they emit a jet of host secretion at the site of attack, and the ejection of secretion is accompanied by a clearly audible explosion. The glands open at the tip of the abdomen into movable turrets that are aimed when the tip of the abdomen is bent toward the attack. The spray contains a mixture of p-benzoquinones.

Quinones are not stored in the reservoir of the gland but rather are produced at the instant of attack. The gland reservoir consists of two compartments. It has been demonstrated as shown by H. Schildknecht, of the University of Heidelberg, Germany, that the inner compartment contains hydroquinones (diphenols), which are precursors to the final defensive quinones. The hydroquinones are mixed with hydrogen peroxide. As soon as a beetle is attacked, some of the content of the inner compartment is passed into the outer compartment. The author compartment is a reaction chamber.

The small outer compartment contains a mixture of enzymes (catalases and peroxidases) that oxidizes and decomposes the hydrogen peroxides to give water and oxygen and also oxidizes the hydroquinones to yield benzoquinone. The oxidation reactions liberate heat and gaseous oxygen. Thomoas Eisner has shown that the temperature of the mixture reaches 100°C. The gaseous oxygen provides that propellent that accounts for the audible "pop" when the glands discharge. Eisner has also used high-speed photography to analyze encounters between attacking ants and bombardier bettles. Thus entire sequence, from ant bite to beetle discharge to ant release, takes place within a second.

Another species known as Darkling beetles, or flour beetles and mealworm beetles, also use quinones in their defensive secretions. Unlike the bombardier beetles, these darkling beetles store the actual quinones, not a precursor, in the reservoir of the glands. But like the bombardier beetles, the darkling beetles protect their secretory cells from quinoid products by a strategic sequence of localized reaction steps. The cells which secrete these defensive glands pass their products to the reservoir via fine cuticular ductules. In these defensive glands there are two distinct populations of secretory cells. The secretory cells that are of particular interest to quinone production are types 2a and 2b, which are arranged in pairs. Each pair is strung in series on a cuticular ductule that leads to the reservoir. Each cell of the pair surrounds a central cavity within which lies the effective ductule.

The secretory product of the cell passes into that extracellular cavity and then through the wall of the different ductule and finally to the reservoir. The structure of cell cytoplasm, cavity, and ductule can be viewed as reaction compartments, arranged in series. The two final steps in quinone production are hydrolysis and oxidation. The reaction steps for quinone production apparently occur in the sequential compartments. In the cytoplasm of the distal cell, phenyl glucoside is stored. The glucoside is hydrolyzed as it is secreted into the cavity of the distal cell, and the resulting diphenol can then pass along through the wall of the ductule. Within the ductule, the diphenol is oxidized to produce the final quionone. The oxidative enzymes (phenolases and peroxidases) are apparently built into the wall of the cuticular ductule. Thus the secretory cells avoid being poisoned by quinones because they are never actually exposed directly to poison.

Another mechanism, to which protects insects from the poison they develop themselves gradually to secrete appropriate enzymatic catalysts and precursors of the allomones into the reservoir. Reservoir fluids often consist of a watery phase and an oily phase. Murray Blum and his colleagues report that in the two-phase secretion of a true bug (*Leptoglossus*), the enzymes are in the aqueous phase and the reactive defensive allomones are in the oily phase.

Like pheromones, defensive secretions are generally mixtures of several substances. The components often differ sharply in their polarity, as in the darkling beeltes, where several relatively polar quinones are mixed with nonpolar hydrocarbons. The mixture is adequate to dissolve and disrupt a variety of permeability barriers at the surface of the integument. Furthermore, a mixture of several different repellent substance contributes to the nonspecific effectiveness of defensive secretions in repulsing many different kinds of predators. In some species, 20 to 40 distinct components make up the defensive fluid.

Defensive Allomones of Social Insects

As soon as the nest of a social insect is invaded, members of the colony pool their resources for collective defense. In some ants and many termites, a special caste, the soliders, is dedicated to nest defense. The secretions they use are doubtly effective repellent to the attackers and alarm signals for nestmates. As shown by endure Quennedey at the University for Dijon in France, the major source in ligher termites is the frontal gland of the soliders in *Cubitermes*, the strong mandibles pinch the attacker (usually an ant), and the frontal gland secretions pour out on the sties of contact. The secretions are somewhat repellent, but in this genus the mandibular pinch is the major defensive act. In *Rhinotermes* there are two caterogries of soldier: a major caste with enormous mandibles and a minor caste with tiny mandibles. The minor caste has a huge frontal gland reservoir that extends back from the head all the way to the abdomen. It secretes a mixture of ketones toxic to many insect species. In the third style of defense, found in nasute termites, the chemical weapon is paramount. The soliders are remarkable in that their mandibles are reduced in size and their heads are huge, elongated snouts-squirting devices for the frontal gland secretion. When the nest wall is broken, soliders advance outward in parallel or in staggered line, precisely spaced by just the distance of antennal contact. A soldier sprays the frontal gland contents when it is bumped. The secretion contains volatile terpenes that are alarm releasers for termites and irritants to the attacker, and other unusual terpenes that are sticky and entangle ants, which are major threat to the lives of termites.

Allomones Promoting Associations

Defensive allomones terminate interspecific associations between potential prey and a diverse array of predator species. In contrast, there are a host of less-studied allomones that promote intimate associations between two paticular species. Many of these associations are mutually aggreable with both partners deriving substantial benefits, while others are parasitic, when only one of the partner gets the benefits. A strange phenomenon is seen in the behaviour of specialized insect speices are "guests" in the nest of social insects.

The alien insect enters an ant nest or termite nest without being attacked and is sometimes even assisted by a host worker. In some cases the volatile body lipids of guest and host are very similar in composition so that the intruder appears to mimic the body odor of the hosts. In other cases the guests placate their hosts with allomones that E. O. Wilson has termed *appeasement substances.* Jacques pasteels, of the University of Brussels,Belgium, has shown that that appeasement substances of rove beetles are produced in a battery of special exocrine glands that are found only in species that are social parasites. Bert Holldobler has observed the process by which a rove beetle guest grains entry to an ant nest. The ant feeds on the appeasement secretion. Next, the ant licks an "adoption gland" on the lateral margins of the beelte's abdomen, and finally it welcomes the beetle into its nest. When inside the nest some guest species behave generously enough, merely soliciting regurgitated food from the workers and exploiting the shelter and stable habitat provided by the hosts. But some guests return the hospitality with a bad turn. Kenneth Hagen, of the University of California, Berkeley, has reported that some lacewing larvae eat their termite hosts.

The lacewing larvae move about freely in the termite galleries. At certain developmental stages the lacewings approach termite workers and wave the tips of their abdomens in front of the termites. The termites are paralyzed by a gaseous allomone, and the larvae then devours the paralysed workers. Allomones also promote symbiosis between insects and micro-organisms. Leaf-cutter ants, certain termites, and bark beetles transport and consume symbiotic fungi. The association are quite specific-each insect species cultures only one or two fungal species. As the fungus grows in ant nests, the cultures remain clear of any contaminating species.

Leaf-cutter ants weed their crops by mechanical and chemical means. Alien spores are picked out of the fungus garden, and allomones, produced by the metapleural gland of the thorax, regulate the species composition of the growing fungi. H. Schildknecht and V. Maschwitz reported that the metapleural secretion has three components: phenylacetic acid, β-indolyacetic acid, and myrmicacin β-hydroxydecanoic acid). Phenylacetic acid is a general-purpose antibiotic that prevents the growth of bacteria and of some weed fungi. Indolyacetic acid, a plant hormone, is thought to promote growth of the symbiotic crop. Myrmicacin inhibits the generation of fungal spores. The speices of fungus normally cultured in colonies of these ants is unaffected by phenylacetic acid or myrmicacin. Both, Bark beetles and ambrosia beetles transport symbiotic fungi within cuticular pits or in large cuticular pockets. The pits and pockets are associated with secretory cells, and cycles of secretary activity correlate with the peak of fungal proliferation in the pocket.

In the case of the southern pine beetles, the secretions apparently nourish the symbiotic fungi and restrict their growth to yeastlike budding rather than production of long hyphal strands. The yeastlike progeny drop out of the transport pocket as the beetle asses through the tunnels and galleries and start new growths of symbiotic fungi that will nourish beetle larvae. In addition, the microflora in the pocket is restricted to two species. Presumably the secretions favor the growth of these two symbionts while excluding all the weed fungi that contaminate the integument of the beetle. Unfortunately we unfamiliar of the chemical nature of allomones that play these role.

The microsymbiont of the association may be an aid to digestion. As shown by the late L. R. Cleveland, of Harvard University, intestinal protozoans (flagellates) help the primitive cockroach *Cryptocercus* to digest its meal of cellulose. The protozoans live in the anaerobic hindgut of the cockroach. In the presence of high concentrations of oxygen, the protozoans die. To grow, the cockroach must molt its entire cuticle, including that of the hingut. If abruptly exposed to the atmosphere together with the shed skin, the protozoans would not survive, and the cockroach would lose its vital digestive parts.

The protozoans save themselves from a molt because they "anticipate" it and enclose themselves in a cyst that excludes poisonous oxygen. The allomone that triggers encystment is the insect the hormone that induces apolysis: the familiar steroid ecdysone. The newly molted cockroach consumes its old cuticle, and once within the anaerobic gut environment, the protozoans become activates again.

An Overview of Chemical Signals

There has been chemical signals between cells since the moment first cells appeared on the earth, and intercellular chemical communication is a distinguishing feature of all major phyla. Hormones, pheromones, and allomones share some common origins in the earlier phase of the evolution of life form on the earth. Exocrine secretions for diverse purposes often have metabolic byproducts, and is not surprising that these products, which are produced only in certain physiological states, might communicate that state to animals of the same species. For phermoones are those allomones that promote associations, selective pressures act on both the producer and the target organism to evolve an unambiguous signal that is detected at low levels. For defensive allomones, slection operates on the producer to make the singal blatant and insulting, while the potential target organisms are at an advantage if they can ignore the insult. Thus many defensive allomones are broad-spectrum reactive protein poisons that are vitally important in the oves of insects.

Same kind carbon chains form the skeletons of the molecules that function as allomones, pheromones, or hormones. The same molecule may play more than one role, as does ecdysone in *Cryptocercus.* Theoretically structural diversity of chemical signals should be almost unlimited, but in practice only a few families of carbon skeletons predominate among pheromones, hormones, and allomones. These include aliphatic straight chains, terpenes and the related steroids, and small ring compounds. There are also a significant number of hormones are peptides, and a significant number of defensive, allomones are complex and unusual poisons.

The explantaion of the widespread molecular conservatism is probably rooted deep in the metabolic pathways common to living systems. All cells have the enzymes for manufacturing certain skeletons. To develop a singal molecule, it is simpler to amend a preexisiting biosynethetic pathway and a preexisting receptor surface than to evolve totally new ones.

All the categories of molecular signals have common characteristics. Mostly pheromones and many allomones are volatile. Many defensive allomones are simple poisons with quite reactive functional groups, while others are complex and unusual toxins (such as percoccinelline) that irritate sensory receptors and are difficult to degrade metabolically. Most pheromone components are scarcely reactive, but their skeletons tend to be rigid and to have asymmetric carbons that determine precisely the shape of the molecule. Both defensive allomones and pheromones are medley, but for very different reasons. Defensive allomones are medley, but for very different reasons. Defensive allomones are medleys so that the producing animal covers as many bets as possible when it plays its defensive hand. Pheromone medleys allows the construction off fever number of indifiniteness signals.